An Expanding World
Volume 18

Plantation Societies in the Era of European Expansion

AN EXPANDING WORLD
The European Impact on World History, 1450–1800
General Editor: A.J.R. Russell-Wood

EXPANSION, INTERACTION, ENCOUNTERS
1 **The Global Opportunity** *Felipe Fernández-Armesto*
2 **The European Opportunity** *Felipe Fernández-Armesto*
3 **The Globe Encircled and the World Revealed** *Ursula Lamb*
4 **Europeans in Africa and Asia** *Anthony Disney*
5 **The Colonial Americas** *Amy Turner Bushnell*

TECHNOLOGY AND SCIENCE
6 **Scientific Aspects of European Expansion** *William Storey*
7 **Technology and European Overseas Enterprise** *Michael Adas*

TRADE AND COMMODITIES
8 **Merchant Networks in the Early Modern World** *Sanjay Subrahmanyam*
9 **The Atlantic Staple Trade** (Parts I & II) *Susan Socolow*
10 **European Commercial Expansion in Early Modern Asia** *Om Prakash*
11 **Spices in the Indian Ocean World** *M.N. Pearson*
12 **Textiles: Production, Trade and Demand** *Maureen Mazzaoui*
13 **Interoceanic Trade in European Expansion** *Pieter Emmer and Femme Gaastra*
14 **Metals and Monies in a Global Economy** *Dennis O. Flynn and Arturo Giráldez*
15 **Slave Trades** *Patrick Manning*

EXPLOITATION
16 **From Indentured Servitude to Slavery** *Colin Palmer*
17 **Agriculture, Resource Exploitation, and Environmental Change** *Helen Wheatley*
18 **Plantation Societies in the Era of European Expansion** *Judy Bieber*
19 **Mines of Silver and Gold in the Americas** *Peter Bakewell*

GOVERNMENT AND EMPIRE
20 **Theories of Empire** *David Armitage*
21 **Government and Governance of Empires** *A.J.R. Russell-Wood*
22 **Imperial Administrators** *Mark Burkholder*
23 **Local Government in European Empires** *A.J.R. Russell-Wood*
24 **Warfare and Empires** *Douglas M. Peers*

SOCIETY AND CULTURE
25 **Settlement Patterns in Early Modern Colonization** *Joyce Lorimer*
26 **Biological Consequences of the European Expansion** *Kenneth F. Kiple and Stephen V. Beck*
27 **European and non-European Societies** (Parts I & II) *Robert Forster*
28 **Christianity and Missions** *J.S. Cummins*
29 **Families in the Expansion of Europe** *Maria Beatriz Nizza da Silva*
30 **Changes in Africa, America and Asia** *Murdo MacLeod and Evelyn Rawski*

THE WORLD AND EUROPE
31 **Europe and Europe's Perception of the World** (Parts I & II) *Anthony Pagden*

Please note titles may change prior to publication

An Expanding World
The European Impact on World History 1450–1800

Volume 18

Plantation Societies in the Era of European Expansion

edited by
Judy Bieber

VARIORUM
1997

This edition copyright © 1997 by Variorum, Ashgate Publishing Limited, and Introduction by Judy Bieber. For copyright of individual articles refer to the Acknowledgements.

Published by VARIORUM
Ashgate Publishing Limited
Gower House, Croft Road
Aldershot, Hampshire GU11 3HR
Great Britain

Ashgate Publishing Company
Old Post Road
Brookfield, Vermont 05036
USA

ISBN 0–86078–506–8

British Library CIP data
Plantation Societies in the Era of European Expansion.
(An Expanding World: The European Impact on World History, 1450–1800: Vol. 18).
1. Slavery–History. 2. Slave trade–Africa–History. 3. Plantations–History. 4. Slave labour–History. 5. Imperialism–Social aspects.
I. Bieber, Judy.
306.3' 62' 0903

US Library of Congress CIP data
Plantation Societies in the Era of European Expansion/ edited by Judy Bieber.
p. cm. – (An Expanding World: The European Impact on World History, 1450–1800: Vol. 18).
1. Plantations–History. 2. Plantation life–History. 3. Slavery–History.
I. Bieber, Judy. II. Series.
HD1471. A3P54   1997   96–40468
306. 3' 49–dc21   CIP

This book is printed on acid free paper.

Printed and bound by Athenaeum Press, Ltd.,
Gateshead, Tyne & Wear.

AN EXPANDING WORLD 18

Contents

Acknowledgements	vii–viii
General Editor's Preface	ix–xi
Introduction	xiii–xxxvi

EUROPEAN ANTECEDENTS TO PLANTION ECONOMIES IN THE AMERICAS: SUGAR AND SLAVES

1 Plantations, Sugar Cane and Slavery
   Sidney M. Greenfield — 1

BRAZIL

2 Indian Labor and New World Plantations: European Demands and Indian Responses in Northeastern Brazil
   Stuart B. Schwartz — 37

3 Slave Families on a Rural Estate in Colonial Brazil
   Richard Graham — 74

4 African Slave Trade and Economic Development in Amazonia, 1700–1800
   Colin M. MacLachlan — 95

SPANISH AMERICA

5 Encomienda, African Slavery, and Agriculture in Seventeenth-Century Caracas
   Robert J. Ferry — 129

BRITISH AMERICA

6 Slaves in Piedmont Virginia, 1720–1790
   Philip D. Morgan and Michael L. Nicholls — 157

7 Plantations, Paternalism, and Profitability: Factors Affecting African Demography in the Old British Empire
   Daniel C. Littlefield — 199

8 A Tale of Two Plantations: Slave Life at Mesopotamia in Jamaica and Mount Airy in Virginia, 1799–1828
   Richard S. Dunn — 215

DANISH CARIBBEAN

9 Slaves and Slave Masters on Eighteenth-Century St. John
   Karen Fog Olwig   249

FRENCH CARIBBEAN

10 Toussaint Louverture and the Slaves of the Bréda Plantations
   David Geggus   266

COMPARATIVE THEORY: CARIBBEAN

11 Freedom and Oppression of Slaves in the Eighteenth-Century Caribbean
   Arthur L. Stinchcombe   285

12 Was the Plantation Slave a Proletarian?
   Sidney W. Mintz   305

Index   323

Acknowledgements

The chapters in this volume are taken from the sources listed below, for which the editor and publishers wish to thank their authors, original publishers or other copyright holders for permission to use their material as follows:

Chapter 1: Sidney M. Greenfield, 'Plantations, Sugar Cane and Slavery', *Historical Reflections/Reflexions Historiques* VI, no. 1 (Alfred, NY, 1979), pp. 85–119. Copyright © 1979 by the Department of History, Alfred University.

Chapter 2: Stuart B. Shwartz, 'Indian Labor and New World Plantations: European Demands and Indian Responses in Northeastern Brazil', *American Historical Review* LXXXIII (Washington, DC, 1978), pp. 43–79. Copyright © 1978 by Stuart B. Schwartz.

Chapter 3: Richard Graham, 'Slave Families on a Rural Estate in Colonial Brazil', *Journal of Social History* IX (Pittsburgh, PA, 1976), pp. 382–402. Copyright © 1976 by Mellon University Press.

Chapter 4: Colin M. MacLachlan, 'African Slave Trade and Economic Development in Amazonia, 1700–1800', in ed. Robert Brent Toplin, *Slavery and Race Relations in Latin America* (Westport, VA, 1974), pp. 112–145. Copyright © 1974 by Greenwood Publishing Group Inc.

Chapter 5: Robert J. Ferry, 'Encomienda, African Slavery, and Agriculture in Seventeenth-Century Caracas', *Hispanic American Historical Review* LX, no. 4 (Durham, NC, 1981), pp. 609–635. Copyright © 1981 by Duke University Press.

Chapter 6: Philip D. Morgan and Michael L. Nicholls, 'Slaves in Piedmont Virginia, 1720–1790', *William and Mary Quarterly* XLVI, no. 2 (Williamsburg, VA, 1989), pp. 211–251. Copyright © by the Institute for Early American History and Culture.

Chapter 7: Daniel C. Littlefield, 'Plantations, Paternalism, and Profitability: Factors Affecting African Demography in the Old British Empire', *Journal of Southern History* XLVII, no. 2 (Houston, TX, 1981), pp. 167–182. Copyright © 1981 by the Southern Historical Association.

Chapter 8: Richard S. Dunn, 'A Tale of Two Plantations: Slave Life at Mesopotamia in Jamaica and Mount Airy in Virginnia, 1799–1828', *William and*

Mary Quarterly XXIX, no. 1 (Williamsburg, VA, 1977), pp. 32–65. Copyright © by the Institute of Early American History and Culture.

Chapter 9: Karen Fog Olwig, 'Slaves and Slave Masters on Eighteenth-Century St. John', *Ethnos* L (Stockholm, 1985), pp. 214–230. Copyright © 1985 by Folkens Museum–National Museum of Ethnography, Stockholm.

Chapter 10: David Geggus, 'Toussaint Louverture and the Slaves of the Bréda Plantations', *Journal of Caribbean History* XX, no. 2 (Bridgetown, Barbados, 1985/6), pp. 30–48. Copyright © 1985/6 by the University of the West Indies.

Chapter 11: Arthur L. Stinchcombe, 'Freedom and Oppression of Slaves in the Eighteenth-Century Caribbean', *American Sociological Review* LIX, no. 6 (Washington, DC, 1994), pp. 911–929. Copyright © 1994 by the American Sociological Association.

Chapter 12: Sidney W. Mintz, 'Was the Plantation Slave a Proletarian?', *Review* II, no. 1 (Binghampton, NY, 1978), pp. 81–98. Copyright © 1978 by the Fernand Braudel Center, State University of New York.

Every effort has been made to trace all the copyright holders, but if any have been inadvertently overlooked the publishers will be pleased to make the necessary arrangement at the first opportunity.

General Editor's Preface
A.J.R. Russell-Wood

An Expanding World: The European Impact on World History, 1450–1800 is designed to meet two objectives: first, each volume covers a specific aspect of the European initiative and reaction across time and space; second, the series represents a superb overview and compendium of knowledge and is an invaluable reference source on the European presence beyond Europe in the early modern period, interaction with non-Europeans, and experiences of peoples of other continents, religions, and races in relation to Europe and Europeans. The series reflects revisionist interpretations and new approaches to what has been called 'the expansion of Europe' and whose historiography traditionally bore the hallmarks of a narrowly Eurocentric perspective, focus on the achievements of individual nations, and characterization of the European presence as one of dominance, conquest, and control. Fragmentation characterized much of this literature: fragmentation by national groups, by geography, and by chronology.

The volumes of *An Expanding World* seek to transcend nationalist histories and to examine on the global stage rather than in discrete regions important selected facets of the European presence overseas. One result has been to bring to the fore the multicontinental, multi-oceanic and multinational dimension of the European activities. A further outcome is compensatory in the emphasis placed on the cross-cultural context of European activities and on how collaboration and cooperation between peoples transcended real or perceived boundaries of religion, nationality, race, and language and were no less important aspects of the European experience in Africa, Asia, the Americas, and Australia than the highly publicized confrontational, bellicose, and exploitative dimensions. Recent scholarship has not only led to greater understanding of peoples, cultures, and institutions of Africa, Asia, the Americas, and Australasia with whom Europeans interacted and the complexity of such interactions and transactions, but also of relations between Europeans of different nationalities and religious persuasions.

The initial five volumes reflect the changing historiography and set the stage for volumes encompassing the broad themes of technology and science, trade and commerce, exploitation as reflected in agriculture and the extractive industries and through systems of forced and coerced labour, government of empire, and society and culture in European colonies and settlements overseas. Final volumes examine the image of Europe and Europeans as 'the other' and the impact of the wider world on European *mentalités* and mores.

An international team of editors was selected to reflect a diversity of educational backgrounds, nationalities, and scholars at different stages of their professional careers. Few would claim to be 'world historians', but each is a recognized authority in his or her field and has the demonstrated capacity to ask the significant questions and provide a conceptual framework for the selection of articles which combine analysis with interpretation. Editors were exhorted to place their specific subjects within a global context and over the *longue durée*. I have been delighted by the enthusiasm with which they took up this intellectual challenge, their courage in venturing beyond their immediate research fields to look over the fences into the gardens of their academic neighbours, and the collegiality which has led to a generous informal exchange of information. Editors were posed the daunting task of surveying a rich historical literature and selecting those essays which they regarded as significant contributions to an understanding of the specific field or representative of the historiography. They were asked to give priority to articles in scholarly journals; essays from conference volumes and *Festschriften* were acceptable; excluded (with some few exceptions) were excerpts from recent monographs or paperback volumes. After much discussion and agonizing, the decision was taken to incorporate essays only in English, French, and Spanish. This has led to the exclusion of the extensive scholarly literature in Danish, Dutch, German and Portuguese. The ramifications of these decisions and how these have had an impact on the representative quality of selections of articles have varied, depending on the theme, and have been addressed by editors in their introductions.

The introduction to each volume enables readers to assess the importance of the topic *per se* and place this in the broader context of European activities overseas. It acquaints readers with broad trends in the historiography and alerts them to controversies and conflicting interpretations. Editors clarify the conceptual framework for each volume and explain the rationale for the selection of articles and how they relate to each other. Introductions permit volume editors to assess the impact on their treatments of discrete topics of constraints of language, format, and chronology, assess the completeness of the journal literature, and address *lacunae*. A further charge to editors was to describe and evaluate the importance of change over time, explain differences attributable to differing geographical, cultural, institutional, and economic circumstances and suggest the potential for cross-cultural, comparative, and interdisciplinary approaches. The addition of notes and bibliographies enhances the scholarly value of the introductions and suggests avenues for further enquiry.

I should like to express my thanks to the volume editors for their willing participation, enthusiasm, sage counsel, invaluable suggestions, and good judgment. Evidence of the timeliness and importance of the series was illustrated by the decision, based on extensive consultation with the scholarly community, to expand a series, which had originally been projected not to exceed eight volumes, to more than thirty volumes. It was John Smedley's initiative which gave rise to discussions as to the viability and need for such a series and he has overseen the publishing, publicity, and marketing of *An Expanding World*. As

General Editor, my task was greatly facilitated by the assistance of Dr Mark Steele who was initially responsible for the 'operations' component of the series as it got under way; latterly this assistance has been provided by staff at Variorum.

The Department of History,
The Johns Hopkins University

Introduction
Judy Bieber

The rise and fall of plantation societies was a phenomenon distinctly associated with European imperial expansion overseas.[1] Although plantation agriculture had existed previously in other regions and eras, the emergence of a widespread 'plantation complex', in which large estates funded through European capital using predominantly slave labour in the production of tropical export crops for European consumption, was largely a New World invention. The kind of slavery associated with New World Plantations differed from its Old World counterparts in Europe, the Mediterranean, Middle East, and Africa. Emergent chattel slavery consisted of a new, more extreme, racially constituted, labour regime.

In the era of European expansion under consideration, (1410–1800) plantations became almost synonymous with sugar and slaves. Slaves, however, were employed in sectors far removed from the plantation sector in the Americas: as artisans, domestic servants, occasional urban wage labourers, prostitutes, etc. Slavery had existed long before the rise of the sugar plantation but in many regions of the world it was an incidental form of labour, limited largely to domestic servants, personal attendants, in the military or as bureaucrats. Slaves could be captured as prisoners of war, enslaved due to religious or cultural difference, and in many cultures, slavery existed as an institution of assimilation. Sugar had been cultivated in the Mediterranean as early as the eighth century, but only in the fourteenth and fifteenth centuries was slave labour widely employed on the estates of Crete and Cyprus. African slaves first laboured on sugar plantations in Portuguese Madeira in the 1450s.[2]

This introductory essay will discuss the institution of slavery as it was understood and practiced in Europe, the Mediterranean, the Middle East and Africa; both the Muslim and European versions of slavery had an impact on the development of the institution in Africa. It also traces the development and expansion of plantation societies which depended upon slavery as a predominant, institutionalized form of labour, and examines some of the trajectories that historical research on plantation societies has taken. A rich and vast body of scholarly work dealing with comparative slavery has emerged since the Second World War. This volume seeks to contribute to the dissemination of that body

---

[1] Philip Curtin, *The Rise and Fall of the Plantation Complex* (Cambridge, 1990).
[2] J.H. Galloway, 'The Mediterranean Sugar Industry', *Geographical Review* LXVII (1977), pp. 177–92.

of research, by bringing together a series of influential essays in a single collection.[3]

Slavery in Europe

The history of slavery in Europe dates back to the Roman Empire. War captives laboured on agricultural estates and were employed in urban professions, as domestic servants, state bureaucrats, healers, educators, artisans, entertainers, concubines, and manual labourers. The working conditions of slaves do not seem to be different from those faced by free laborers. Slaves were subject to both discipline and protection, much like any other dependent member within the patriarchal Roman family. Their basic needs were met and slaves enjoyed certain basic rights by law: subsistence, freedom from excess abuse, the right to worship and to marry. Freedom through manumission was relatively frequent through gift or self purchase. Although most slaves held out for possible manumission, slave resistance, flight and revolt were constant factors.

Slavery in the Roman Empire began to gradually diminish in the beginning of the Christian era. The pax romana (31 BC–AD 14) led to a decline in war captives and the law forbade the enslavement of Roman subjects. Although Rome collapsed, the eastern part of the Empire endured for many more centuries. In the mid 500s the Byzantine ruler Justinian codified Roman law in the legal code that bears his name. This initiative was to have repercussions in the development of New World slavery. Roman law was adopted by many emerging European states or grafted onto other legal systems. It included specific, detailed slave codes and offered legal precedents for the development of Atlantic slavery in the sixteenth century.

Slavery persisted after the fall of Rome under the Visigoths and Vandals and continued during the reign of Charlemagne (AD 800). A series of invasions following Charlemagne's death: Vikings from the north, Magyars from the east and Muslims from the south, led to fragmentation and decentralization, encouraging the emergence of feudalism between the tenth and twelfth centuries. Slaves became integrated into a system of peasant production in which dependents contributed labour to the lord's lands and a portion of their own harvest in exhange for access to a plot of land.

By 1400, European slavery was waning with the exception of Spain and Portugal. Muslims who had conquered and settled southern Spain and Portugal

---

[3] The literature on comparative slavery is vast indeed. Joseph C. Miller's, *Slavery: A Worldwide Bibliography, 1900–1982* (New York, 1985) contains 5,177 entries. The reader who wishes to learn more about slavery and the slave trade is encouraged to consult the volumes of this series by Patrick Manning, *Slave Trades, 1500–1800* (1996); Colin Palmer, *From Indentured Servitude to Slavery* (1998); Helen Wheatley, *Agriculture, Resource Exploitation, and Environmental Change* (1997); and Kenneth Kiple, *Biological Consequences of the European Expansion* (1997).

had kept the institution alive, importing both Middle Eastern and black African slaves. This situation changed with the beginnings of Portuguese maritime expansion initiated with the conquest of Ceuta in 1415. From this base on the North African coast, the Portuguese had hoped to tap into the Moroccan and trans-Saharan trade. When this venture proved to be unsuccessful, the Portuguese reoriented their efforts and began to descend by sea along the West African coast to try to circumvent the Berber trading networks which controlled overland trade. They set up fortified trading posts along the West African coast and on offshore archipelagos, setting up 'factories' which traded European goods, cloth and copper for gold and slaves which were then used to purchase spices in India.

Through these commercial activities on the West African coast and the Atlantic islands, small numbers of African slaves began to enter Spain and Portugal in the 1440s. Relatively few were imported; in Portugal 1,000 slaves per year entered Lisbon between 1440–1530, at most comprising 2–3 per cent of the total population. Most slaves worked in urban areas as domestics, artisans and in service industries and were not used in plantation agriculture. In Castile, Seville became another enclave of black slavery. African slaves were set to similar tasks, although they were prohibited membership in the craft guilds. Valencia of the kingdom of Aragon also introduced black slaves, although its slave population consisted largely of Muslim war captives, foreign Jews, enslaved criminals and debtors, and even Canary islanders. About 10 per cent of Iberian slaves were manumitted annually, typically through self-purchase.[4] Black slavery differed from the lot of white commoners in degree of servitude only; in terms of diet, clothing, shelter, medical care and working conditions, slaves and poor whites faced similar constraints.

It was in Madeira, however, where the link between African slavery and sugar was forged. Europeans had become acquainted with sugar production in the eleventh century during the Crusades and began to grow and market slave-produced sugar in Palestine and Cyprus by the thirteenth century. In the Mediterranean, sugar had followed the spread of Islam, from Syria, Palestine, and Egypt to Morocco, Northern Africa and finally present-day Greece, Sicily and southern Spain and the Algarve by the beginning of the fifteenth century. Crete and Cyprus came under Venetian rule between the thirteenth and fifteenth centuries, and it was in these Christian European islands where the availability of slave labour combined with Italian merchant capital enabled a new type of agricultural production, more efficient and rational than feudalism, to emerge.

---

[4] For more on European and Iberian slavery see William D. Phillips, *Slavery from Roman Times to the Early Transatlantic Trade* (Minneapolis, 1985), especially chapter 8, and A.C. de C.M. Saunders, *A Social History of Black Slaves and Freedmen in Portugal, 1441–1555* (Cambridge, 1982).

During the fifteenth century, both the Spanish and the Portuguese had explored, and laid competing claims to the Canary islands, Madeira, and the Azores, known collectively as the Atlantic Islands. In 1479, Portugal and Castile signed the treaty of Alcaçovas, giving the Canaries to the Spanish and ceding the remaining islands to the Portuguese. Madeira proved to be the Portuguese island most suitable for cane cultivation. Its forests were quickly and indiscriminately razed and slaves began to enter in the mid 1450s, both African and Guanche, the indigenous inhabitants of the Canary islands. Sugar was also introduced into the Azores, the Cape Verdes and São Tome, but the climate and lands were less favorable and these islands were mostly devoted to grain, fruits, grazing, and the re-export of slaves. Spain experimented with cane in the Canary islands, using the native Guanches as slave labour, but production was short-lived, entering into decline by the 1530s.

The emergence of sugar plantations in the Atlantic Islands marked the transition between old world and new world slavery. In sub-Saharan Africa, the Mediterranean, and Europe, slavery had been an incidental form of labour recruitment, never the dominant form. As Portugal and Spain came to dominate, regulate, and expand the slave trade to the Americas, the institution of slavery was transformed to fit the needs of plantation societies. Peoples of African descent were subjected to a new variety of slavery, unprecedented in harshness. The eminent Africanist scholar, Philip D. Curtin, has argued persuasively that it 'became a new kind of institution, where the bundle of property rights in the masters' hands was far more elaborate, all inclusive and permanent than its model in Europe'.[5] This transition was complete by 1650.

Slavery in Africa

The European model of slavery differed markedly from its counterparts in the Islamic Middle East and in sub-Saharan Africa. Slavery was a vital institution in North Africa and the trans-Saharan trade exceeded the transatlantic in both duration and volume. An estimated 17,000,000 slaves were transported from black Africa to the Islamic world between 650–1905, compared to the transatlantic trade which lasted from the 1490s–1860s and carried about 11.5 million slaves.[6] Islamicist John R. Ellis has emphasized the importance of slavery in the region, stating:

---

[5] Philip D. Curtin, 'Slavery and Empire', in eds. Vera D. Rubin and Arthur Tuden, *Comparative Perspectives on Slavery in New World Plantation Societies*, Annals of the New York Academy of Sciences, vol. 292 (New York, 1977), p. 9.

[6] Philip D. Curtin, *The Transatlantic Slave Trade: A Census* (Madison, 1969). Curtin's figures have stood the test of time. See Paul E. Lovejoy, 'The Volume of the Atlantic Slave Trade: A Synthesis', *Journal of African History* XXIII (1982), pp. 473–501.

Slaves of African origin formed a vital thread in the living lines of economic production in the Near and Middle East and formed the cord of economic activity in Islamic Africa itself. Slaves sustained the salt pits and date palms of desert societies; they worked the spice plantations of the East African littoral – became the porters and placemen in the trans-Saharan trade; and they constituted the entourage – the veritable wealth and currency – of the notables of Islamic societies... they stand as 'marks of favor' in the gift of the strong, a kind of investment 'hedge' against the vagaries of the nomadic orb.[7]

The justification for enslavement under Islam, like that which would be adopted by European Christians, was religious conversion. Many slaves were prisoners of holy wars of religious expansion. The mechanism of slavery, for the first generation, was typically a violent act. Self-enslavement to repay debt or prevent starvation also occurred but was rare. Slavery was conceived as a form of religious tutelage and as a means of conversion and cultural assimilation. It was not a permanent condition but a temporary status which dwindled over successive generations.

Under Islam, manumission was relatively common, especially for slave concubines who bore children for their owners. A slave woman could not be involuntarily sold if she bore her owner a child; moreover, that child would receive its freedom. Such women could even produce heirs to ruling dynasties. Slave status passed through the father, not the mother, contrary to slavery based on Roman law. Women and children had the greatest chances for assimilation into free society. Male captives entered into military, government administration and domestic service. Commonly, male servants were castrated; the contrast between the eunuch and the female concubine highlights the slave's lack of control over sexuality and reproduction.

Islam penetrated sub-Saharan Africa through religious wars (*jihads*) and raids, and through long distance trade. Trading networks also contributed to the rise of large scale, interconnected states in the sahel and savannah belts. As sub-saharan Africa was incorporated into the Islamic fold, African kings and notables began making pilgrimages to Mecca, opening up a two-way slave trade.

Many scholars believe that the institution of slavery that emerged in black Africa borrowed from a Muslim model. Paul Lovejoy, for example, has conceived African slavery as an incorporative mechanism consistent with a lineage or domestic mode of production.[8] To elaborate, small-scale African societies were organized according to hierarchies determined by age, sex, and kinship. Elders dominated access to resources, especially women, who were important both in (agricultural) production and reproduction. Wealth and status was measured in people, not land or material possessions.

---

[7] John Ralph Willis, *Islam and the Ideology of Enslavement* (London, 1985), p. vii.
[8] Paul Lovejoy, *Transformations in Slavery: A History of Slavery in Africa* (Cambridge, 1983).

In sub-Saharan Africa, mechanisms of slavery included war, punishment for a serious crime or social transgression, and debt. Slaves served an important social function as kinless dependents at the bottom of the hierarchical heap. In Mali, for example, slaves were preferred as soldiers and bureaucrats because they were believed not to have outside loyalties. The acquisition of slave women was especially important in order to expand one's lineage. Kinless women who required no brideprice produced loyal offspring with no ties to competing relatives. These children would be gradually incorporated as full members of society, although some social stigma could linger for generations. The high value placed on women's productive and reproductive capabilities made Africans reluctant to sell them to European traders; hence one reason for the 2:1 predominance of males over females in the transatlantic trade.[9]

Slavery in Africa changed with the advent of the European slave trade. When European slave traders began to procure human captives in the fifteenth century, the costs to African societies were relatively low. Initially, European ideas about slavery were not internalized and a Muslim version (lineage slavery) continued. A pre-existing supply of domestic slaves filled the need. As demand grew, Africans shifted from a 'political' model of slavery, to an 'economic' one, in which wars began to be waged for the purpose of profit and the capture of prisoners-of-war which could be sold externally.[10] Market changes caused an increase not only in number of slaves but also the uses to which such slaves were put within and without Africa. Warfare and conflict increased in order to meet the ever-expanding demand of plantation societies in the New World.

The expanding transatlantic trade was key to the transformation of slavery *as an institution* which was essential to the economic functioning of society, rather than a marginal or incidental social relation which did not significantly effect the economy. This transition occured not only in European-dominated societies, but was internalized in Africa when the external slave trade came to a halt in the nineteenth century.

Transition to the Americas

Making the abundant lands and mineral resources of the Americas profitable for Europeans depended on capital investment and labour. Various labour options were available, but slavery quickly became the preferred solution for plantation owners. Indentured servitude proved to be impractical for a number of reasons. In Spain, the masses were earning decent wages and had little incentive to relocate. Portugal

---

[9] Joseph C. Miller has highlighted the shortage of women in the slave trade from the supply side point of view in *Way of Death: Merchant Capitalism and the Angola Slave Trade, 1730–1830* (Madison, 1988).

[10] Philip Curtin, Steven Feierman, Leonard Thompson and Jan Vansina, eds., *African History*, 5th impression (London and New York, 1984), pp. 221–4.

was underpopulated and migrants opted for Asia over Brazil. The English colonies did experiment with indentured servitude initially, but the risk of escape was high. In the Americas, European migrants preferred to strike out on their own than subject themselves to unremitting field labour and social subservience.

The next logical alternative was the appropriation of Native American labour. What would become the Americas had an estimated indigenous population of 57,300,000.[11] In the Caribbean, the Spanish attempted to enslave the peaceful Arawaks and the more fierce Carib Indians. Both groups succumbed to European diseases, violent assaults, and overwork and were virtually extinct by 1600. On the mainland the Spanish found more advanced Indian societies organized into complex city states – the Mexica (Aztec), Maya, and Inca civilizations predominating. Large populations of tribute-paying settled farmers in the Valley of Mexico and the Andean highlands could be profitably exploited by both the Spanish and their indigenous imperial predecessors. Although these areas also suffered demographic decline, they were able to recover over time.

Surviving indigenous peoples were subject to various forms of coerced labour: outright slavery gave way to royal grants of Indian labour known as the *encomienda* until pressure from religious activists like Bartolomé de las Casas forced its demise by 1550. These mechanisms were followed by obligatory rotational labour under the *repartimiento, mita*, and tribute payment systems. Indians worked in all sectors, from urban occupations, domestic servitude, trade, mining and diversified agriculture. Large scale, monocrop, agriculture was limited to a few areas such as the sugar plantations of Santo Domingo and Oaxaca and the cocoa plantations of Venezuela.

Unlike the Spanish, the Portuguese did not find settled indigenous populations in Brazil, but rather, semi-nomadic peoples who subsisted by swidden agriculture supplemented by hunting and gathering. The only immediately profitable resource found was brazilwood, which produced a reddish purple dye which commanded high prices in European markets. As Stuart Schwartz shows in his essay, 'Indian Labor and New World Plantations' (see chapter 2 below), Indians cut and transported this wood to the coast in exchange for barter items such as steel axe heads, fish hooks and other technically advanced items which would make their lives easier. The Tupí-speaking Indians were willing to engage in trade but not to subject themselves to plantation agriculture. The groups which practiced settled agriculture designated this activity as women's work while men were hunters, land clearers and warriors.

---

[11] Scholars still debate the size of the precontact indigenous populations of the Americas due to the fragmentary nature of historical documentation and the archaeological record. The above estimate comes from William M. Denevan, ed., *The Native Population of the Americas in 1492* (Madison, 1976), p. 291, cited in James Lockhart and Stuart B. Schwartz, *Early Latin America: A History of Colonial Spanish America and Brazil* (Cambridge, 1983).

Once the Portuguese crown realized that Brazil had little potential as a trading post colony, it opted for settlement, much along the lines of the pattern laid out in the Atlantic Islands. Donatory grants were extended to members of the minor nobility and the imperial bureaucracy. To make the land profitable, labour was needed. The Portuguese in Brazil used the concept of 'just war' to justify the enslavement of Indians, preying on tribes who had allied themselves with the French and 'rescuing' prisoners-of-war who had been captured by allegedly cannibalistic groups.[12] For the Indians, the only alternatives were flight or conversion to Christianity, acculturation to European norms, and settlement in one of the Jesuit *aldeias*, proto-peasant villages where Indians were introduced to the joys of settled agriculture.

In both Spanish and Portuguese America, indigenous peoples suffered massive demographic decline due to exposure to previously unknown European disease. Smallpox, measles, and influenza had devastating effects, in some areas, mortality rates of up to 90 per cent. Disease would be one of the critical factors which led Europeans to exploit African slave labour;[13] Africans, in addition to having been exposed to the same disease pool as Europeans also had extra immunities against tropical diseases that Europeans lacked. In addition to being immunologically superior, Africans also had advanced metal working and mining skills and were accustomed to large scale agriculture.

Africans were readily identifiable because of their race and were therefore more easily controlled and less able to escape. American slavery became unique in the history of the institution both by linking race to the servile condition as well as being unprecedented in its cruelty, oppression, and intensity. Peoples of African descent unfortunately hold the unique distinction of being victims of this new form of institutionalized slavery.

The Portuguese and the Spanish were the first European powers to embrace African slavery on a large scale. Over the course of the transatlantic trade, an estimated total of 1.55 million slaves arrived in Spanish America and 3.5 million were sold in Brazil. Columbus brought black slaves and sugar cane cuttings to the Caribbean on his voyages, and the first New World sugar plantations emerged on the island of Hispaniola. Mervyn Ratekin, in his essay 'The Early Sugar Industry in Española', demonstrates the early challenges that the

---

[12] Peter Hulme and Philip Boucher persuasively make the contention that the cannibalistic reputation of indigenous Caribs and Tubinambas may have been invented or distorted by European conquerers. See Peter Hulme, *Colonial Encounters: Europe and the Native Caribbean, 1492-1797* (London, 1986) and Philip P. Boucher, *Cannibal Encounters. Europeans and Island Caribs, 1492-1763* (Baltimore and London, 1992).

[13] See Kenneth Kiple, *The Caribbean Slave: A Biological History* (Cambridge, 1984); Alfred Crosby, *The Columbian Exchange: Biological and Cultural Consequences of 1492* (Westport, 1972); and Philip D. Curtin, 'The Epidemiology of the Slave Trade', *Political Science Quarterly* LXXXIII (1968), pp. 190–216.

plantation sector faced in the Spanish Caribbean (see *An Expanding World*, vol. 7, ch. 10).

On the mainland, slaves filled the gap between conquerers and vanquished Indians. For example, in regions dominated by nomadic tribes like Zacatecas (New Spain), imported labour was needed to work the silver mines. African slaves were most heavily employed in mining, textile workshops (*obrajes*), cattle ranches, and plantations.[14] They laboured in lesser numbers as domestic servants, artisans and petty traders. Sugar plantations were introduced in the late 1520s and flourished in the southern lowlands around Veracruz and Oaxaca. By 1533, Africans were being imported specifically to work on the plantations. Most seventeenth-century plantations were small, averaging 20–40 slaves, but estates could exceed 200 captives. Most slaves laboured in the fields but a few practiced subsidiary skilled occupations associatated with the running and maintenance of the mill. Cocoa plantations also flourished in the lowlands of New Spain and in coastal Venezuela, the subject of Robert Ferry's essay (see chapter 5 in this volume).

In the viceroyalty of Peru, African slaves played an active role in the conquest and served as military auxiliaries during the period of the civil wars. The mountainous terrain and difficulties of overland and coastal transport via Panama did not favour the emergence of large scale plantation sectors. African slaves were used in small-scale agriculture and ranching, in the service industries, on public works, and as domestics, artisans, fishermen, construction workers, and muleteers. They did not labour in the silver mines because they adapted poorly to the cold and high altitude, but did pan gold in the lowlands of present-day Colombia.[15] Agricultural estates based on slave labour did emerge after 1600, dominated by religious orders.[16] At its height, the slave population in Peru reached 100,000, comprising 10–15 per cent of its population, compared to New Spain's peak of only 35,000; less than 2 per cent of its viceregal population.

The donatory lords of Brazil began to make the transition to black slavery as early as the 1550s and it was well underway by the 1580s. By the 1580s, northeastern Brazil boasted more than 100 sugar mills, producing two-thirds of all American sugar and effectively undercutting the sugar plantations of Hispaniola. By then, roughly one-third of the plantation labour force was African; by 1600 they comprised half of all sugar workers and by 1650, the transition to African slave-produced sugar on the coast would be nearly complete.[17] Colin

---

[14] The standard source on slavery in New Spain is Colin A. Palmer, *Slaves of the White God: Blacks in Mexico, 1570–1650* (Cambridge,1976).

[15] Frederick Bowser, *The African Slave in Colonial Peru* (Stanford, 1979).

[16] Nicholas P. Cushner, 'Slave Mortality and Reproduction on Jesuit Haciendas in Colonial Peru', *Hispanic American Historical Review* LV, no. 2 (1975), pp. 177–99.

[17] Herbert S. Klein, *African Slavery in Latin America and the Caribbean* (New York, 1986).

MacLachlan shows, however, in his essay, 'The African Slave Trade and Economic Development in Amazonia, 1700–1800', that indigenous labour still occupied a crucial place in the regional economy of the Amazon as late as 1750, due to the continued availability of indigenous labour, and the high transportation and mortality costs of shipping slaves inland (see chapter 4 in this volume).

By 1650, 350,000 Africans had arrived in Spanish America and 250,000 in Brazil. In comparison, the French and British colonies had absorbed maybe 25,000 more. British colonists only turned to African labour in the mid-seventeenth century because they could not afford the overpriced captives sold by Dutch smugglers who had managed to evade the Portuguese and Spanish slaving monopoly. As England developed into a dominant maritime power, however, it was able to enter into the slave trade itself. Barbados and Jamaica became dominant sugar producers, and the Cheasapeake and tidewater Virginia focused on tobacco plantations in the seventeenth century.

The Dutch occupation of the Brazilian northeast from 1630–1654 undercut Brazilian supremacy as a sugar producer, especially after the Dutch withdrew and began to establish plantations of their own. Dutch capital and technical know-how benefited other nations as well. British Barbados, for example, enjoyed a clean slate advantage. It was able to invest in the most modern machinery and methods while Brazil depended on outmoded mills operating at reduced efficiency. The French and Danes followed suit in the eighteenth century, developing competitive planter societies in the West Indies. The Caribbean was thus transformed into a viable economy, moving beyond its original function as a strategic base from which to trade with the mainland.

The Decline of Slave Plantations in the Americas

The late eighteenth and early nineteenth centuries represent a period of broad social, political and economic change. Decades of revenues from the production and sale of tropical exports and New World mining had provided the capital accumulation necessary to fund the Industrial Revolution in Europe. Enlightenment thought challenged the colonial mercantilist system. Free trade and expanding markets overseas were necessary to absorb the increase in manufacturing within Europe. New political ideas advocating representative government (for the propertied and literate) and legal equality for all men began to take hold. The French, North American, and Haitian Revolutions, followed by the independence struggles in Spanish and Portuguese America represented tangible manifestations of these changing ideas.

The Haitian revolution (1791–1804) also served as a cautionary tale of the dangers inherent in the brutal exploitation of slaves. The French colony of St Domingue had become the premier sugar island in the latter part of the eighteenth century. The conflict was grounded in social divisions between slaves, free blacks, free mulattoes, *petit blancs* (lesser or petty whites) and *grand blancs* (the dominant planter and administrative class). The revolution began among the white elite in 1789 when they demanded representation in the Estates General in France and set up local assemblies which excluded from membership those owning fewer than 20 slaves. The *petit blancs* then revolted against this elite, starting a civil war. Then well-to-do mulattoes became involved by sending representatives to Paris to fight for equality under the Rights of Man. Each faction convinced themselves that they were the true supporters of the Revolution in France.

The whites maintained the upper hand until 1791, when slaves in the north rose up and destroyed numerous plantations. Both whites and mulattos sought to bring slaves over to their side, but in continuing to ignore black needs allowed the slaves to take over the struggle and make it their own. Shifting alliances of whites, mulattoes, and slaves prevailed until one of the slave leaders, Toussaint Louverture gained an advantage. The demographic superiority of the black slave population, which outnumbered both the whites and free coloureds by a factor of twenty, would eventually prove to be decisive.[18] Independence was secured by one of Toussaint's allies in 1804, although formal international recognition would take decades to achieve. The revolt made tangible what the planter class had long feared, that the numerically superior enslaved population could not only destroy the plantation complex itself, but also abolish slavery and declare political independence.

Slave resistance was hardly a new phenomenon in the New World. Runaway slave communities in the Americas did not take long to emerge after the introduction of Africans. Although most such communities enjoyed an ephemeral existence, those that held out long enough against military pressure could and did negotiate treaties with dominant societies for formal recognition, amnesty, and political autonomy. The quintessential maroon society was the Brazilian *quilombo*, Palmares, which endured from 1605 to 1695, achieved a population in excess of 20,000, created a diversified, nearly self-sufficient economy, political stability, and held out against decades of sustained Portuguese attack. Other forms of resistance, both passive and violent, were endemic wherever plantations could be found.[19]

The Haitian Revolution had a profound effect upon slaveholding societies throughout the Americas. As long as blacks outnumbered European whites, which was the case in the New World until 1840, the planter elite had reason to fear.[20] New World anxieties combined with the rapidly changing ideological context in Europe resulted in intellectual and political challenges to the institution of slavery.

---

[18] In 1787, whites numbered 24,000; *gens de couleur*, 20,000; and slaves, 408,000. Thomas O. Ott, *The Haitian Revolution, 1789–1804* (Nashville, 1973).

[19] See the edited volume by Richard Price, *Maroon Societies: Rebel Slave Communities in the Americas* (Baltimore, 1979); and the synthesis by Eugene Genovese, *From Rebellion to Revolution: Slave Revolts in the Making of the Modern World* (Baton Rouge, 1979).

[20] Curtin, *Plantation Complex.*

Slavery did not fit within the changing world order for a number of reasons. Slaves did not earn wages and therefore could not become consumers of European manufactured goods. Economic theorists began to argue that slavery was inefficient, that slaves were incapable of learning complex tasks, and that plantation slavery itself was becoming unprofitable. Although contemporary scholars have since disproved many of these assumptions, they nonetheless served as persuasive tools of propaganda at the time.[21] The growing strength of evangelical Christian sects in England, such as the Quakers and Methodists, added a moral dimension to attacks on the slave trade and slavery.[22]

Great Britain outlawed the slave trade in 1807 and eliminated the institution of slavery in its colonies in 1834. Many of the Spanish American colonies followed suit during their independence struggles in order to win blacks over to the rebel side. The British imposed cessation of the trade through unilateral decrees and bilateral treaties, often demonstrating flagrant disregard of international law in the process. In the Americas, islands like Cuba and Puerto Rico, which enjoyed a clean slate advantage, held out longer than did islands like Jamaica and Barbados where the land was rapidly becoming depleted. Cuba and Brazil were the last holdouts. Philip D. Curtin has argued that the failure of the Democratic Revolution in Cuba and Brazil was partially responsible for the continuation and expansion of the plantation complex there in the nineteenth century.[23] Brazil became a constitutional monarchy in 1822 and became a Republic in 1889, not coincidentally a year after Princess Isabel abolished slavery. Cuba abolished slavery in 1886, but remained a colony of Spain until 1898.

Ironically, the British-led abolition of the slave trade and the eventual emancipation of slaves in the Americas led to an extension of the 'plantation complex' to new areas. In the nineteenth century, the plantation system shifted from the Caribbean to untapped regions like Natal, Réunion, and Mauritius, in South and East Africa and the South Pacific and Hawaii. Plantations diversified beyond sugar to encompass coffee and bananas in Central America; cotton and rice in the U.S. South; cocoa, coffee, rubber, and cotton in Brazil, and tea in India, to name just a few examples. Instead of slaves, European planters predominantly used various forms of coerced labour, contract labour of Asians and the subsidized labour of immigrant Europeans.

---

[21] Seymour Drescher, *Econocide: British Slavery in the Era of Abolition* (Pittsburgh, 1977); David Eltis, *Economic Growth and the Ending of the Transatlantic Slave Trade* (New York, 1987); and Barbara L. Solow and Stanley L. Engerman, eds., *British Capitalism and Caribbean Slavery: The Legacy of Eric Williams* (Cambridge, 1987).

[22] See Roger Anstey, *The Atlantic Slave Trade and British Abolition* (London, 1975); and Betty Fladeland, *Abolitionists and Working Class Problems in the Age of Industrialization* (Baton Rouge, 1984).

[23] Philip D. Curtin, *Plantation Complex*, p. 156.

In Africa, the abolition of the external trade and the rise of legitimate trade served to internalize the slave market and the 'plantation complex'. Prior to 1700, slaves had not been used extensively in large-scale agriculture in Africa.[24] The closing of the external trade, however, did not halt slave generating activities like warfare. The moving frontier of Islam as typified by leaders like Samori Ture in Senegambia and Uthman dan Fodio in Northern Nigeria created new sources of slaves and the administration to use them effectively. African slaves came to labour on palm oil, cocoa and rubber plantations in West Africa and on clove plantations on the Omani Arab-controlled islands of Zanzibar and Pemba.[25]

Historiography: Scholarship on Comparative Slavery

This brief historical overview of what Philip D. Curtin has dubbed the 'plantation complex' may give the impression that the development of plantation colonies by Europeans overseas was a uniform process. Within the parameters outlined here, however, significant variations exist. European nations differed in their legal codes governing slavery, in the organization of their plantations, the treatment of slaves, demographic patterns, and a host of other variables.

Interest in comparative slavery expanded after the publication of Frank Tannenbaum's *Slave and Citizen: The Negro in the Americas* in 1946. Both Tannenbaum and Stanley Elkins, who published *Slavery: A Problem in American Institutional and Intellectual Life* ten years later, emphasized the 'moral personality' of the slave, contrasting differences between Protestantism and British common law with Catholicism and Roman legal traditions to explain cross-cultural variation in slave treatment. The Portuguese, Spanish and French had pre-existing metropolitan legal codes which contained regulations for slavery. British law did not, however, and so slave legislation had to be invented on the spot in its colonies. It was argued that Iberian slavery was less harsh than British slavery, because the more inclusive legal and ecclesiastical legislative protectionism benefitted slaves.

Sidney Mintz (see chapter 12 in this volume) has redefined the question, asking not whether antecedents to slavery existed in the metropole, but rather whether or not they were transferred effectively and enforced systematically in the Americas.[26] Mintz argues that the extent of capitalist development in the Americas determined the relative power of the planter class relative to the metropole, and consequently, their ability to shape policies affecting slave labour. Also crucial was the degree to which representative government was possible at

---

[24] John Fage, 'Slaves and Society in Western Africa, c. 1445–c. 1700', *Journal of African History* XXI, no. 3 (1980), pp. 289–310.

[25] Curtin, *Plantation Complex*, conclusion.

[26] 'Review of Stanley M. Elkins' *'Slavery'*, *American Anthropologist* LIII (1961), pp. 579–87.

a local level. In this volume (see chapter 11), Stinchcombe adopts a position similar to that of Sidney Mintz, emphasizing the strength of local government and as determinants of 'freedom' in the eighteenth-century Caribbean.

David Brion Davis has been yet another proponent of materialist approaches and he has discouraged the comparison of legal systems.[27] Frederick Bowser has applied a similar critique to slave studies done in Spanish America, advocating research which goes beyond the formal legal codes of the Siete Partidas, the *Recopilaciones*, and the Spanish slave code of 1789.[28] Robert Conrad's *Children of God's Fire*, for example, provides ample evidence that the legal rights of slaves were honored more in breach than in practice. Court proceedings in the Americas provide some of the richest data concerning the inner world of slaves and the living law which shaped the boundaries of their existence. A synthesis between the law as practiced in the Americas with the law as constructed in the metropole(s) might enable us to mesh local realities with structural conditions.

Formalist legal studies on the institution of slavery fell into disfavour as scholars recognized the disparity which often existed between legal codes and actual practices. In the 1970s, a new school of comparative studies emerged, this one based on computerized analysis of quantitative data, the so-called 'cliometric school', launched by Robert W. Fogel and Stanley Engerman, in their book, *Time on the Cross*. This approach quickly fell into disfavour for its manipulation of statistical data at the expense of other forms of evidence. In this regard, Richard S. Dunn has commented:

> Quantification itself is surely not at issue; the historian who wishes to interpret slave records cannot get very far without employing techniques of aggregative analysis. What is at issue is the cliometricians' habit of counterfactual hypothesis, their manipulation of synthetic figures extrapolated from mathematical models, and their certitude that by such tactics they can 'correct' the 'errors' of previous interpreters'.[29]

Dunn's essay in this volume (see chapter 8) represents an attempt to move beyond the distortions of the cliometric approach. His piece uses aggregation of quantitative data over time from two plantations to reconstruct the social

---

[27] 'A Comparison of British America and Latin America', in *The Problem of Slavery in Western Culture* (Ithaca, 1966).

[28] Frederick P. Bowser, 'The African in Colonial Spanish America: Reflections on research achievements and priorities', *Latin American Research Review* VII, no. 1 (1972), pp. 77–92.

[29] See his essay in this volume: 'A Tale of Two Plantations: Slave Life at Mesopotamia in Jamaica and Mount Airy in Virginia, 1799–1828', *William and Mary Quarterly* XXXIX, no. 1 (1977), here p. 216. A more sustained critique may be found in Herbert G. Gutman, 'The World Two Cliometricians Made', *Journal of Negro History* LX (1975), pp. 54–227, reprinted as *Slavery and the Numbers Game: A Critique of Time on the Cross* (Urbana, 1975).

dimension of slave life. Although the study falls slightly beyond the periodization of this series, it was included for its theoretical and methodological significance.

The cliometric school, despite its flaws, has led to a greater use of quantitative and demographic data in comparative slave studies. The study of disease, morbidity and mortality rates across regions has demonstrated that in most plantation societies, the slave population suffered imbalanced sex ratios and did not reproduce itself naturally, providing clear evidence of the physical harshness of slavery. The U.S. South was anomalous in this regard in that it enjoyed rates of natural increase comparatively early. Other demographic exceptions include Barbados which adopted a conscious policy of favoring female imports, and eighteenth century Minas Gerais, Brazil, which made a transition from labour intensive mining to a more diversified and less exploitative economy based on mixed agriculture, ranching, craft production, and internal trade.[30]

Kenneth Kiple has attempted to enter the inner world of slaves through analysis of the systemic factors of diet and disease. Kiple recognizes an intellectual debt to Philip D. Curtin, commenting on Curtin's discovery that slaves in the U.S. South enjoyed high rates of natural increase while their Caribbean counterparts experienced demographic decline,

> This revelation had, among other things, the effect of redirecting scholarly attention from the comparative institutional and economic advantages and disadvantages of a particular slave system for the slave toward confronting directly fundamental questions of life, death and physical well being.[31]

Comparative slavery also began to borrow from social history, examining previously under-utilized sources such as baptism, marriage and death records; household censuses; court proceedings, notary documents including wills and contracts; and manumission records in order to attempt to reconstruct the inner lives of the slaves as well. The subject of manumission, elaborated upon by Tannenbaum in his comparative legal study, has been turned upside down by scholars such as Katia de Queirós Mattoso who consciously has criticized broad structural variables such as Church, State, and capitalism. In *To Be a Slave in Brazil*, she explores the pychological conflict experienced by slaves in opting for the possibility of manumission. Assimilating to white values in order to increase one's chances meant alienation from Afro-Brazilian slave society, resulting in perpetual cultural alienation. For slave owners, manumission served an effective form of social control and the possibility of an added return on one's initial investment.

---

[30] Hilary McD. Beckles, *Natural Rebels: A Social History of Enslaved Women in Barbados* (London, 1989); and Roberto B. Martins, 'Growing in Silence: The Slave Economy of Nineteenth-Century Minas Gerais, Brazil', Ph.D. Diss. (Vanderbilt University, 1980).

[31] *The Caribbean Slave: A Biological History* (Cambridge, 1984), p. 105.

Research on manumission is also connected to a greater scholarly concern about the experience of free people of colour in slave societies. The emergence of free people of colour challenged the bipolar categories of black and white. They faced legal and economic marginality in most plantation societies. Tensions amidst this caste could lead to violent conflict, embodied in the Tailors' Revolt in Brazil in 1798 and most dramatically, in the Haitian Revolution. The treatment and opportunities open to manumitted slaves say a great deal about the definition of 'freedom' and the openness or rigidity of particular slave societies. Historians have explored corporate groups created by slaves and free coloureds including religious brotherhoods and mutual aid societies.[32] Interest about the preservation of African-American religious beliefs, rituals, and culture has also expanded.

Some scholars have attempted to recreate the intimate details of slave existence by focusing on gender, family, and the household economy. Much attention has been drawn to the 'peasant breach' discussed by Sidney Mintz in his essay (see chapter 12 in this volume), 'Was the Plantation Slave a Proletarian?' Planters in the Caribbean and Brazil often gave their slave families usufruct rights to small plots of land which were unfit for sugar cultivation. Slaves worked these provision grounds one day a week and on holidays. They were to feed themselves by their own labours but could market the surplus to gain cash. In Jamaica, slave markets became so extensive that they eventually controlled 20 per cent of the coin in circulation.[33] Karen Fog Olwig (see chapter 9 in this volume) also reveals the existence of active slave marketing on the island of St John in the eighteenth century.

This relative 'freedom' and autonomy given to slaves can also be seen in a more negative light. Marietta Morrissey, in her study, *Slave Women in the New World: Gender Stratification in the Caribbean*, focuses on the household as the nexus of non-capitalist and capitalist production. The slave household economy, a 'precapitalist' site of production and reproduction, was embedded within world capitalist relations of production. Intensification of plantation production could mean erosion of the slave's domestic and economic autonomy, if the owner chose to substitute the functions of individual households and garden plots with collectivized housing, child care, cooking and provisioning. Her study provides

---

[32] Some representative studies are: David W. Cohen and Jack P. Greene, eds., *Neither Slave nor Free, the Freedmen of African descent in the Slave Societies of the New World* (Baltimore, 1972); Ira Berlin, *Slaves Without Masters: The Free Negro in the Ante-bellum South* (New York, 1974); Jerome S. Handler, *The Unappropriated People: Freedmen in the Slave Society in Barbados* (Baltimore, 1974); and A.J.R. Russell-Wood, *The Black Man in Slavery and Freedom in Colonial Brazil* (New York, 1982).

[33] Sidney W. Mintz, *Caribbean Transformations* (Chicago, 1974); and Sidney W. Mintz and Douglas Hall, 'The Origins of the Jamaican Internal Marketing System', *Yale University Publications in Anthropology* LVII (1960), pp. 1–26.

a gendered elaboration to Mintz's question, 'Was the Plantation Slave a Proletarian?'

This volume represents an attempt to capture some of the complexity and richness of the comparative study of slave plantation societies created by Europeans and African migrants overseas. Surprisingly, selection of the essays was difficult, not due to an overabundance of material, but rather, through a shortage of article-length studies which fit into the periodization of this series (1450–1800). There are several facets to this deficiency of published sources. Although monographs on plantation societies abound, published articles do not. Scholarship also becomes more abundant as one progresses in the historical continuum to the present. Finally, the extension of plantation systems to European colonies and former colonies in Africa, India, Southeast Asia and the Pacific was largely a nineteenth and twentieth century phenomenon and therefore falls out of the range of our study.[34] So too does the expansion of coffee plantations in nineteeth-century Central America and Brazil which employed a variety of semi-coerced forms of labour in its production for export. Very little work exists outside the confines of the Americas for the fifteenth to eighteenth centuries in any language, much less those European languages contained in this series. If nothing else, this volume serves to demonstrate the lacunae existing in the scholarly literature on plantation systems during the era of European expansion.

Finally, although academic journals still continue to produce enormous amounts of scholarship on *slavery*, research on *plantations* appears to be on the wane. A lively theoretical debate concerning the differentiation of the plantation and hacienda in Spanish America intrigued scholars throughout the 1960s and early 1970s.[35] Interest in the slave plantation as a unit of analysis seems to have peaked in the late 1970s with the publication of Vera Rubin and Arthur Tuden's comprehensive and authoritative edited volume, *Comparative Perspectives on Slavery in the New World*. Symptomatic of this declining interest is the fact that the journal, *Plantation Society in the Americas*, founded in 1979 with the intention of focussing on the plantation as a unit of analysis, has published only eight issues to date. The inaugural volume stated: 'Plantation Society hopes, lastly, to

---

[34] James L. Watson, ed., *Asian and African Systems of Slavery* (Berkeley, 1980); Suzanne Miers and Igor Kopytoff, eds., *Slavery in Africa: Historical and Anthropological Perspectives* (Madison, 1977); and Curtin, 'Plantation Complex', conclusion.

[35] Eric Wolf and Sidney W. Mintz, 'Haciendas and Plantations in Middle America and the Antilles', *Social and Economic Studies* VI (1957), pp. 380–412; Shane J. Hunt, 'La economia de las haciendas y plantaciones en America Latina', *Historia y cultura* IX (1975), pp. 7-66; James Lockhart, 'Encomienda and Hacienda: the evolution of the Great Estate in the Spanish Indies', *Hispanic American Historical Review* XLIX, no. 3 (1969), pp. 411-29; Jay R. Mandle, 'The Plantation Economy: An Essay in Definition', *Science and Society* XXXVI, no. 1 (1972), pp. 49-62; Robert G. Keith, ed., *Haciendas and Plantations in Latin American History* (New York, 1977).

foster the development of studies of the Europeanization of the New World. The commendable trend of recent decades towards greater interest and sophistication in Afro-American studies has not been matched by similar attention to the role of Europeans in the plantation world'.[36] Had scholars taken up the challenge offered in *Plantation Societies*, this volume might have been very different.

The essays in this volume trace the development of slave plantations from their Mediterranean antecedents to the transition to free labour in the Americas in the nineteenth century. Sidney M. Greenfield, in his essay, 'Plantations, Sugar Cane and Slavery', outlines the trajectory of sugar cultivation, technology and production from the medieval Middle East to the era of European maritime expansion (see chapter 1 in this volume). In doing so, he poses two fundamental questions:

> where, when and how did the plantation system first develop; and how did it come to be introduced into tropical America so as to become the central institution around which society and culture was to be organized to replace what had been found and destroyed?[37]

Greenfield demonstrates that plantation societies came to serve not only as sources of wealth for expanding nation states but as a means of settling new territories that were of strategic value.

The extension of slave plantations from the Atlantic Islands to Portuguese America is detailed in the articles by Schwartz, Graham and MacLachlan. Schwartz addresses the transition from indigenous to African labour in the northeastern sugar regions. Graham examines a late-eighteenth century census from a mature estate to demonstrate that slave owners implicitly recognized the humanity of the slaves by enumerating them in family units. MacLachlan also deals with the transition from Indian to black slavery, in this case in the Amazon. He shows that the inability of Portuguese administrators to recognize that the slave plantation model was not a universal model for economic success led to the adoption of an economic policy doomed to failure.

The shift from indigenous to African labour is also addressed in two selections dealing with colonial Spanish America. Mervyn Ratekin describes the shaky implementation of the first sugar plantations in early sixteenth-century Española and the quick rise of the local planters to local political and economic dominance (see *An Expanding World*, vol. 7, ch. 10). Ferry details the emergence and capitalization of cacao plantations on coastal Venezuela. Pre-existing stands of trees located on coastal *encomiendas* produced enough capital for the

---

[36] *Plantation Society in the Americas* I, no. 1 (1979), p. 6. See also the introductory essay in the same volume by Carl N. Degler, 'Plantation Society: Old and New Perspectives on Hemispheric History', pp. 9–14.

[37] Greenfield, 'Plantations', p. 85.

encomenderos to purchase slaves and weather the transition from indigenous to African labour.

The remaining essays deal with coastal North America and the Caribbean. Many of them employ quantitative methodology to draw conclusions about slave existence. These articles bear out the generalization that slaves in mainland North America experienced natural increase. Littlefield argues that this demographic pattern was due in part to a greater willingness for North American planters to purchase female slaves. In South Carolina, it took rice farmers five years to regain their capital investment on their slaves so they could not afford to work them to death as sugar planters in the Caribbean did. Moreover, seasoned, acculturated, trained slaves could be sold for significantly more than raw Africans. Taking good care of one's slaves was a sound investment. Morgan and Nicholls found that eighteenth century Piedmont Virginia, despite being a frontier area, achieved parity in the sex ratios of its slaves and positive rates of increase relatively quickly, again due to a conscious policy of importation of females.

In contrast, Geggus examines the demographic profile of two plantations of the northern plains of St Domingue immediately prior to the Haitian Revolution (see chapter 10 in this volume). His findings demonstrate a slave society dominated by Africans characterized by imbalanced sex ratios, a low fertility rate, and endemic illness. Dunn's findings are mixed. In his comparison of two fourth-generation plantations, Mount Airy in Virginia and Mesopotamia in Jamaica, he found natural increase at the former and decrease at the latter. The Mount Airy slaves demonstrated a more diversified occupational structure and a less intensive work regime, contributing to higher rates of reproduction (see chapter 8 in this volume). Finally, Karen Fog Olwig shows that Danish slaveowners, although they ideologically demonized their slaves, nonetheless recognized their humanity by caring for sick and permanently disabled slaves, permitting marriage both on and off the plantation, and passing relatively humane slave codes. Her analysis is not explicitly demographic but uses local legislation and court proceedings to reconstruct slave life, kinship, and race relations (see chapter 9).

The final two selections construct broader theoretical arguments. Stinchcombe identifies a series of variables to determine the relative harshness of slave systems in the Caribbean (see chapter 11). These include the importance of sugar, the degree of political and economic organization of the planter class, the relative influence of metropolitan legal systems, and the timing of the 'frontier' period. If the peak of the frontier came before 1750, then planters could have the opportunity to develop political and social institutions which would narrowly restrict the liberties enjoyed by slaves. The two extremes would be represented by the Dutch Antilles which were devoted largely to commerce and expended relatively little effort to restrict slaves and Barbados which converted to sugar early and restricted options for slaves such as access to provision grounds and manumission.

Sidney W. Mintz attempts to reinsert slave plantations within Marx's theory of capitalism. Mintz asserts,

> ...it is not analytically most useful to define either 'proletarian' or 'slave' in isolation, since these two vast categories of toiler were actually linked intimately by the world economy that had, as it were, given birth to them both, in their modern form.[38]

He divides the historical evolution of New World plantations into five phases; in only one of these, from 1650–1750, did 'pure slavery' predominate. Mintz concludes that slaves do not offer their labour as a commodity; they are themselves commodities or a form of capital. Nonetheless, in the marketing of provision ground surpluses, they could potentially sell the fruits of their labour. Mintz reveals the ambiguity of the plantation slave within the expanding world economy and the transition to a rural proletariat in the nineteenth century.

The plantation systems that emerged in European colonies had far reaching effects. Tropical exports such as sugar, tobacco and cotton provided the capital necessary to fund the Industrial Revolution. A relatively small elite planter class made money through extreme human exploitation and suffering, although profits were not disproportionate, compared to other investment opportunities available. The human cost, however, was extremely high.

Hopefully, these essays avoid one of the early pitfalls of comparative slavery, determining that some systems were 'better' than others. These studies demonstrate that plantation societies were exploitative and inhumane, and if value judgements need be made, one should talk of varying degrees of cruelty, not relative benevolence. The creation of plantation societies by Europeans resulted in the forced migration, death, and suffering of millions of Africans. The slave trade also stimulated endemic violence and depopulation within African societies. Finally, institutional slavery as it emerged in the Americas was racially constructed and has left a legacy of racism and discrimination that continues to plague former European colonies to the present day. We must ask whether the wider availability of cheaper coffee, sugar, tobacco, and cocoa in European markets was worth the price in capital investment and human lives.

---

[38] See chapter 12 in this volume, 'Was the Plantation Slave a Proletarian?', p. 321.

Select Bibliography

Historiographical

Davis, David Brion, *The Problem of Slavery in Western Culture* (Ithaca, 1966).
Elkins, Stanley, *Slavery: A Problem in American Institutional and Intellectual Life*, 3rd edn. (Chicago, 1956).
Engerman, Stanley L., and Genovese, Eugene D., *Race and Slavery in the Western Hemisphere: Quantitative Studies* (Princeton, 1975).
Fogel, Robert W., and Engerman, Stanley, *Time on the Cross* (Boston, 1974).
Freyre, Gilberto, *The Masters and the Slaves*. Samuel Putnam, trans., 2nd edn. revised (Berkeley, 1986).
Genovese, Eugene, *The Political Economy of Slavery* (New York, 1968).
Gutman, Herbert, *Slavery and the Numbers Game: A Critique of Time on the Cross* (Urbana, 1975).
Patterson, Orlando, *Slavery and Social Death: A Comparative Study* (Cambridge, 1982).
—, *The Sociology of Slavery* (London, 1967).
Stampp, Kenneth, *The Peculiar Institution* (New York, 1956).
Tannenbaum, Frank, *Slave and Citizen: The Negro in the Americas* (New York, 1946).
Williams, Eric, *Capitalism and Slavery* (London, 1964).

General and Comparative

Curtin, Philip D., *The Rise and Fall of the Plantation Complex* (Cambridge, 1990).
Deerr, Noel, *The History of Sugar*, 2 vols. (London, 1949–50).
Galloway, J.H., *The Sugar Cane Industry: An Historical Geography from its Origins to 1914* (Cambridge, 1989).
Emmer, Pieter C., ed., *Colonialism and Migration: Indentured Labour Before and After Slavery* (Dordrecht, 1986).
Genovese, Eugene, *From Rebellion to Revolution: Slave Revolts in the Making of the Modern World* (Baton Rouge, 1979).
Hall, Gwendolyn M., *Social Control in Slave Plantation Societies: A Comparison of St. Domingue and Cuba* (Baltimore, 1971).
Kiple, Kenneth F., *The Caribbean Slave: A Biological History* (Cambridge, 1984).
Klein, Herbert S., *African Slavery in Latin America and the Caribbean* (New York, 1986).
—, *Slavery in the Americas: A Comparative Study of Virginia and Cuba* (Chicago, 1967).
Miller, Joseph C., *Slavery: A Worldwide Bibliography, 1900–1982* (White Plains, 1985).
Mintz, Sidney, *Caribbean Transformations* (Baltimore, 1976).

—, *Sweetness and Power: The Place of Sugar in Modern History* (New York, 1985).
Morrissey, Marrieta, *Slave Women in the New World: Gender Stratification in the Caribbean* (Lawrence, 1989).
Phillips, William D., *Slavery from Roman Times to the Early Transatlantic Trade* (Minneapolis, 1980).
Price, Richard, *Maroon Societies: Rebel Slave Communities in the Americas* (Baltimore, 1979).
Rubin, Vera and Tuden, Arthur, eds., *Comparative Perspectives on Slavery in New World Plantation Societies* (New York, 1977).
Saunders, Kay, ed., *Indentured Labour in the British Empire, 1834–1920* (London, 1984).
Solow, Barbara L., ed., *Slavery and the Rise of the Atlantic System* (New York, 1991).
Watson, James L., ed., *Asian and African Systems of Slavery* (Berkeley, 1980).

Histories According to Region

Africa

Cooper, Frederick, 'The Problem of Slavery in African Studies', *Journal of African History* XX (1979), pp. 103–125.
Kopytoff, Igor, and Miers, Suzanne, eds., *Slavery in Africa: Historical and Anthropological Perspectives* (Madison, 1977).
Lovejoy, Paul, *Transformations in Slavery: A History of Slavery in Africa* (Cambridge, 1983).
Manning, Patrick, *Slavery, Colonialism, and Economic Growth in Dahomey, 1640–1960* (Cambridge, 1982).
Robertson, Claire, and Klein, Martin A., eds., *Women and Slavery in Africa* (Madison, 1983).

Brazil

Conrad, Robert, *Children of God's Fire: A Documentary History of Black Slavery in Brazil* (Princeton, 1983).
Dean, Warren, *Rio Claro: A Brazilian Plantation System, 1820–1920* (Stanford, 1976).
Marchant, Alexander, *From Barter to Slavery: The Economic Relations of Portuguese and Indians in the Settlement of Brazil, 1500–1580* (Baltimore, 1942).
Mattoso, Katia Queiroz, *To Be a Slave in Brazil* (New Brunswick, 1985).
Schwartz, Stuart, *Sugar Plantations in the Formation of Brazilian Society* (Cambridge, 1985).

Stein, Stanley J., *Vassouras: A Brazilian Coffee County, 1850–1900* (Cambridge, 1957).

British Caribbean

Beckles, Hilary McD., *Natural Rebels: A Social History of Enslaved Women in Barbados* (London, 1989).
Bush, Barbara, *Slave Women in Caribbean Society, 1650–1838* (Kingston, 1990).
Craton, Michael, *Testing the Chains: Resistance to Slavery in the British West Indies* (Ithaca, 1982).
Dunn, Richard S., *Sugar and Slaves: The Rise of the Planter Class in the English West Indies, 1624–1713* (Chapel Hill, 1972).
Goveia, Elsa, *Slave Society in the British Leeward Islands at the End of the Eighteenth Century* (New Haven, 1965).
Higman, Barry, *Slave Population of the British Caribbean, 1807–1834* (Baltimore, 1984).
Sheridan, Richard B., *Sugar and Slavery: The Economic History of the British West Indies, 1623–1775* (Baltimore, 1974).
Schuler, Monica, *"Alas, Alas, Kongo": A Social History of Indentured Immigration into Jamaica, 1841–1865* (Baltimore, 1980).
Solow, Barbara L., and Engerman, Stanley L., eds., *Caribbean Slavery and British Capitalism* (New York, 1988).

French Caribbean

Cauna, Jacques, *Au temps des isles à sucre: histoire d'une plantation de Saint-Domingue au XVIIIe siècle* (Paris, 1987).
Debien, Gabriel, *Les Esclaves aux Antilles Francaises, XVIIe–XVIIIe Siècles* (Basse-Terre, 1974).
Hall, Gwendolyn M., *Africans in Colonial Louisiana: The Development of Afro-Creole Culture in the Eighteenth Century* (Baton Rouge, 1992).
Martin, Gaston, *Histoire de l'esclavage dans les colonies françaises* (Paris, 1949).

Spanish America

Barrett, Ward, *The Sugar Hacienda of the Marqueses del Valle* (Minneapolis, 1970).
Fraginals, Manuel Moreno, *The Sugar Mill: The Socioeconomic Complex of Sugar in Cuba, 1760–1860*. C. Belfrage, trans. (New York, 1976).
Knight, Franklin, *Slave Society in Cuba during the Nineteenth Century* (Madison, 1970).
Lombardi, John V., *The Decline and Abolition of Negro Slavery in Venezuela, 1820–54* (Westport, 1971).

Scarano, Francisco A., *Sugar and Slavery in Puerto Rico: The Plantation Economy of Ponce, 1800–1850* (Madison, 1984).
Scott, Rebecca J., *Slave Emancipation in Cuba: The Transition to Free Labor, 1860–1899* (Princeton, 1985).

U.S. South

Fox-Genovese, Elizabeth, *Within the Plantation Household: Black and White Women of the Old South* (Chapel Hill, 1988).
Gutman, Herbert, *The Black Family in Slavery and Freedom, 1750–1925* (New York, 1976).
Genovese, Eugene, *Roll, Jordan, Roll: The World the Slaves Made* (New York, 1972).
Kulikoff, Allan, *Tobacco and Slaves: The Development of Southern Cultures in the Cheasapeake, 1680–1800* (Chapel Hill, 1986).
Morgan, Edmund S., *American Slavery, American Freedom: The Ordeal of Colonial Virginia* (New York, 1974).

Dutch and Danish Caribbean

Hall, Neville, and Higman, B.W., *Slave Society in the Danish West Indies: St. Thomas, St. John, and St. Croix* (Baltimore, 1992).
Lamur, Humphrey, *The Production of Sugar and the Reproduction of Slaves at Vossenburg, Suriname, 1705–1863* (Amsterdam, 1987).
Olwig, Karen Fog, 'West Indian Research in Denmark', *Plantation Society in the Americas* III, no. 2 (1993), pp. 51–62.

1
Plantations, Sugar Cane and Slavery
Sidney M. Greenfield

When Christopher Columbus accidentally landed in the tropical New World while trying to find a westward route to Asia, the region already was inhabited by peoples he erroneously called Indians. Within little more than a century of their discovery, however, the aboriginal peoples of the Circum-Caribbean — and their socio-cultural systems — had disappeared. The broad region stretching from the Atlantic shores of northeast Brazil in the south, along the coastal lowlands of northern South, Central and southern North America into the southern United States to the north, and including the islands in the Caribbean Sea, had become colonial dependencies of the competing nation-states of Western Europe; and plantation societies, producing commerical crops to be sold in the markets of Europe, worked by a labour force composed primarily of enslaved Africans, had been established as what was to be the institutional form that was to dominate society and culture in the area for centuries to come.

The Europeans to discover and to conquer the tropical New World, and to establish plantation societies there, came almost exclusively from the Atlantic fringe at the western extreme of the continent. It is significant to observe that slave plantations such as were to be established in tropical America were unknown in this part of the Old World prior to the discovery of America. They were to be found, however, in the islands in the Mediterranean Sea and in Madeira in the Atlantic Ocean. They were not invented as a response to the conditions of the New World tropics. Instead, the slave plantation was brought to tropical America as an integrated system by its discoverers and first settlers. The questions to which this paper is addressed are: 1) where, when and how did the plantation system first develop; and 2) how did it come to be introduced into tropical America so as to become the central institution around which society and culture was to be organised to replace what had been found and destroyed?

Charles Verlinden has traced many of the European customs and institutions that had served as precedents for the establishment of colonial society in the New

---

[1] Gratitude is expressed to the National Science Foundation and to the Graduate School of the University of Wisconsin-Milwaukee for the financial support that made the research on which this paper is based possible. All opinions presented, however, are those of the author and not of the sponsoring organisations.

World.[2] My intent in this paper is to follow his lead, but more specifically to trace the combination of separate features that when put together became the institution of the slave plantation as it was to be known in the New World. I shall argue, on the one hand, that there had been a developmental continuity between the plantations that first appeared in the eastern Mediterranean in the early twelfth century and those to be found in the Atlantic Ocean in the fifteenth and in tropical America in the sixteenth, seventeenth and eighteenth centuries. On the other hand, however, I shall maintain that there was something new added from the fifteenth century on that made the plantations in the Atlantic Ocean and in tropical America something different from their Old World predecessors. That which was new was not part of the plantation itself. Instead, it was the larger socio-cultural setting of which the plantation was to become a part that was to change. Specifically, the Portuguese to establish plantations in Madeira and later in Brazil were part of an expanding nation-state. The same was to be true of the English, the French, the Dutch and the Spanish who set up plantations on the Portuguese model in the New World. This was not true, however, of the Normans, Venetians, Genoese and others who founded the plantations in the eastern Mediterranean. As parts of nation-states the plantations in the Atlantic Islands and in the Americas were to take on new functions that were to make them something different from their Old World predecessors.

The sugar plantations of Cyprus, Crete and the other islands in the Mediterranean, as we shall see, were sources of wealth for their founders and owners, as both autonomous kingdoms and as colonies of Italian city-states. The plantations in Madeira, the Azores, the Cape Verdes, and São Tomé, however, not only were sources of wealth, but also a means of settling new territories that became part of, and of strategic value to the Portuguese nation in the implementation of its foreign policy. That is, the plantations not only were of value in their own right, they also served as a model for colonising new regions that had to be settled for other reasons.

The plantations of Madeira, as we shall see, developed independently of Portuguese national authority. A handful of settlers, stimulated primarily by Genoese merchants, taking slaves in the Canary Islands, established a successful plantation society as the Portuguese government was embarking on a policy of expansion and trade.[3] In the pursuit of that policy the government was to respond to peoples and places discovered primarily in terms of their commerical potential.

---

[2] Charles Verlinden, *The Beginnings of Modern Civilization* (Ithaca, New York: Cornell University Press, 1970); *Les Origines de la civilisation atlantique* (Paris and Neuchâtel: n. pub., 1966); *Précédents médiévaux de la colonie en Amérique* (Mexico City, n. pub., 1954); "Les Origines coloniales de la civilisation atlantique: Antécédents et types de structure," *Journal of World History* I (1953): 378-398.

[3] Sidney M. Greenfield, "Madeira and the Beginnings of New World Sugar Cultivation and Plantation Slavery: A Study in Institution Building," in *Comparative Perspectives on Slavery in New World Plantation Societies*, Vera Rubin and Arthur Tuden, eds., Annals of the New York Academy of Sciences, No. 292 (New York: The New York Academy of Sciences, 1977), pp. 536-552. Hereafter cited as "A Study in Institution Building."

Consequently, where local populations were producing and/or trading in commodities that could be sold at a profit in Europe, *feitorias* (factories), centres of trade, were established. Alternatively, where local production and trade were not organised, but the Portuguese saw the potential for profitable commerce, monopolies were awarded to individuals and companies to exploit specific resources. And where no possibilities for profit could be seen, entire regions were disregarded and left alone.

On occasions, however, lands and peoples that otherwise would have been passed by were found to be necessary for the protection of the nation and the implementation of its policies of expansion and trade. In such cases the government had to settle (or resettle) these territories, which were made a part of the national domain, with people loyal to the crown. And given the state of the royal treasury, it had to be done at but minimal cost. The means, as we shall see, especially during the second half of the fifteenth and first half of the sixteenth centuries, was to lead to the establishment of plantations in the islands of the Atlantic Ocean, in the Gulf of Guinea, and on the coast of Brazil.

The Azores and Cape Verdes were uninhabited groups of islands when discovered. Both, however, were situated in strategic locations with respect to wind systems and ocean currents. Portugal's control of the sea lanes in the Atlantic — and later to Asia — were to depend on her ability to secure the islands as bases.

The plantation system was to be the key to establishing settlements that were to make these islands loyal Portuguese dependencies. That the plantations themselves were not always successful, in that many never produced crops of commercial value, was insignificant in the larger scheme of things. They made it possible for Portugal to secure a strategic part of her now enlarged national territory. Furthermore, especially in the Cape Verdes, the experimentation associated with the attempt to establish plantations resulted in adaptations in flora, fauna and human culture that eventually were to make life in the tropics possible for the Portuguese.[4]

From Madeira the plantation also was exported to São Tomé and to Príncipe in the Gulf of Guinea. It also was introduced into Brazil in the early sixteenth century when French Huguenots threatened Portugal's claim to the sparsely inhabited region. When discovered, the native peoples of Brazil produced little of value to the commercially minded Portuguese. Not until after the introduction of sugar cane and the plantation was the Brazilian territory to become a valuable part of the empire. First, however, the native inhabitants had to be displaced, made slaves and/or replaced by imported Africans. Only then was Brazil to become a source of wealth to be envied and then copied by the other expanding nation-states of Europe.

The main structural features of the plantations established in the tropical New World may be defined as follows. A plantation was: a) a unit of land in a tropical

---

[4] Antonio Carreira, *Cabo Verde. Formação de uma Sociedade Escravocrata (1460-1878)* (Lisboa: Imprensa Portuguesa, 1972); Orlando Ribeiro, *Aspectos e Problemas da Expansão Portuguesa* (Lisbon: Junta de Investigações de Ultramar, 1962).

of sub-tropical region; b) on which a commercial crop (or crops) was (were) raised; c) for sale in the markets of Europe — for a profit; d) in which labour was supplied by slaves of non-European (and primarily African) origin; e) who, along with the land and other means of production, were owned and managed by Europeans; f) who, in turn, were subject to the sovereignty of a European nation-state.

Sugar cane was to be the first crop the plantation was to be created to produce. The story of its diffusion from the Mediterranean basin, where the first plantations were established, to the New World shall be told in six parts. In Section I the introduction of the crop into the Mediterranean and the establishment of the first plantations is examined. Section II shifts the focus to continental Portugal to explore the national development of the first society to incorporate the plantation as part of a policy of national expansion. Section III then examines the settlement of the island of Madeira and the development of its sugar plantations in the fifteenth century. Section IV then places Madeira and its plantations within the framework of Portugal's national expansion to show how the new form was to become a model for the settlement of other, strategic areas. Section V then examines the experimental efforts to transfer the system to the Azores, the Cape Verdes and to São Tomé. The paper ends with a brief section examining its transfer to and establishment in Brazil.

I

The earliest accounts available refer to sugar cane as a form of bamboo from which honey could be made without the intervention of bees.[5] It was native to southeast Asia from where it was carried westward to India, where it most probably was domesticated. Then it continued its westward migration to Asia Minor, where in the Persian Gulf the techniques for the conversion of the liquid obtained from crushing the plants into sugar were developed.[6]

Europeans first learned of sugar when Alexander the Great's generals brought some back after an expedition to India in 327 B.C. It then was planted in Arabia, Egypt and in western Asia, but never in sufficient quantities to make it commercially valuable. This would have to await the birth of Islam, after which its commercial cultivation was brought to the Mediterranean in the wake of the expansion of the new religion.

In Moslem areas, it is interesting to note, sugar cane was to be grown and processed almost exclusively by free labour, although slavery was an integral part of life in the expanding Moslem world. Slavery, of course, had had a long history in the Mediterranean and in Asia Minor. Its practice, dating back to ancient Greece and beyond, was to continue in Christian Europe down through the Middle Ages.[7] Its origins can be traced back at least to classical Greece and

---

[5] Henrique Gomes de Amorim Parreira, "História do Açúcar em Portugal," *Estudos de História da Geografia da Expansão Portuguesa* (Lisboa: Junta das Missões Geográficas e de Investigações do Ultramar, 1952), VII, 1: 15.

[6] W. Heyd, *Histoire de commerce du Levant au Moyen-Âge* (Amsterdam: Adolf M. Hakkert, 1959), 2: 680.

[7] Verlinden, *The Beginnings of Modern Civilization*.

probably beyond.[8] In Imperial Rome commercial crops had been grown by slaves taken in battle, on estates that shared many of the features of the plantations of the New World. By the third century, however, the Roman slaves were being transformed ito free *coloni*, while the estates on which they toiled were in the process of becoming the self-sufficient manors of the European Middle Ages.[9] For ten centuries to follow then, from before the fall of Rome until the beginnings of the expansion in Europe in the fifteenth century, slaves were not used in the production of agricultural commodities. Instead, free peasants raised the crops that were consumed directly or indirectly by those who had feudal rights in the lands on which they were grown. Markets and commerce also virtually had disappeared, with the natural economy of the manor dominating the society of medieval Europe. Then in the eleventh century a commercial revival was to begin, highlighted by the growth of the Italian city states and the towns of Germany and other parts of central Europe. A major stimulus for the re-establishment of trade and the development of the European commercial economy to follow was to be the military conquest of the eastern Mediterranean by combined Christian forces from Europe at the end of the eleventh century.

Shortly after its birth, Islam was to spread eastward across southern and eastern Asia and westward across North Africa. In its march westward it was to come in direct conflict with Christianity, which it was to displace first on the southern shores of the Mediterranean and then in the Iberian peninsula. In the centuries to follow, however, Christian military might from the north was to increase as disputes between rival caliphs were to diminish the military ability of Islam.

In the year 1095, at the initiative of Pope Urban II, Christian forces from northern and western Europe began the first of what were to be several Crusades, military ventures whose original purpose was to regain from Moslem control lands formerly Christian, and material relics associated with the birth of Christianity.

As its diverse and competing leaders planned and organised the First Crusade (1096-1099), little time and energy could be spared to ponder the problems they would have to face should the venture turn out to be a success. A host had to be recruited, mobilised, transported from Europe to the Holy Land, supplied with arms, provided with food and organised into a force that could take on the champions of Islam. But when victory did come, its long range, and at the time unanticipated, consequences were to be of far greater significance than contemporaries could appreciate. For it was from the Crusader states in the Holy Land that a major stimulus for the recommercialisation and growth of the economy of Europe was to come; and, with respect to the subject of this paper, it was in the

[8] Moses I. Finley, *Slavery in Classical Antiquity* (New York: Barnes and Noble, 1968); *The Ancient Economy* (Berkeley: Univ. of California Press, 1973), ch. 3; William L. Westerman, *The Slave Systems of Greek and Roman Antiquity* (Philadelphia: The American Philosophical Society, 1955).

[9] Max Weber, "The Social Causes of the Decay of Ancient Civilization," Eugene D. Genovese, ed., *The Slave Economies, Historical and Theoretical Perspectives* (New York: John Wiley and Sons, 1973), p. 1.

Holy Land that Europeans for the first time were to have direct contact with sugar cane and its production.

The European response to victory in the Holy Land was the establishment of feudal states such as the Kingdom of Jerusalem, the Principality of Antioch, and the Counties of Edessa and Tripoli. The model of organisation followed was that of Europe, with its feudal hierarchy and natural economy. But as Joshua Prawer has observed, "The crusaders probably created the only feudal society in a developed economy ...."[10] The economy of the Levant at the time was based on commerce and trade. The new feudal lords, therefore, to quote Prawer again, "introduced the use of money into their feudal system."[11]

Direct contact with the peoples of the Levant, among other things, was to expose the Europeans to material commodities previously unknown, or little known to them. Among these was sugar cane, perhaps the most important crop being produced commercially at the time in the conquered lands.

"In the course of the wearisome marches in Palestine during the First Crusade," Verlinden tells us, "the warriors of the West came to know the sweetness of what one chronicler called 'this unsuspected and inestimable present from Heaven.' "[12]

In Jerusalem and Acre, according to Prawer,

> *the European colonist had his first fill of spices and condiments. Sugar, a medicine for many generations to come (because of its benign taste as well as its price) was consumed directly by sucking the marvellous* cana mellis *(sugar cane) or as syrup. Fulk of Chartre during the First Crusade does not find words to describe the marvels of this unknown plant when he encountered it for the first time* ....[13]

"The crusaders," as Prawer tells us elsewhere, "seized upon the importance of this plant for Europe.... They consequently preserved the old centers of production and expanded the industry elsewhere."[14]

The cultivation of sugar cane had flourished in the Holy Land since the ninth and tenth centuries in the Plain of Acre, around Tyre, Sidon and in the Jordan Valley. "The Crusaders," writes Benveniste, "quickly learned the economic value of this branch of agriculture. It is not surprising that, a short time after they had settled in the country, they began to grow sugar cane and soon became experts in the processes of sugar manufacture."[15]

According to Prawer, however, the crusaders themselves did not move on to the land. Therefore, for the most part, they left the production of sugar — and most other crops — to the local inhabitants while they remained, with but minor

---

[10] Joshua Prawer, *The Latin Kingdom of Jerusalem: European Colonialism in the Middle Ages* (London: Weidenfeld and Nicolson, 1972), p. 66.
[11] Ibid.
[12] Verlinden, *The Beginnings of Modern Civilization*, p. 18.
[13] Prawer, p. 517.
[14] Ibid., p. 364.
[15] Meron Beneveniste, *The Crusaders in the Holy Land* (Jerusalem: Israel Universities Press, 1970), p. 253.

exceptions, in the cities as administrators and bureaucrats.[16]

The crusaders also did not develop the commerce of the crop themselves. This was done by merchants from Venice, Genoa, Pisa and the other Italian city-states who had followed closely and at times even participated in the several crusades. When Venice, the most important and powerful of the city-states, whose ships were familiar visitors to Alexandria and other parts of the Levant, was slow to react to developments in the new Latin states, the younger cities, whose ships already plied the coasts of western Italy and of southern France, entered the breach in search of new markets and secure sources of supply.

Agreements, Prawer tells us, soon were signed

> between the Latin Kingdom and the Pisans and Genoese, granting the latter spoils and extensive privileges in future conquests. A little later, Venice, which commanded the strongest fleet, threw in her lot with the new crusader states.[17]

As a result, the Italians came to have increased access to goods from the East and an interest in the sugar industry established by the crusaders.

The products available to the Italian merchants in the Levant were to become a major stimulus for the continuing expansion of trade and the growth of the economy of Europe. It is of interest to note at this point, however, that of all the exotic items to which Europeans were exposed in the Holy Land, and for which a demand developed on this continent, the only one whose production they came to master was sugar cane. The others they obtained in trade, which meant that their future profit was to depend on their having regular sources of supply. A Christian presence in the Holy Land provided the Italian merchants with access to supplies. To protect themselves, however, the merchants also developed contacts with Moslem and other traders able to provide them with goods, most of which were carried overland to the Levant from India and eastern Asia. For sugar, however, the European merchants now could rely on Christian producers.

But the European presence in the Levant, to the detriment of the merchants and the economic growth of the continent, was to be short lived. By the beginning of the thirteenth century the Ottoman Turks were mobilising armed assaults that in the course of the next few centuries were to give them control not only of the Levant and Byzantium, but of much of eastern Europe as well. They began by driving the Christians out of the Holy Land; and in time they succeeded in closing off all of the eastern Mediterranean to European access.

At first the merchants were able to find new sources of supplies for goods from the East. But by the middle of the fourteenth century, even these had dried up. Sugar, however, continued to be available since the Europeans themselves were producing it. In this respect, the history of sugar was to differ significantly from that of the other commodities from the East to which Europeans had been exposed, and for which a demand had developed on the continent. But to produce sugar from the cane plant the Europeans were to need tropical or sub-tropical land

---

[16] Prawer.
[17] Ibid., p. 19.

with abundant water, along with a substantial labour force to clear the fields, plant and harvest the crop, grind the cane and make the juice into sugar.

As the Ottoman Turks began their assault on the eastern Mediterranean, the Europeans established the production of sugar elsewhere. Within a decade of the fall of Acre, for example, sugar cane was being raised on the Norman-controlled island of Cyprus.[18] There, for reasons we shall get to in a moment, what probably were the first plantations were established.

According to Riley-Smith, when Guy de Lusignan, whose heirs were to rule the island kingdom, first landed in 1192 after purchasing the sub-tropical territory from the Templars, "he found much vacant land, some belonging to the old public domain, but some newly deserted by those who had fled the troubles of the past year."[19] Riley-Smith then tells us that Guy repopulated the island "with colonists from the Latin East, creating about three hundred fiefs for knights and two hundred for men-at-arms, besides making grants to burgesses and others."[20] In describing the structure of the new society in Cyprus on the following pages he tells us that as in Latin Syria, "the Franks occupied the upper reaches of society without much altering the Greek agricultural system; but unlike Syria," he adds, "the lords held demesne land and demanded *corvées* of their peasants."

The production and manufacture of sugar had been introduced into Cyprus by the Arabs after they had captured the island in the seventh century.[21] Then, after declining, it was re-established under Lusignan hegemony at the initiative of Venetian merchants and entrepreneurs. This was done, of course, in cooperation with the Norman and other feudal lords.[22]

Sugar on Cyprus, Riley-Smith tells us, was planted on land that was treated as demesne.[23] Consequently, the landowners could have it cultivated by corvées recruited from the local peasants. The population of the island, however, was relatively sparse; and it was to be reduced further in the following century by wars and plagues. The manpower that could be recruited to raise sugar cane, therefore, was not sufficient for the large scale commercial production of the crop.

The native population at first was supplemented by immigrants from the mainland who feared their new Moslem overlords. But when added to the natives their numbers still were not sufficient to produce sugar on a large scale. The problem of a labour force to grow sugar cane on the demesne lands of Norman and

---

[18] Heyd, 2: 7-9; Verlinden, *The Beginnings of Modern Civilization*, pp. 18-20.

[19] Jonathan Riley-Smith, *The Knights of St. John in Jerusalem and Cyprus, 1050-1310* (London: Macmillan St. Martin's Press, 1967), p. 104.

[20] Ibid.

[21] Verlinden, *The Beginnings of Modern Civilization*, p. 19.

[22] Estates appeared near Limosso and Baffo, along with sugar mills. The richest plantations, however, were along the southern coast, as for example the royal domains of Lemva, Paphos, Aschelia and Kuklia, the lands of the Venetian Cornaro family near Piscopi, those of the Bishop of Limassol, of the Catalan Ferrer family, and the Hospitalers' monastery near Kolossi. Venetian merchants then purchased the crops for sale in Venice and Western Europe—see ibid.

[23] Riley-Smith, *The Feudal Nobility and the Kingdom of Jerusalem, 1174-1277* (London: Macmillan St. Martin's Press, 1973), p. 46.

Venetian producers on Cyprus was resolved by the importation of slaves.

Slavery, as has been noted, had existed in Europe since the time of classical Greece, and had continued as a practice in the Christian West through the Middle Ages. It had had its origins in the taking of prisoners of war. Captives had been made the slaves of their captors.

As Christianity had spread, pagans and others who resisted conversion, as we shall see in greater detail below, were assumed by Church definition to be declaring war against the faith. Consequently, forceful efforts to save them by Christians were defined as the conduct of a just war. An "aggressor" captured in such as "just war" could be enslaved by his captor.[24]

The slaves that were to be put to work on the sugar estates of Cyprus — and later on Crete and the other labour-short islands of the Mediterranean where the crop was to be introduced in the following century — were "Greeks, Bulgarians, Turkish prisoners of war, and Tartars brought from the shores of the Black Sea."[25]

In the thirteenth century then, on the Norman island kingdom of Cyprus we find sub-tropical land holdings owned by Europeans on which sugar cane is being raised, to be manufactured and sold at a profit in Venice and in Western Europe, by corvée and slave labour native to the island or taken as prisoners in Greece, Bulgaria, Turkey and on the shores of the Black Sea. The basic elements of the slave plantation have already been combined. Galloway, in his recent survey of the Mediterranean Sugar Industry, therefore, can conclude that:

> *As the organization of the Mediterranean sugar industry evolved, the antecedents of plantation agriculture can be recognized. The cultivation of sugar in many parts of the Mediterranean employed forced labor, at first corvée and later also slave labor. The link between sugar cultivation and slavery which was to last until the nineteenth century became firmly forged....*[26]

In the thirteenth century Cyprus was an independent kingdom ruled over by a Norman line. In time, however, it was to become a colony of Venice. Hence Galloway, taking a broad view, is able to add that:

> *In addition to forced labor, there were other harbingers of plantation agriculture in the Christian-ruled lands of the Eastern Mediterranean. For example, a colonial relationship was established between the primary producing, cane-growing areas and the metropolitan, manufacturing, and refining centers of Europe. Crete and Cyprus became colonies....*[27]

From Cyprus the cultivation of sugar cane was introduced, also under the initiative of Italian merchants, into Crete, Rhodes, and some of the other islands

---

[24] Maurice Keen, *The Laws of War in the Late Middle Ages* (London: Routledge and Kegan Paul, 1965), pp. 137, 156.
[25] J.H. Galloway, "The Mediterranean Sugar Industry," *The Geographical Review*, No. 67, 2 (1977): 190.
[26] Ibid.
[27] Ibid.

in the eastern Mediterranean. Where labour was in short supply slaves also were introduced and the crop raised on what in every sense were plantations. Sugar cane also was introduced, under the order of Emperor Frederick II, into Sicily after the Norman capture of the island,[28] although it had been grown there by the Moors since the eleventh century. From Sicily, as we shall see, the crop was to be introduced into Madeira in the early fifteenth century.[29]

At the end of the fourteenth century sugar was being raised throughout the Mediterranean. Islamic centres of production were concentrated on the shores of the Levant, in the Nile delta, and in pockets across North Africa to Morocco and north into southern Spain. Christians, meanwhile, were producing the crop on Cyprus, Crete, Sicily and on some of the smaller islands. In the half century to follow, however, the Ottoman Turks were to overrun most of the Christian possessions in the eastern Mediterranean. Shortly thereafter they were to close off Western access to the Levant, which was to mean to products from India and the East also.

European supplies of sugar were in danger as the Turks threatened the island producers. However, sugar was to be only one of the many items Europe would have to do without unless new sources of supply could be found. As the fifteenth century dawned, therefore, Europeans were to be in search both of new lands in which to produce sugar and of new routes to the East in order to obtain many other commodities on which both personal fortune and continuing economic growth were to depend. The story of sugar, a commodity that first came from the East, was to differ from that of other condiments, spices and Asian products because the Europeans had learned to produce it themselves. In the frantic search for a sea route to Asia in the fifteenth century, the production of sugar, however, was to be carried, along with Europeans, into and then across the Atlantic Ocean. At the time, however, changes were taking place in the political and social organisation of Europe that were to affect significantly the future of the plantation. To appreciate these changes and their implications a few words must be said about the nation-state in general, and about the Portuguese nation in particular.

The major universal properties of the nation-state, as Crawford Young has recently observed, are its territoriality and sovereignty.[30] A nation is a territorial unit whose government is sovereign within the national territory. Sovereignty Young defines as "the untrammelled perogative of the state to the exercise of authority within its territory and over the population within its frontiers."[31] To enforce its sovereignty, which is expressed in the form of laws, governments depend upon organised police forces. They also must have a military to protect

---

[28] Heyd, 2: 686.
[29] In contrast with Cyprus and Crete, Sicilian sugar was grown by Europeans with free labour. Unlike the other islands, the population of Sicily was relatively dense when it was taken from the Moors. Consequently, large scale commercial production of sugar was possible without the introduction of slaves. See Giovanni Rebora, *Un'impresa Zucceriera del Cinquecento* (Napols: n. pub., 1968).
[30] Crawford Young, *The Politics of Cultural Pluralism* (Madison, Wisconsin: The Univ. of Wisconsin Press, 1976), p. 68.
[31] Ibid.

and defend the national boundaries. Nation-states then have governments that pass laws that are enforced by their police. They also have militias to protect both the boundaries and the regime that exercises sovereignty within the territory. The final element that characterises the modern nation-state is a system of taxation to support the regime and to pay the police and the military.

In a world of nation-states, such as we find in our own era, the borders of one end where those of the next begin. But this was not always the case. When the territorial state first came into being there were few other national units − with militias − to constrain the expansion of the earliest to adopt the form. And this was especially true with respect to the lands and peoples outside of Europe.

When Portugal assumed the leadership of a Europe in search of a sea route to the Orient in the fifteenth century, she was one of the first nation-states, having already reached her present day continental limits. But as she moved out into the Atlantic, and then beyond, her subjects came in contact with lands and peoples − heretofore unknown in Europe − not organised in nation-states and without militaries to defend them. As a result, with legitimacy provided by the Church and the Papacy, both the lands and the peoples became a part of an expanding Portuguese national society. This state of affairs was to have numerous unanticipated consequences, only one of which is of concern to us here. I refer to the problem of securing those lands that had nothing of value to offer the commercially-minded Lusitanians. Other things being equal they would have preferred to leave them alone and proceed in search of more rewarding domains. But at times they could not. The territories now were a part of Portugal and often of value politically and/or militarily. Although they offered little to attract merchants, settlers and others needed to secure them in the name of the Portuguese state, as they could not be abandoned.

The means used by the Portuguese state to secure and settle these lands was to be the plantation. As we shall see, sugar cane was introduced into previously uninhabited Madeira at the very beginning of the expansion. Within half a century, and decades before the Cape of Good Hope was rounded and the New World discovered, and independent of the political turmoil that was to preceed the espousing of trade, conquest and the search for a sea route to Asia as national policy, successful and profitable plantations were established in Madeira. The plantation then, with Madeira as the model, became the means by which islands and continents were to be settled and incorporated politically into the expanding Portuguese state.

In some cases the plantations were to become economically viable, while in others they were not. Either way, however, they served the political objectives of settling and securing new sections of the national territory. The plantation, previously an end in itself measured in economic terms, was to become a political tool to be established, however, because of its commercial potential. In this way the islands in the Atlantic and off the coast of Africa came to be settled as plantations, with the pattern repeated later across the Atlantic in Brazil. To examine this process we turn first, however, to the national unification of Portugal and to the struggle for leadership and a foreign policy in the fifteenth century.

II

Portugal first appeared as an independent state at the end of the eleventh century when Count Henri of Burgandy, son-in-law of Sancho IV, King of Leon and Castile, and his wife the Countess Tarasia, ceased to fulfil their feudal obligations while continuing to govern the people of the territory. Their son, Afonso Henriques, then was to pursue the matter of independence with the Pope. After thirty-five unsuccessful years Pope Alexander III in 1179 A.D. "solemnly recognized Afonso as king and his state as a kingdom...."[32] Even prior to this, Afonso Henriques had begun to expand his kingdom under the aegis of the crusades.

Much of the Iberian Peninsula had been under Moslem control since its conquest in the eighth century. When Pope Urban II had proposed the idea of a crusade in the eleventh century he had planned a campaign at the western as well as the eastern end of the Mediterranean. A full-scale crusade in Iberia never materialised. However, when Christian princes there requested it, northern knights on the way to the Holy Land were given permission to stop off and help in what came to be known as the *Reconquista*. On no less than six occasions, Christian forces from the north were to assist a ruler of Portugal and his army in the capture of Moslem held regions. But in contrast with the disposition of conquests in the Holy Land, where new Christian states came into being, in Iberia they became part of an expanding Portuguese state.

By the middle of the thirteenth century the territory that was to be modern continental Portugal had been unified as an independent state under the descendants of Henri and Tarsia. As the new lands were being incorporated into the prevailing feudal system, however, a revolutionary transformation was in process. The market principle was introduced into the society in the eleventh and twelfth centuries.[33] Soon thereafter production was to be geared first to local markets and then, more significantly, to foreign trade. In addition, commercial development was to be stimulated by external forces.

The Portuguese nation occupied a strategic position with respect to the commercial and economic revival of Europe. Located on the Atlantic Ocean just beyond the Straits of Gibraltar and the entrance to the Mediterranean Sea, its sheltered harbours were natural ports of call for the small vessels of Italian and other merchants carrying trade goods from the Levant to England, France and the Germanies. From these merchants the Portuguese learned both of the products of the East and of the profits to be earned from their trade.

Within a short time a national bourgeoisie was to appear that soon was to compete openly with the traditional landed nobility not only for the ear of the king but for control of the polity. By the time the first major dynastic crisis occurred at the end of the fourteenth century, the bourgeoisie and their allies were able to determine the line of royal succession and the future direction of national policy.

In 1383 King D. Fernando I (1367-1383), the only legitimate son of King D. Pedro I (1357-1367), died without a male heir. The crown, in theory, then was to

---

[32] Antonio H.D. Oliveira Marques, *History of Portugal from Lusitania to Empire* (New York: Columbia Univ. Press, 1972), 1: 43.
[33] Ibid., I: 94.

have passed to his daughter D. Beatriz. Beatriz, however, at the time was the wife of Juan I, King of Castile. The wedding clauses, according to Oliveira Marques, "clearly entrusted the regency and the government of the realm to the queen-mother Leonor Teles until a son or a daughter was born to Beatriz."[34] Leonor, however, who represented the interests of the traditional landed nobility, was hated by the people and opposed openly by the bourgeoisie. To further complicate matters, Juan I of Castile, most probably in response to the strong opposition of Leonor and her lover, Count João Fernandez Andeiro, a Galician nobleman, invaded Portugal. A victory most probably would have ended Portugal's existence as an independent state.

Leonor Teles and Andeiro had the support of most of the landed aristocracy. They were opposed primarily by the middle and lower ranks of the bourgeoisie, led by the Master of Avis, João, an illegitimate son of King Pedro I. João, it appears had invited the Castilian monarch to enter Portugal in fear of the situation in which he and his followers found themselves.[35] He then changed his mind, assuming leadership of the opposition to both groups. First he himself killed Andeiro and forced the queen-mother to flee for her life. He then proclaimed himself regent and defender of the realm. As such he organised the resistance against the Castilian forces.

The war to follow went through three main phases. First, while the Castilians beseiged Lisbon, João's forces, under the leadership of his constable, Nuno Álvarez Pereira, defeated them in the south. Then, in perhaps the most famous battle in Portuguese history, Pereira defeated superior Castilian forces at Aljubarrota where he was aided by English archers and, quite probably, advisers. Finally, a full treaty with England brought the Duke of Lancaster, a pretender to the throne of Castile, to the peninsula. Juan then was forced to withdraw from Portugal in order to protect his own kingdom.[36]

The *Côrtes*, a body representative of all of the estates of the kingdom[37] then was convened in Coimbra in 1385. By means of election the Master of Avis was placed on the throne as D. João I, with his primary support coming from the bourgeois representatives of the municipal councils.[38] Foreign recognition soon followed.

During the Great Schism Portugal had changed sides on several occasions. Regent João had switched again from the Pope of Avignon to the Pope of Rome. In return, his vows as a clergyman were quickly cancelled, enabling him to marry Philippa, daughter of John of Gaunt, Duke of Lancaster. This established the House of Avis which was to rule Portugal until 1580.

As the fifteenth century dawned, then, the government of the unified

[34] Ibid., 1: 126.
[35] Ibid., 1: 126-127.
[36] Ibid., 1: 127.
[37] Marcelo Caetano, *Subsidies para a História das Cortes Medievais Portuguesas*, No. 15 (Lisboa: Revista da Faculdade de Dereito da Universidade de Lisboa, 1963).
[38] Oliveira Marques, 1: 124ff.; Antonio Sergio, *Sobre a Revolução de 1383-85* (Lisboa: Cadernos da Seara Nova, 1946); Joel Serrão, *O Carácter Social da Revolução de 1838* (Lisboa: Cadernos da Seara Nova, 1946).

Portuguese nation-state was in the hands of a new king who had been supported and placed on the throne by the representatives of commercial and mercantile interests. As a troubled Europe still torn by the Hundred Years War[39] and the Great Schism[40] was to respond to the expansion of the Ottoman Turks and the closing off of the eastern Mediterranean to the trade upon which her economic growth was dependent, Portugal, under the House of Avis, gradually was to rise to a position of leadership. The first step was to be the landmark capture of Ceuta, the Moslem stronghold across the Straits of Gibraltar in North Africa, in 1415. For the first time Christian forces had taken the offensive, conquering lands traditionally Moslem, rather than recapturing what had been taken previously from Christians.

Although invited to participate in the councils of Europe and to contribute to resolving the major problems of the day, Portugal was not yet ready to take the lead in what in retrospect were to be called the expansions and discoveries. First the problem of a national foreign policy raised by the capture of Ceuta had to be resolved.

After the initial excitement had worn off, the Portuguese realised that Ceuta itself was insignificant to them. The question was what they should do next. Two alternatives were presented. One was a policy of conquest in which national forces would attack other towns and cities in the hinterland of Morocco and elsewhere in the Moslem world. The other, by contrast, emphasised trade and commerce, with conquest limited to a tactic in an overall strategy of commercial expansion. The next half a century and more was to be spent in turmoil as the supporters of the opposing positions battled each other to determine which of them would control the government and decide what was to be the foreign policy of the nation.

Before turning to the details, we first must return to the story of sugar cane and plantations. While the battle for the direction Portuguese foreign policy was to take was being waged, the island of Madeira was being settled and sugar cane plantations established. The plantations, and the slave trade upon which they were to depend, developed independent of the at times bloody struggle over foreign policy. When the latter finally was resolved, however, the plantation system was to be a valuable instrument in the pursuit of the policy adopted.

III

Although known about previously by Europeans, the Madeira Islands, composed of Madeira, Porto Santo, Deserta and the Selvagens, were uninhabited when Portuguese seamen landed there sometime shortly after the victory of Ceuta.

The islands, of volcanic origin, are located in the Atlantic Ocean off of the northwest coast of Africa. Although situated between 32°7'50" and 33°7'50" north latitude, they are warmed, especially near the coast, by the Gulf Stream and winds emanating from the Azorean anticyclone.

---

[39] Peter E. Russell, *The English Intervention in Spain and Portugal in the Time of Edward III and Richard II* (London: Oxford University Press, 1955).

[40] Julio Cesar Baptista, "Portugal e o Cisma do Ocidente," *Lucitania Sacra* 1: (1956) 65-203.

The main island of Madeira, which measures some 36 miles in length and almost 14 miles in width, with an area of a little over 400 square miles, is mountainous and rugged. The mountains, which rise precipitously from the sea, climb to approximately 2,000 meters in height. The landscape is broken and there are but few relatively flat areas, mostly along the coast. The interior of the island is dominated by sharply rising peaks that are difficult to ascend.

The soils are generally good, especially along the coast. However, the abundant moisture carried in the winds that come mostly from the south-southwest does not condense until it reaches the higher elevations. Hence, most of the precipitation occurs above 1,300 meters, with the water then cascading down the mountain sides to form streams and rivers that then flow forcefully into the ocean. To plant crops in the warm, fertile soils near the coast the water must be controlled and brought to the area under cultivation. In brief, the agricultural potential of the island at the time of its settlement was considerable, provided that an irrigation system could be established.

According to the chroniclers, two squires of the household of Prince Henry sighted the islands when blown off course by a storm in 1419.[41] When they returned home they are reported to have asked their lord for permission to establish a settlement. With the permission granted, in May of 1420 the squires, João Gonçalves Zarco and Tristão Vaz Teixeira, along with Bartolomeu Perestrelo, departed with somewhat less than one hundred colonists.[42] Elsewhere, I have argued that the most likely motivation that can be attributed to the settlers was their desire for social mobility.[43]

The ships, supplies and provisions for the venture are reported to have been provided by Prince Henry, who was to become the lord proprietor *(donatário)* of the islands. He then divided the main island of Madeira into two captaincies, Funchal and Machico, with Porto Santo left as one — the smaller islands never were to be settled. Zarco was appointed captain of Funchal, Tristão captain of Machico and Perestrelo captain of Porto Santo.

The captains then were empowered to distribute the lands in their jurisdictions as *sesmarias*[44] to their followers. Although the original documents no longer exist, from later records we may reconstruct that holdings were given to the eligible heads of households in parcels that ran from the ocean to some unspecified point in the mountainous interior. Natural markers, such as rivers, streams and hills, then divided one holding from the next.

The recipient of a land grant then held it in perpetuity, provided that it was in cultivation within five years of receipt. If cultivated the land then could be sold, given, and/or inherited as private property by the beneficiary and his heirs. If not,

---

[41] Gomes Eanes da Zurara, *Crônica de Guiné*, Segundo o ms. de Paris, modernizada, introdução, notas, novas considerações e glossário de José de Bragança (Lisbon: Livraria Civilização-Editora,, 1973), 83; João de Barros, *Ásia, Primeira Década* (Coimbra: Imprensa da Universidade, 1932), Book 1, Chs. 2-3; Damião de Goes, *Crônica do Príncipe D. João* (Lisboa: n. pub., 1724), Ch. 8.

[42] Oliveira Marques, 1: 152.

[43] Greenfield, "A Study in Institution Building."

[44] Virginia Rau, *Sesmarias Medievais Portuguesas* (Lisboa: n. pub., 1946).

ownership reverted to the crown, to be assigned by the captain to another settler. In this way the land in the islands was made available to the colonists.

The lord proprietor, as noted, had provided the colonists with the materials and supplies for founding their settlement. These included, besides domestic animals, seeds, such as wheat, to be planted and raised for their subsistence. First, however, the land had to cleared of its dense virgin forest cover. This was accomplished with the aid of fire.[45] The ash, when added to the fertile soils, produced a land of considerable agricultural potential. But before crops could be raised, especially in the relatively flatter coastal areas, the water abundant in the mountainous interior had to be brought to the coast.

The Portuguese settlers already were familiar with irrigation techniques like those that were adopted in Madeira.[46] Their implementation was to be extremely difficult, however. First, the relief of the island was to make the construction of irrigation works extremely hazardous.[47] Second, the tiny colony lacked manpower in sufficient numbers to perform such an extraordinary feat. But irrigation works were constructed. A letter by King D. João II dated 1493, for example, refers to a document by his grandfather, King D. João I, regulating the use of water in the island.[48] The obvious implication is that some time prior to the death of D. João I in 1433 irrigation works had been constructed in Madeira and regulations had been set out for their use.

It is highly unlikely that the European settlers in Madeira built the irrigation works themselves. Instead, as I have suggested previously,[49] the *levadas*, as the irrigation ditches were called,[50] were constructed by slaves taken in the Canary Islands. The slaves then were present when sugar cane was introduced into the island. Given the sparse, socially mobile European population, the slaves taken in the Canary Islands became the labour force for the early plantations.

At the time Madeira was being settled, the Portuguese also were contesting jurisdiction over the nearby Canary Islands. A century earlier a Castilian nobleman had "procured a grant..., with the title of King, from Pope Clement VI, upon condition that he would cause the Gospel to be preached to the natives".[51] By the

---

[45] Barros, Book 1, Ch. 3; Gaspar Frutuoso, *As Saudades da Terra*, Book 2 in História das Ilhas do Porto Santo, Madeira Deserta e Selvagenms, manuscrip do Século XVI amotado por Alvara Rodrigues de Azevedo (Funchal, Madeira: Tipografia Funchalense, 1873), p. 64.

[46] Jorge Dias and Fernando Galhano, *Aparelhos de Elevar a Água de Rega*, contribuição para o Estudo de Regadio em Portugal (Porto, Portugal: Junta da Provincia de Douro-Litoral, 1953).

[47] Maria Lamas, *Arquipélago da Madeira* (Funchal, Madeira: Editora Eco, 1956); Padre Fernando Augusto da Silva, "Levadas," in Padre Fernando Augusto da Silva and Carlos Azevedo de Mennezes, eds., *Elucidaro Madeirense*, 3d ed. (Funchal, Madeira: Tipografia Esperanca, 1965).

[48] José Lucio de Arquivo da Camara, *Municipal do Funchal* (Funchal, Madeira: n. pub., n.d.), 1: 207.

[49] Greenfield, "A Study in Institution Building."

[50] Silva, "Levadas."

[51] Juan de Abreu de Galindo, *The History of the Discovery of the Canary Islands* (London: n. pub., 1764), pp. 1-2.

beginning of the fifteenth century the condition had not been satisfied, since the natives had not been converted to Christianity. The Portuguese, therefore, laid claim to the Canaries by trying to satisfy the condition Castile had been unable to meet.[52]

As I have stated elsewhere:

> *The natives of the Canary Islands had resisted the efforts of the European Christians to subjugate and convert them. To make matters worse they fought off the European intruders who, in their eyes, were attacking their homeland. But to the Europeans, resistance to the advance of the "true faith" was the equivalent of a declaration of war against all of Christendom. Consequently, all future relationships between Europeans and Canary Islanders, in the eyes of Europeans, were placed in the legal framework of parties engaged in war. Furthermore, for the Europeans the war was of a special kind referred to as a "just war." Roman law (the law of the Roman church that is) had laid down that persons captured in a just war became the slaves of their captor and their goods the property of the victor.*
>
> *The immediate consequence was that all Canary Islanders taken as captive by Europeans, no matter how they were captured, became the legal slaves of the Europeans.*[53]

In 1415 the Portuguese had launched an unsuccessful attack on Grand Canary island. Then in 1424, just four years after the colony in Madeira had been settled, Fernando de Castro led a force of 2,500 infantry men and 120 horses against Grand Canary, Palma and Tenerife.

The leaders of the Madeira settlement, for the most part, were knights in the household of Prince Henry, trained more in the arts of war than in agriculture. Most had served with their lord in Ceuta in 1415. Since Henry had organised the assault on the Canaries, and de Castro's forces more than probably stopped in Madeira before proceeding to the Canaries, it is more than likely that at least a number of the Madeiran settlers participated in the campaign of 1424.[54]

Although the assault of 1424 ended in failure, we may safely assume that captives were taken, as they had been in previous raids and were to be taken in later raids on the Canaries.[55] Instead of selling them into slavery on the continent, as was common practice, the Madeirans most probably took home with them small numbers of captives. The Canary Islanders, however, were uniquely suited

---

[52] João Martins da Silva Marques, *Descombrimentos Portugueses — Documentos para a Sua História* (Lisboa: Instituto para a Alta Cultura, 1944), 1: 291-346.

[53] Greenfield, "A Study in Institution Building," p. 542.

[54] Although there is no direct evidence to support this contention, Sarmento reports a document dated 1425 in which the king thanks the settlers of the island for the help they have given to the forces of Portugal. Since there were no other campaigns in which the settlers of Madeira could have participated at the time, it is safe to conclude that he was referring to de Castro's attack on the Canaries. Alberto Artur Sarmento, "Madeira e Canárias," *Fasquias da Madeira*, Diaria de Noticias (Funchal, Madeira: n. pub., 1935).

[55] D.J. Wolfel, "La Curia Romana el la Colonia de España en la defensa de los Aborígines Canarios," *Anthropos* 25 (1930): 5 and 6:1011-1083.

to the needs of the Madeiran colonists. They were excellent physical specimens conditioned to scaling precipitous mountains and leaping gorges and ravines.[56] Raised to defend themselves in islands similar to Madeira in their geography, they were ideal labourers for the construction of the irrigation works that were a prerequisite to the development of agriculture in Madeira.

Between 1420 and 1446 there were at least four major and several minor assaults by Portuguese forces on the Canary Islands. Madeirans more likely than not participated in all of them. In addition, although there are no records to document it, the enterprising leaders of the island colony probably conducted raids of their own in the nearby Canaries, taking home additional captives who were utilised in the construction of *levadas*.

With the irrigation works completed and slaves in the island, cereals then were planted with surpluses exported to the continent.

We are not sure exactly when sugar cane was first planted in Madeira. Most of the sources claim that it was introduced from Sicily at the instigation of Prince Henry.[57] Some authorities, however, doubt this since the crop already was being raised in Portugal. Still others, like Duarte Leite, believe that the Genoese, already present as merchants in Lisbon and Madeira, introduced the crop on their own from the Mediterranean.[58]

The *Cartas de Doacaō* that formalised the rights and duties of Zarco (1450), Tristaō (1440) and Perestrelo (1446) make no reference to sugar cane, although taxes are specified for other crops.[59] The first of the chroniclers, Zurara, who wrote in 1452, says almost nothing about the value of the sugar being produced.[60] Three years later, however, the Venetian Cadamosto reports an annual production in excess of 6,000 *arrobas*.[61] And by the following year Madeiran sugar already is being exported to Bristol.[62]

Although sugar cane most probably was being planted in Madeira long before the middle of the fifteenth century, the commercial value of the crop was not to be significant until sometime after 1450. In the three decades or so between the founding of the colony and the first export of valuable quantities of sugar, a society was being formed by the upwardly mobile would-be nobility in the island

---

[56] Galindo, p. 6.
[57] Luis de Cadamosto, "Le Navegazioni Atlantiche di Luis de Cadamosto," *Descobrimentos Portugueses*, João Martins de Silva Marques, ed., trans. from Italian to Portuguese by Giuseppe Carlo Rossi (Lisbon: Instituto para a Alta Cultura, 1944), suplemento ao Vol. 1: 164-248; Jeronimo Dias Leite, *Descobrimento da Ilha da Madeira e Discurso da Vida e Feitos Dos Capitães da dita Ilha* (Coimbra: n. pub., 1947), p. 19; Frutuoso, As Saudades da Terra, p. 431.
[58] Duarte Leite, "Os Primeiros Açúcares Portugueses," *Coisas de Varia* História (Lisboa: n. pub., 1941).
[59] Silva Marques, 1.
[60] Zurara.
[61] Cadamosto.
[62] Noel Deerr, *The History of Sugar* (London: Chapman and Hall, 1949-50), 1:100; Maria de Lourdes de Frietas Ferraz, "O Açucar e a sua Importância na Economia Madeirense," *Geogr*áfica, 25 (1971): 30-38, 78-88.

who were incorporating slaves taken in the Canary Islands to perform the drudgerous tasks that a lower class peasantry had performed on the continent. Since there had been no native population on the uninhabited island, the difficult and unpleasant tasks, such as building *levadas* and then planting and harvesting crops, to be shunned by a late medieval European elite — aspiring or arrived — had been assigned to the slaves.[63]

As society and economy in Madeira developed then, the slave component was to increase in number and importance. And as sugar cane came to be of considerable value in the second half of the century, the colonists had their dream come true. With the wealth obtained from the sale of their sugar crop they soon were to be recognised by the continental authority as a part of the new nobility.

But as a plantation society, based on slave labour, was being fashioned and elaborated as an institutional form on the island of Madeira, a struggle for a foreign policy that was to shape the national destiny of Portugal was being waged on the continent. Before continuing with the development and diffusion of the new plantation form, therefore, we must return to the continent and to the struggle between the supporters of conquest and those advocating trade as the basis for national policy.

IV

Two parties had appeared in Portugal in the years following Ceuta, each championing one of the possible directions foreign policy might take. Those advocating conquest and crusades were led by Prince Henry and the Count of Barcelos, an illegitimate son of the king who had married the daughter of the popular Nuno Álvarez Pereira. Their opponents, on the other hand, had as their leader Prince Pedro, supported by his younger brother Prince João.

The aging king would not take sides on the issue that was to divide his sons and his kingdom for half a century after his death. But he did curb the enthusiasm of the pro-crusade, pro-conquest faction and there were no new crusades undertaken. With his death in 1433, however, the advantage was to swing temporarily in favour of the supporters of crusades.

The new king, D. Duarte, at first was cautious and hesitated, but then gave in to what Oliveira Marques refers to as the war party.[64] The result was the disaster at Tangiers in 1437.[65] Badly routed and surrounded by the enemy, the Portuguese, in order to escape, were forced to give up as a hostage the king's younger brother, D. Fernando. On their return home, the war party pressed for another expedition. But before they could succeed, King D. Duarte died.

D. João I's oldest son and heir had been a sickly child who never had been expected to outlive his father, let alone succeed to the throne. D. Pedro, the second son, by contrast, was strong and healthy, and possessed of extraordinary abilities. In fear of Duarte not surviving, or if so, not being able to marry and to conceive an heir, D. João I had both boys prepared as his replacement on the

---

[63] Greenfield, "A Study in Institution Building."
[64] Oliveira Marques, 1:131.
[65] Domingos M.G. dos Santos, *D. Duarte e as Responsibilidades de Tanger (1433-1438)* (Lisboa: Comissão Executiva de V Centenario da Morte do Infante D. Henrique, 1960).

throne. But Duarte did live; and not only did he succeed his father as king, he married D. Leanor of Aragon, by whom he had two children. When he died only five years after his father, however, the worst of the old king's fears came close to being realised. D. Duarte's heir, the future D. Afonso V, was but six years of age. A regency would be necessary until he came of age.

In his will D. Duarte had left his widow both as guardian of the future king and as ruler *(regador)* of the kingdom.[66] But when the *Côrtes* met to formally declare the boy king, they took the position that only they, and not the deceased king, had the right to appoint the regent.

When the members of the war party closed ranks behind the claims of the queen-mother, the opposition rallied behind Prince Pedro. Prince Henry then tried to mediate between the factions.[67] But when the queen-mother refused the compromise, the representatives of the bourgeois advocates of trade and commercially oriented discoveries who dominated the *Côrtes* moved to turn everything over to D. Pedro.[68] To avoid what might have developed into civil war, the supporters of the queen-mother advised her to accept D. Pedro as regent and Defender of the Realm, on the condition that she be permitted to remain as the young king's guardian. The advocates of discoveries and trade had won the round.

During the eight years D. Pedro was to serve as regent (1439-1447) the expansions and discoveries began. By 1440, for example, Portuguese mariners had reached Senegal. A year later Nuno Tristão had arrived at Cabo Branco. Two years after that Nuno da Cunha and Gonçalo de Cintro were to estabish the first *feitoria* (factory), or trading post on the African coast. In 1444 the Azores were discovered, as were several islands off Africa. A year later the Portuguese had reached Gambia and the Rio do Ouro. In 1446 four of the islands in the Cape Verde group were discovered and the Portuguese were down to the Rio Grande. By the end of D. Pedro's regency 198 leagues along the west coast of Africa had been explored and the Azores and the Cape Verdes had been discovered and were being colonised. In addition, D. Pedro had induced his brother Prince Henry, the lord proprietor of Madeira, to grant the settlers tax concessions to stimulate the production of crops.[69]

But the victory was to be short-lived. In 1446 the regent, in response to the pressures exerted by his opponents, was to convene the *Côrtes* in order to turn the government over to the boy king. When that body refused to let him, however,

---

[66] Ruy de Pina, *Cronica de D. Duarte* (Lisbon: Biblioteca de Classicos Portugueses, 1901), 1:17.

[67] He proposed that the country be run by a troika composed of D. Pedro, the queen-mother, and the Count of Arrioles (the son of the Count of Barcelos, who now was the Duke of Braganza, representing the traditional nobility) until D. Afonso attained his majority. Pedro was to be regent, with the title of Defender of the Realm. D. Leanor was to be guardian of the young king, while Arrioles was to be responsible for the administration of justice. See Ibid., 1:34; Gaspar Dias de Landim, *O Infante D. Pedro, Crônica Inédita* (Lisbon: Biblioteca de Classicos Portugueses, 1892), 1:42-43.

[68] Ibid., 1:45-46.

[69] Silva Marques, 1:400.

insisting that because of Afonso's youth and inexperience D. Pedro should continue to govern,[70] the queen-mother and her war party supporters then intensified their efforts, now directed personally against the regent. Therefore, when the government eventually was turned over to the young king a few years later he was well on the way to being convinced that his uncle had mistreated his mother, in addition to mishandling the affairs of state. Early in his reign then the young monarch was to banish his uncle from court. Still not satisfied, however, the leaders of the war party now sought further vengeance against the regent. At the instigation of the Duke of Braganza the king ordered D. Pedro to disarm his castles. When the regent at first refused, Braganza organised the royal forces to enforce compliance. And when Braganza was refused passage on his estates demanded in the king's name, the regent was branded a traitor. Forced by Braganza to defend himself, D. Pedro was killed in battle by the royal forces at Alfarrobeira on February 20, 1449. With the leader of the opposition gone, the war party was back in control.

During the early part of the reign of D. Afonso V the expansions and discoveries were slowed almost to a standstill and new crusades in north Africa were organised. Moreover, the king himself was to lead many of the wars of conquest. But in the long run, the victory of the war party was to be but a last gasp as the king himself was to permit the pendulum to swing back in the direction it was to remain in the centuries to follow. The new leader who was to redirect Portugal's foreign policy back to commercial expansion and trade where it was to remain was to be the crusader king, D. Afonso's son, the future D. João II.

While still regent D. Pedro has arranged the marriage of his royal nephew to his own daughter, D. Isabel. In spite of their opposition to him, the new leadership permitted the wedding to take place. After bearing two children, however, the young queen died suddenly at the tender age of twenty-three. The king never married again. At first he consoled himself with further crusades, but in later years he tired and even tried to renounce his crown in order to enter a monastery.

The young prince and princess were raised primarily by servants of their deceased mother and maternal grandfather. And as he grew to maturity the future D. João II demonstrated a combination of ability, leadership and responsibility that endeared him to his father who had tired very quickly of the affairs of state. As the boy matured, therefore, D. Afonso came to rely on him in the administration of the government. Then, when he was nineteen, to "occupy himself in some good and worthy exercise, thereby to sharpen his wits and give him practice of how kings and princes negotiate and govern and dispatch business . . . ," he was entrusted by his father, "with the affairs of Guinea . . ."[71] From 1474 on, then, D. João was in charge of Portugal's activities in the south Atlantic.

Having previously been convinced of the merit of the policy that had been advocated by his maternal grandfather, the regent D. Pedro, the prince redirected

---

[70] Landim, 2:151.
[71] Elaine Sanceau, *The Perfect Prince* (Porto, Portugal: Editoria Livaria Civilização, 1959), p. 62.

the energies of the state in support of discoveries and the search for the sea route to the Orient. For the remainder of the reign of his father (1474-81), and while king himself (1481-95), D. João II relentlessly pursued the discoveries and the rounding of Africa.

The eventual objective of the advocates of discoveries and commercial expansion, it must be remembered, was to reach India and the Orient, the source of the trade goods that had stimulated the growth of the European economy. In the years D. João II was in command of Portuguese affairs in the south Atlantic everything was done to prepare the way to achieve that goal; and with the full mobilisation of the energies of the kingdom while he was on the throne, Africa was to be rounded and India reached before the end of the century. Although Vasco da Gama was not to arrive in India until three years after the death of the Perfect Prince, as D. João II has come to be known, it was to his energy, ability and determination that the achievement is owed.

After pushing down the coast as far as the Kingdom of the Congo (present day Angola), a fleet was fitted out in 1487, under the command of Bartolomeu Dias, whose avowed purpose was to reach India by rounding Africa. Although Dias never did reach India, he did round the Cape of Good Hope and the southern tip of Africa. Furthermore, he returned home the following year to instruct others as to how it could be done again. Then, in 1498, after several setbacks including the death of D. João II, da Gama was to complete successfully the first sea voyage to India. The flood of wealth and the power and prestige that accrued to the kingdom then was to end the internal struggle that had divided the nation. The merchants, mariners and traders finally had won a decisive victory and the war party never was to challenge them again.

As the Portuguese had moved sporadically down the coast of Africa and into the Atlantic Ocean in the fifteenth century before reaching Asia, three forms, or patterns of colonisation and settlement were to be noted,[72] each of which is to be understood in relationship to the commercial goals of the expansion and discoveries.

What the proponents of discoveries and trade had hoped for all along was the sea route to Asia so that they could establish themselves and trade for the items that would bring them wealth when sold on the markets of Europe. Unable to round Africa for the better part of a century, however, they sought other places where resident populations already were producing commodities for which there was a demand in Europe. Where these were found, the Portuguese set up *feitorias* in which trade was to be conducted.[73] The model for these trading posts, which were set up along the African coast, and later on in India, the Far East and the New World, were the *feitorias* that had been established in cases centuries before in

---

[72] Viera Higno, "Panoramica Geral da Cultura Portuguesa na Época Henriquina," *Boletim do Instituto de Angola,* 13 (1960):5-31.
[73] Manuel Múrias, "A Política da Feitorias," *Boletim Geral das Colonias,* 7 (1936):131, 151-178; Rau, *Feitores e Feitorias: "Instrumentos" do Comércio Internacional Português no Seculo XVI* (1966) Brotéria LXXXI.

Flanders,[74] Antwerp, Bruges and other parts of northwest Europe.

But not all of the places discovered in the long search for the all water route to Asia were inhabited by peoples already producing commodities for which there was a demand in Europe. Instead, Portuguese mariner-merchants at times discovered lands that were inhabited by peoples not producing anything of commercial value. Although their first inclination might have been, and often was, to push on in the hope of finding something better further on, they often observed something that if developed could be of value to them. In such cases the crown, which formally owned all lands discovered, would grant exclusive rights to an individual or company formed to develop the resources in question. In return for the concession, which was a monopoly, the entrepreneur or his company would pay the royal treasury a percentage, usually a fifth, of the profit to be earned from the development of the concession. The earliest example of this form of colonial exploitation, and the model for its later use, was the company formed in Lagos in the Algarve to exploit the monopoly Prince Henry had obtained from his father to fish in the Ocean Seas.

When the lands to be developed were uninhabited, however, the exploitation of their resources required that a settlement be established, which ment recruiting colonists. The difficulty of finding settlers in an expanding economy with a small population was to make this at times an investment not considered worth the cost by the recipient of the concession. But, as we shall see, the often uninhabited islands in the Atlantic Ocean and off the coast of Africa were of importance to the crown for reasons other than their profit potential. Not only were they now a part of the national domain, at times they also were of strategic importance in the defence and management of the national policy of commercial expansion and overseas trade. Although of marginal value commercially, they were vital to the overseas program of the state.

In fairness to the monarchs of Portugal in the fifteenth and early sixteenth centuries, it is doubtful that they had the resources needed to have settled uninhabited islands and territories. Therefore, they delegated concessions and privileges to private persons (and/or institutions)[75] in the hope of having them settled in the name of the crown.

Where the possibilities for profit appeared clear, Portuguese and other merchants and entrepreneurs in the service of the king were quick to exploit royal monopolies and concessions. But where the chances for material gain were unlikely, as was the case in some of the strategically important but poor islands and mainland territories discovered, there was little to motivate both entrepreneurs and the potential settlers they would have to recruit.

One solution to this problem, as we shall see, was to be the introduction of

---

[74] Anselmo Braacamp Freire, *História da Feitoria Portuguesa de Flandres*, Archivo Histórico Portugues.

[75] Perhaps the best known example of the delegation of royal privileges to an individual and/or institution in return for undertaking ventures on behalf of the crown was Prince Henry. As Grand Master of the Order of Christ, he was able to draw on its resources to exploit concessions and to found settlements on authority from three different kings, first his father, then his brother, and finally his nephew.

sugar cane. First produced in quantity in Madeira, sugar continued to be in short supply in Europe. The prospect of establishing its productin in other parts of the expanding kingdom held out the hope both of wealth and of further reward from a grateful monarch.

As the discoveries continued, therefore, the system of producing sugar on plantations manned by slave labour that had started in the eastern Mediterranean after the fall of the Latin Kingdoms, and which had been re-established and further developed in uninhabited Madeira, was introduced into otherwise worthless territories by a series of entrepreneurs who hoped to profit while also settling lands that were important to the crown and its implementation of the policy of commerical expansion.

In the process of securing and expanding the national patrimony then, sugar, the one crop the Europeans had learned to produce themselves during the conquests that led to the commercial revitalisation of the continent, was introduced, along with the plantation system and slavery, as the Portuguese secured their lines of supply and communication in their efforts to discover a sea route to the Orient. In this way the sugar plantation and plantation slavery came to be an integral part of the administrative, political and social system that was disseminated throughout the world with the expansions of the fifteenth, sixteenth and seventeenth centuries. For some examples of the spread of sugar cane cultivation, the plantation and plantation slavery we turn next to the discovery and occupation of the islands of the Azores, the Cape Verdes and São Tomé. As we shall see, both the crops produced and the system for their production changed as colonists adapted to new and diverse ecological settings. We may think of this retrospectively as a series of experiments in which the Portuguese used the experiences to adapt and refine the plantation system as a form of colonial settlement.

V

The Azores are a group of nine volcanic islands located in the central north Atlantic due west of the Iberian Peninsula. São Miguel, the largest island — slightly larger than Madeira — along with Santa Maria and the tiny rocks called the Fornigas, constitute what are referred to as the Eastern Azores. Terceira, Graciosa, São Jorge, Pico and Fayal then form the Central Azores, while Corvo and Flores, several hundred miles to the northwest, form the Western Azores.

When exactly the uninhabited islands first were discovered is not really known, although considerable speculation has been put forth on the matter.[76] It appears, however, that seven of the islands, the eastern and central groups, were known by the early years of the regency of the Infante D. Pedro. A strong effort then was made by his administration to settle the islands, especially São Miguel and Santa Maria. But with his death in 1449, and the return to power of the war party, interest declined as did colonising activity. Then, about a quarter of a century later, after the commercially oriented future D. João II was given responsibility for the "affairs of Guinea," a second effort resulted in the successful occupation and colonisation of the archipelago.

---

[76] T. Bentley Duncan, *Atlantic Islands* (Chicago: Univ. of Chicago Press, 1972), pp. 11-12.

The early history of the Azores, as Duncan has summarised it, "was associated with two other factors: (1) the introduction of the caravel for long-distance Atlantic navigation; and (2) the Portuguese push down the West African coast."[77]

"The two-masted, lateen-rigged, high-boarded caravel," he elaborates, "could sail close-hauled, six points off the wind . . . . It was fairly swift and nimble, but more important, it could withstand the heavy pounding of the Atlantic sea. Furthermore, it was better adapted to navigation in the Atlantic than the ships previously used by the Portuguese. The caravel, " he concludes, "made navigation to the Azores, and down the coast of Africa, cheap and practicable."[78] But still more important for our purposes, it made the Azores an integral part of the Portuguese overseas venture.

Ships returning from the African coast south of Senegal and Cape Verde had to battle strong head winds if they chose to stay close to the shores of the continent. A much easier route, it soon was discovered, was to swing out to sea from Senegal and beyond, forming a great semicircle that passed through the Azores on its way to Portugal. The islands of the Azores, as Duncan emphasises, then could become normal ports of call for ships choosing this safer route home from Africa and beyond.[79] But first the islands had to be settled and made secure in the name of the crown.

The earliest records indicate that the first Portuguese to land in the Azores left domestic animals to breed as a food supply for later settlers. This practice was to be repeated often elsewhere.

In 1439 Prince Henry became the lord proprietor *(donatário)* of the islands under a grant from his nephew, the young king D. Afonso V — but actually from his brother D. Pedro, the regent. According to the terms of the grant Henry was given permission to settle the seven islands that already had been discovered.[80] In a document dated 1443, D. Pedro, also in the name of the king, confirmed Henry's appointment of Gonçalo Velho Cabral, a knight of the Order of Christ, as his commander or captain in the islands.[81] The document also granted the colonists "a five year exemption from paying tithe and portage on the produce they exported from the islands to Portugal."[82]

The regent took a personal interest in the colonisation of the Azores, owning himself at one point the island of São Miguel.[83] Documents dates 1453 inform us that he had hoped to settle it with convicts, a plan that had been raised and rejected in the settlement of Madeira and was to be tried again in the Cape Verdes and São Tomé.[84]

---

[77] Ibid., p. 13.
[78] Ibid.
[79] Ibid., p. 14.
[80] Silva Marques, 1:401.
[81] Ibid., 1:425.
[82] Verlinden, *The Beginnings of Modern Civilization*, p. 221; Silva Marques, 1:425.
[83] Ibid., 1:452.
[84] Ibid., 1:517; 2:344.

According to Magalhães Godinho following the suggestion of Duarte Leite,[85] he also was responsible for the early introduction of sugar cane into the islands.[86] Velho Arruda, for example, refers to a mestre Antonio Catalão (referred to as Curvelo by Magalhães Godinho), one of the first settlers, "who came to the islands to teach the cultivation and industry of sugar cane."[87] Furthermore, Zurara makes reference to Gonçalo Velho, the captain of the island, receiving half the sugar produced in São Miguel.[88]

It appears, therefore, that sugar cane had been introduced into both São Miguel and Santa Maria, the first two of the Azores settled, quite likely in the hope of establishing in them plantations whose profits would help in attracting additional settlers who would secure the islands in the name of the crown, thereby integrating them into the transportation and communication systems of the expanding kingdom. But the regent, it must be remembered, was killed in 1449; and with his death the war party was back in control of the nation and the expansion and discoveries tapered off. Interest in the Azores also declined as did the efforts devoted to their occupation. When Prince Henry died in 1460, therefore, there were but several small colonies each in São Miguel and in Santa Maria which had not increased, and perhaps had even decreased, during the previous decade. Furthermore, they were not to grow significantly for the next quarter of a century.

When Gonçalo Velho died, the captaincies of São Miguel and Santa Maria passed to his nephew João Soares. Although Soares settled in Santa Maria, he did little to develop the island. In 1474, however, he sold São Miguel to Rui Gonçalves da Camara. Rui, the second son of João Gonçalves Zarco, captain of Funchal in Madeira, had served the king in Africa. Under his leadership, and with support from D. João, the population of São Miguel increased substantially and its economy took form. He started by bringing with him many residents of Madeira to whom he gave land grants in São Miguel. By the time of his death in 1497, five *vilas* (chartered, self-governing municipalities) had been established on the island, and there were other flourishing communities at Povoação and Lagôa. But sugar cane, it appears, was not to become the basis of the economy. Instead, wheat was raised — and exported to Portugal — and before the end of the century the blue-dye pastel (woad) was introduced to become the backbone of the island's economy for the next century and a half.

Sugar cane, however, eventually was to be reintroduced into São Miguel, but not until about 1540,[89] long after it had been tried in the Cape Verdes, São Tomé,

---

[85] Leite, p. 461.
[86] Vitorino Magalhaes Godinho, *Os Descobrimentos da Economia Mundial* (Lisbon: Editorial Arcádia, 1963-65), 2:450. See also *Documentos Sobre a Expansão Portuguesa* (Lisbon: Edições Cosmos, 1943-56), 1, Ch. 7.
[87] Manuel Monteiro Velho Arruda, *Coleção de Documentos Relativos ao Descobrimento e Povoamento dos Açores* (Ponto Delgado, Azores: Oficina de Artes Graficas, 1932), 135.
[88] Zurara, Ch. 83.
[89] Frutuoso, *Suadades da Terra*, Book 4 in Ilha de São Miguel (Ponta Delgado, Azores: Tipografia do Diario dos Açores, 1926), Vol. 2.

and was being introduced into Brazil. But by then the eastern islands already were an integral part of the successful commercially oriented overseas venture of Portugal; and the central and western islands also were well on their way. Long before this, however, the Portuguese were to discover other uninhabited islands in which sugar cane, plantations and slavery were to be experimented with in an effort to settle them and to integrate them into the expanding kingdom.

The Cape Verdes are composed of ten islands and some half a dozen islets located off the coast of West Africa between 22° 42' and 25° 22' west longitude and 14° 48' and 17° 13' north latitude. Boa Vista, the easternmost island, is approximately 300 miles west of Senegal, and some 500 miles southeast of Santo Antão, the westernmost island. The archipelago is composed of two groups. Santo Antão, São Vicente, São Nicolau, Sal, Boa Vista and Santa Luzia to the north form what are called the *Barlavento,* or Windward group, while Brava, Fogo, Santiago and Maio to the south form the *Sotavento,* or Leeward group. Santiago, the largest island, is somewhat larger than Madeira, but slightly smaller than São Miguel. The entire archipelago is situated due east of the Lesser Antilles.

Although the Portuguese mariner Dinis Dias reached the Cape Verde Peninsula, Africa's westernmost projection into the Atlantic, in 1441, the then uninhabited islands off the coast were not to be sighted for almost another fifteen years. And then it was not to be a Portuguese who was to discover and settle them, but rather an Italian sailing under the Portuguese flag. Although the Venetian Cã da Mosto reports sighting several of them when blown off course by a storm on his second voyage in 1456, and the Portuguese Diogo Gomes also claimed their discovery, it was the Genoese, Antonio da Noli, sailing in the employ of Prince Henry, who is credited with the official discovery in 1455.[90]

Prior to his death Prince Henry, as he had done previously in Madeira, divided the largest island into two captaincies. The southern half he gave to da Noli as reward for the discovery, while the northern half was given to Diogo Afonso, a squire in his household who had served him as *contador* in Madeira. In his will the Cape Verdes, along with Madeira and the Azores, passed to his nephew and adopted son, D. Fernando. Fom a document dated 1462, the first official source to provide us with information on the settlement of the Cape Verdes, we learn of the king's confirmation of D. Fernando as lord proprietor, along with the confirmation of da Noli and Afonso as captains of Santiago.[91] Afonso, however, never settled in the island and there was little development in his captaincy. Antonio da Noli, in contrast, settled in Ribeira Grade along with his brother and a nephew. The charter of privileges granted by the king on June 12, 1466 informs us that the colonisation of Santiago had begun some time in 1462.

With a charter and official permission for settlement, da Noli's major problem was to find settlers. Along with his brother and nephew, and a handful of members of the households of Prince Henry and D. Fernando, he founded a small settlement at Ribeira Grande, which soon was augmented by a group of Flemish

---

[90] Verlinden, *The Beginnings of Modern Civilization*, pp. 163ff.
[91] Carreira.

adventurers, some reprieved convicts, and a number of Jews and/or new Christians — beginning to feel the repression that was to blossom into the Inquisition at the end of the century.

According to Carreira, the first settlers probably subsisted on the wild cattle that had been left to breed by the discoverers of the island.[92] Shortly thereafter, he maintains, they began to produce crops for their own consumption. This was followed, he adds, by the "introduction of sugar cane and the setting up of *engenhos* to produce *aguardente* (sugar cane brandy) and moscovite sugar."[93] Amorim Parreira tells us that the cane plants most probably were introduced from Madeira.[94] Valentim Fernandes, writing in the early sixteenth century, refers to sugar cane being produced early in Santiago, but he does not give us a date.[95] Based on these sources Duncan concludes that "as in Madeira, Genoese seamen and merchants, backed by Genoese capital, tried to turn Santiago into a rich sugar-producing area."[96]

But sugar cane never was to thrive in Santiago. Although sufficiently warm, the island, along with other islands in the archipelago, was much too dry; and without sufficient water, sugar will not prosper. Instead, as we shall see, Santiago was to develop as a purveyor of slaves both for its own use and for sale in the other islands and later in the New World where the plantation system was to spread. And in the process Santiago itself was to become a plantation society.

The Portuguese had been taking slaves in tropical Africa since the voyage of Antão Gonçalves and Nuno Tristão in the early 1440s. In search of trade goods, and to augment their tiny population that was in need of a labour force, da Noli and his followers went off to the nearby Guinea coast. The slaves he brought back increased the population of his captaincy, while providing his settlers with a replaceable labour force. And they also came to be a commercial commodity that was to be as valuable as sugar cane.

Needless to say, a market already had developed for slaves in the metropole and in recently settled Madeira and the Azores in the Atlantic. The first slaves used in agricultural production, as we have seen, to grow sugar cane in Madeira, had been taken in the Canary Islands. But these islands, which are estimated to have had no more than 100,000 inhabitants when first discovered,[97] were close to being decimated by the end of the fifteenth century. They could not continue to provide labourers for the growing number of plantations created in the wake of the Portuguese expansion. But Africa could; and the settlers of Santiago in the Cape Verdes were to seize on that fact in the development of their island's economy. First, however, they satisfied their own needs; and with the aid of slave labour they settled the interior of their island, experimenting with other crops while using the plantation as their form of organisation.

[92] Ibid., p. 17.
[93] Ibid.
[94] Amorim Parreira, p. 56.
[95] Ibid.; Magalhães Godinho, *Os Descobrimentos da Economia Mundial*, 2:452
[96] Duncan, p. 19.
[97] Magalhães Godinho, *A Economia das Canárias nos Séculos XIV e XV*, Revista da Historia 10.

Although never able to find a crop that could equal sugar cane as a potential source of wealth, thriving cotton plantations and a cotton textile industry did develop. But the finished product never was to be sold in Europe. Instead, it was traded along the Guinea coast, becoming one of the preferred items accepted in exchange for slaves.[98] The plantations of Santiago therefore produced a commodity whose commercial value was to be realised only secondarily, from the profits earned from the sale of slaves.

In the growth of their colony, with the incorporation of the African slaves into the population and their use as a labour force on the plantations and elsewhere, a unique situation developed in Santiago that was to become commonplace in the centuries to come in the tropical New World. A small group of whites, who came from and were loyal to an expanding European nation-state, dominated a growing economy manned by a superior number of black slaves imported from continental Africa. And as the Europeans mated with their increasing number of African slaves, a mulatto population was created. In brief, Santiago was to become the prototype of a plantation society inhabited by a small number of whites, served by a large number of blacks, to which were added in each generation a growing number of mulattos.

In the charter of privileges granted the settlers of Santiago by King D. Afonso V, permission was given for them to trade in both goods and slaves along the Guinea coast, with the exception of Arguin where a *feitoria* had been established.[99] Furthermore, once they had paid the royal duties on the goods obtained, they were exempted from additional taxes otherwise imposed on items sold in Madeira, the Azores, the Canaries and the metropole.[100] In brief, they could trade in slaves. But in 1469 the king leased all trade on the Guinea coast, including that in slaves, to Fernão Gomes for a period of five years for a fee of 200,000 *reais* per year. The consequence of the contract, which gave Gomes a royal monopoly, made the most lucrative activity of the colonists in Santiago illegal. Although they could continue to obtain slaves for their own use, they were prohibited from engaging in the one activity that had brought them a profit. In defiance of royal autority, therefore, they went off surreptitiously to the African coast and from there to the Azores, Canaries and Madeira to trade illegally in slaves.

As a racially mixed, plantation society was being created on the island numerous European settlers moved off to the African coast and then upriver to trade in slaves and other merchandise. There they formed alliances — both commercial and sexual — with African women which brought them a privileged position in local commerce. These *lançados*, or *tangomãos*, as they were called,[101] with their African consorts, also were to produce mulatto offspring who, with their creole language and a toehold in both cultures, were to dominate trade in slaves and other commodities as intermediaries between successive

---

[98] Carreira.
[99] Ibid.
[100] Ibid., pp. 22ff.
[101] Ibid., pp. 47ff.

European powers and the leaders of West African states for centuries to come.[102]

In the meantime, however, a viable colony had been founded in Santiago that was to provide one more link in the chain that was to enable the Portuguese to continue down the African coast and then around it to the Orient by the end of the century. Although never becoming a plantation society based on sugar cane, as had Madeira before and as were São Tomé and Brazil after, Santiago and the Cape Verdes developed as the first racially mixed society spawned by Europe in the tropics. As such it was to serve as a model for later colonising efforts in the New World.[103] But as this new dimension was being worked out, the Portuguese were moving into the Gulf of Guinea where further experimentation was to take place.

São Tomé is a volcanic island of 325 square miles located in the Gulf of Guinea approximately 100 miles west of Cape Gabon at 6° west longitude at the equator. It was first discovered in 1470 by Pero Escobar and João de Santarem. Its importance was to be its location at the convergence of northerly and southerly wind systems that made it a natural turning point for vessels sailing the African coast. That is, northerly winds would carry southbound ships to its position where southerly winds could be picked up to take them into the Atlantic where the Prevailing Westerlies then would carry them back to Portugal. In spite of this, however, little was done to settle the island for more than a decade. But then, after D. João II succeeded to the throne and was making a major effort to round Africa and reach Asia in a programme of commercial expansion, São Tomé was to be settled.

To establish a colony in the uninhabited island D. João resorted to the system that had been applied so successfully by his predecessors in Madeira, the Azores and the Cape Verdes. In 1485, therefore, he named João de Paiva of Óbidos as *donatário* and captain of the island.[104]

Paiva left for São Tomé in 1486 with a small group of settlers, "either from the homeland or from Madeira."[105] But the colony was not to prosper. Paiva himself died in 1490, to be followed just three years later by his successor, João Pereira — who Garfield assumes married Paiva's daughter.

By this time, however, the strategic location of the island had become clear to the crown as it was coming closer to rounding the African continent. Therefore, D. João was to make a more determined effort to have São Tomé settled. In 1493 he transferred its ownership to Alvaro de Caminho, "a gentleman of the Royal Household who had served in the navy and in 'Guinea.'"[106] Although receiving greater privileges than had Paiva and Pereira, Caminho faced the same problem that they, along with Zarco, Tristão and Perestrelo in Madeira, Velho in the Azores and Noli and Afonso in the Cape Verdes, had faced. He had to recruit

---

[102] Ibid., pp. 62-73.
[103] Ribeiro.
[104] Actually of half of the island. The other half, also on the model of Madeira and Santiago, was given to Paiva's daughter, with its administration to be vested in whomever she might marry. See Robert Garfield, "A History of São Tomé Island, 1470-1655," (Ph.D. dissertation, Northwestern University, Ann Arbor, Michigan, 1971).
[105] Ibid., p. 5.
[106] Ibid., p. 7.

settlers to augment the tiny group already there and to develop the island.

According to the terms of his charter, he was given permission to take condemned prisoners from the jails of the mother country. But more important in this case, he also was allowed to take with him to São Tomé some 2,000 Jewish children, many of whom had only recently arrived in Portugal with their families after escaping repression against them in Castile.[107] The children, mostly between the ages of two and ten, were baptised and then sent off as part of the São Tomé colony.

In his charter Caminho also was given permission to take slaves on the nearby African coast. The settlement in São Tomé that he eventually established, therefore, was composed of the small group brought by Paiva to which Caminho added prisoners and others recruited in Portugal, the Jewish children, and the blacks brought as slaves now from the Congo. The slaves, of course, were to be the labour force for the Europeans.

In the original charter given to Paiva there is a clause reserving one-fourth of the sugar to be produced on the island for the king. By 1485, therefore, the crown had assumed that sugar cane would be introduced into the uninhabited island to be the basis of its economy. The Europeans to settle then were assumed to be the future plantation owners and managers, while the blacks brought as slaves from Africa would be their labour force. In this way, it was assumed, that a colony would be established at the expense of the captain-*donatário*, from which he would profit, as would his colonists and the royal treasury, all while securing for the crown a strategically located island.

Sugar cane, we are told was introduced in São Tomé by its first settlers who in 1485 brought it from Madeira, along with specialists in its cultivation and production.[108] These specialists, Garfield adds, "included not only Portuguese, but also Genoese and Sicilians who had been employed in sugar-cultivation in that island."[109]

In contrast with the arid Cape Verdes, and more like Madeira, São Tomé had an abundance of rain and numerous swift, flowing streams. It also had rich volcanic soils, was tropical, and on the north side had extensive flat lands. Given the introduction of sugar cane, specialists to teach its production, European settlers given land in *sesmaria*, and a labour force from Africa, it was to be just a matter of time before the island was to become a rich plantation colony. Although it was to take some time for the plantations to prosper, by 1495 we have the first

---

[107] Portugal, at the time, was assumed to have had some one hundred thousand Jews in a population of little more than a million. With the repression in Castile and Granada that number was being increased to a point at which the king came to see it as a potential threat to the Catholicity of the country. See José Lucio de Azevedo, *História dos Cristãos Novos Portugueses* (Lisbon: Livraria Iassica, 1921), p. 21 and Ch. 5.

[108] Amorim Parreira, p. 57; Garfield, p. 47.

[109] Ibid.; see also Anonymous, *Viagem de Lisboa à Ilha de S. Tomé* (Lisbon: n. pub., 1555), p. 57; Francisco Tenreiro, *São Tomé, Um Exemplo de Organização do Espaço* (n.p.: Junta da Investigações do Ultramar, 1951), p. 70.

reports of sugar from São Tomé arriving in Antwerp.[110] In the meantime, however, the island was occupied, to become one of the many in an expanding network of strategically located island bases.

In a series of fits and starts that was to cover more than half of the fifteenth century, then, we have seen a series of Portuguese governments use sugar cane and the model of the plantation as the carrot, so to speak, that motivated private individuals to undertake and finance themselves the settlement of uninhabited islands in the Atlantic Ocean and in the Gulf of Guinea. In what in retrospect has been likened to a series of experiments, the plantation system, now combining African slaves under the authority of European settlers in a racially mixed society, producing sugar cane and other commercial crops, spread as island after island was integrated as part of the expanding kingdom. In only some of the islands did sugar cane plantations prosper, however, and provide their founders with the wealth they had hoped for. But overall, sugar cane and the plantation did enable the government of Portugal, once it had committed itself to the policy of commercially oriented expansion, to have settled, at the expense of private citizens, island bases that gave her control of the South Atlantic and made possible the rounding of Africa and trade in the East.

VI

Within two years of the return of Vasco da Gama from the first successful sea voyage to India, an impatient King D. Manuel I, who had succeeded D. João II in 1495, organised a fleet of thirteen ships to return around Africa and establish trade and a Portuguese presence in Asia. After sailing past the Canary and Cape Verde Islands, the fleet, under the command of Pedro Álvares Cabral, swung far into the Atlantic Ocean to the west. It has been claimed by some that Cabral was just following the advice given him by Vasco da Gama to facilitate the rounding of the Cape of Good Hope. Others, however, maintain that Cabral deliberately crossed the Atlantic to lay claim to land that had been previously discovered. In any event, on April 22, 1500 land was sighted and Cabral officially laid claim to what he called Vera Cruz, soon changed to Santa Cruz. And so Brazil was officially discovered.

After but brief contact with the land and its inhabitants, Cabral dispatched a ship to inform the king of his discovery. He then proceeded with the rest of his ships and men to India, never to return to the continent he had discovered. His undertaking, however, had enabled Portugal to lay claim to a portion of the New World.

By the time Portugal had resolved her internal struggle and had become fully committed to commercial expansion down the coast of Africa and beyond to Asia, a rival had appeared seeking the same objective in the Far East. Castile, which had merged with Leon and was united with Aragon by royal marriage, had gained in military stature. As she then took the lead in finally expelling the Moors

---

[110] Amorim Parreira, p. 58; Edmund von Lippmann, *História do Açúcar*, trans. into Portuguese by Rodolfo Coutinho (Rio de Janeiro: Instituto do Açucar e do Alcool, 1941-42), 2:374.

from the peninsula, and unifying Spain, she also challenged Portugal's hegemony over the lands and waters of the Atlantic and the coast of Africa. Then a short war between the Iberian neighbours ended in 1479 with the signing of the Treaty of Alcaçovas. Portugal had agreed to relinquish its long standing claim to the Canary Islands in exchange for Castile renouncing all claims to the other islands in the Atlantic and to the African coast. When the treaty was endorsed two years later by papal bull, it appeared that Portugal had gained a monopoly both in the Atlantic and in Africa and Asia — when she was to reach it. But in 1492 that monopoly in the Atlantic — and by implication in Asia — was broken by the voyage of Christopher Columbus. The Genoese, sailing in the employ of Ferdinand and Isabel, it must be remembered, had claimed to be seeking a westward route to Cathay and the spice islands. The Catholic Monarchs then appealed to the papacy for rights in Asia based on Columbus' voyage.

To avoid conflict between the most powerful Christian nations of the period, the Pope "proceeded to divide the unknown and newly discovered portions of the globe between them. His decision," as Poppino has summarised it, "reflecting scant knowlege of the geography of the known world, provided for an imaginary line from pole to pole one hundred leagues west of the Azores and the Cape Verde Islands."[111] To the east of this line Portugal would exercise monopoly rights while to the west Castile would be supreme.

The Portuguese, however, found this unacceptable. Under threat of war with Spain they insisted that the line be drawn further to the west. After extensive negotiation it was reset at 370 leagues to the west of the Cape Verde Islands. The agreement was signed with papal endorsement in 1494 as the Treaty of Tordesillas. Although there is no evidence to demonstrate prior knowledge of Brazil by the Portuguese, such knowledge would explain their insistence on the second line to mark the respective spheres of domination. Under the new agreement the east coast of South America fell within the Portuguese half of the globe.

Pero Vaz da Caminha, the scribe who had accompanied Cabral, sent back to D. Manuel a lengthy report describing the land and the people that had been found in Brazil. But since he had stated what others to follow him later confirmed, that the natives were little advanced and had little of commercial value, and furthermore that there was no evidence of either spices or precious metals, the crown and the merchant community were to have little interest in the New World. Besides, there was business to attend to in India where untold riches and treasures in jewels, silks and spices were to be gained in trade and tribute. Little attention, therefore, was to be paid to the New World for the first three decades of the sixteenth century. Although occasional voyages were made, and *feitorias* established along the coast — to obtain the dye-producing tree whose name was to be given to the colony — it is questionable whether Brazil ever would have been settled and integrated into the overseas Portuguese domains had it not been for the interest shown in it by the French.

France, like the other nation-states of Europe, had been excluded from direct

---

[111] Rollie E. Poppino, *Brazil: The Land and People* (New York: Oxford Univ. Press, 1968), p. 43.

overseas commerce and settlement by the Tordesillas treaty. The French king, however, protested, refusing to accept the monopolies awarded to Portugal and Spain. Although not strong enough to challenge the Iberian powers openly, and not wishing to test papal sanction, he, like his English counterpart soon after, would send out merchant-adventurers to trade in and/or to plunder in the domains of the Iberians. In addition, he gave permission to Protestant dissenters to settle in the New World. By 1530, therefore, France and French settlers constituted a serious threat to Portugal's hold on Brazil.

The crown had responded to the French presence in its New World domains by sending troops on four occasions between 1516 and 1530, to sweep the "pirates" from the sea. But in spite of these Portuguese efforts, permanent French settlements continued to be present on the continent.

A further incentive had been provided the Portuguese crown to secure Brazil by the Spanish experience in the Andes and in the Valley of Mexico. Since gold and silver had been found in the Spanish part of the continent, why should there not also be precious metals on the Portuguese side? But to even search for them the French had to be driven from the continent.

In contrast with the peoples to occupy the lands the Portuguese had conquered in Africa and Asia, and the Spaniards had subjugated in the highlands and in Mexico, the natives of Brazil were not organised politically. Consequently, they could not be used, as had been the others, to combat European and other interlopers. The only way to secure Brazil, therefore, was to have it settled by a population loyal to the Portuguese crown. The question was how?

In 1530 King D. João III, who had succeeded D. Manuel I in 1521, sent Martim Afonso de Souza to clear the French from Brazil, to establish *feitorias* in which to trade with the natives, to search for precious metals, and to establish a colony. Afonso did plant the first Portuguese colony in America. But on his return the king decided to resort to what Merêa refers to as the by now traditional solution to the problem of settlement.[112] He divided the continent into twelve captaincies over which proprietorship was given to loyal subjects who had served him. In return for royal privileges, the captains agreed to settle their lands with colonists at their own expense while paying the royal treasury a percentage of the profits to be earned — and the precious metals discovered. The model of Madeira, which had worked successfully in the Azores, the Cape Verdes and in São Tomé, was to be transferred to Brazil. And as in the Atlantic and in the Gulf of Guinea, sugar cane was to be the catalyst that would transform a worthless — to the merchants and settlers of Portugal — wilderness into a rich and successful colony.

Sugar cane had been introduced into northeast Brazil sometime before the captaincy system was established. In 1526, for example, duty had been paid to the Casa da India on sugar from Pernambuco. But under the leadership of Duarte Coelho, who was to receive the Captaincy of Pernambuco, it was to be firmly established on an extensive basis, to spread up and down the entire coast.

In general the captaincy system is regarded as a failure. In only two cases were

---

[112] Paulo Merêa, A. Solução Tradicional da Colonização do Brasil," in Carlos Malheiro Dias, ed., *Historia da Colonização Portuguesa do Brasil*, Porto, Portugal: Litografia Nacional, 1921), pp. 167-83.

permanent settlements established. But these were to be sufficient to attract further settlers and to secure the continent for Portugal.

In Pernambuco, as we have just noted, Duarte Coelho brought settlers who planted sugar cane. The plants, along with the techniques for cultivation and processing, were brought from Madeira and São Tomé, from where many of the colonists also came. But in contrast with the small mountainous islands in the Atlantic and in the Gulf of Guinea, Pernambuco, and the Brazilian coast in general, was immense and flat. It also had sufficient rainfall. The average captaincy measured 50 leagues along the coast and extended inland to the imaginary Tordesilla line — wherever that was. Consequently, the *sesmarias* given to the settlers were larger and even better suited to sugar cane than had been those in the islands. The plantations of Pernambuco, therefore, were to be larger and more productive than anything that had been known before.

Brazil also was to differ from the islands in another way. Whereas the latter had been uninhabited when discovered, the continent was not. But to transform it into sugar plantations on the model of Madeira and the other islands, the inhabitants had to be either eliminated and replaced, or transformed into slaves. The second alternative was to be tried first. The Bandeirantes, as the descendants of Martim Afonso's colonists on the São Paulo plateau came to be known, organised an economy around the capture and enslavement of the natives of the interior who then were sold to the settlers of the northeast. But the Indians proved themselves to be inefficient and difficult workers. Therefore, the plantation owners in the northeast turned to what by now had become the accepted source of agricultural labour. As had their predecessors in Santiago and in São Tomé, they turned to Africa where they found an inexhaustible supply of more efficient workers. Within a short time northeast Brazil, like the Cape Verdes and São Tomé, was to become a racially mixed, plantation society in which black Africans, as slaves, toiled for white Europeans, to produce commercial crops for sale and profit in Europe.

Sugar, the plantation and plantation slavery had arrived in the New World. But given Europe's insatiable demand for sugar, the plantations of northeast Brazil soon were to earn wealth equal to that which had been obtained in the Asian trade. By the end of the sixteenth century Brazil was to become one of the richest and most oppulent colonies the world had ever known.

From Brazil, of course, the system that had gradually been put together over the course of more than three centuries was to be carried by the Dutch to the West Indies. After first capturing the sugar rich Brazilian colony of Pernambuco, and then being expelled, the Dutch introduced sugar cane, the plantation and plantation slavery into the islands of the Caribbean Sea. In the hands of yet other Europeans to enter into competition with Portugal and Spain the system then spread throughout the New World, to replace what had existed when the Europeans first arrived. The plantations that had started in the eastern Mediterranean, and had been incorporated by the Portuguese state as part of a policy of commercial expansion for the settlement of politically important but of minimal commercial worth regions, had been transferred to tropical America to become the institution, with its racially mixed populations of freemen and slaves, that was to dominate the shape of society in the region for centuries to come.

2
Indian Labor and New World Plantations: European Demands and Indian Responses in Northeastern Brazil

Stuart B. Schwartz

IN RECENT YEARS INTEREST in the early years of European conquest and colonization in the New World has undergone remarkable resurgence. No longer concerned solely or primarily with the effect of this process on subsequent national histories, scholars have increasingly turned their attention to the role that the New World colonies played in the creation of an Atlantic—even a world—economic system. This resurgence of interest has resulted in lengthy and often heated debates on the nature of colonization in the sixteenth and seventeenth centuries and on the dominant mode of production which prevailed in these colonies. Recently, Immanuel Wallerstein has argued that the Americas, as a peripheral zone of capitalist expansion, experienced certain forms of coerced labor such as chattel slavery and the *encomienda*, both of which were necessary forms of colonial exploitation that permitted the formation of a surplus sufficient to make such colonial ventures worthwhile.[1] A Brazilian scholar, Fernando Novais, has even suggested that European merchants seeking high returns on investments were responsible for imposition of the Atlantic slave trade and that the slave trade created African slavery in the

---

I would like to express my thanks to Richard Graham, David Sweet, Consuelo Pondé de Sena, Dauril Alden, and Stanley Engerman for helpful suggestions and criticisms. Part of the research for this article was done with the aid of grants from the American Council of Learned Societies (1974–75) and the University of Minnesota Graduate School (1972). The following abbreviations have been used: *ABNR* (*Anais da Biblioteca Nacional do Rio de Janeiro*); AGS (Archivo General de Simancas); AHU (Arquivo Histórico Ultramarino, Lisbon); ANTT (Arquivo Nacional da Tôrre do Tombo, Lisbon); BA (Biblioteca da Ajuda, Lisbon); BI (Biblioteca do Palacio da Itamaraty, Rio de Janeiro); BNL (Biblioteca Nacional de Lisboa); Cart. *Jesuitas* (Cartório dos *Jesuitas*); *DH* (*Documentos Históricos da Biblioteca Nacional do Rio de Janeiro*, 120 vols. [1928–]); *DHA* (*Documentos para a História de Açúcar*, 3 vols. [Rio de Janeiro, 1954–63]); *HCJB* (Serafim Leite, ed., *História da Companhia de Jesus no Brasil*, 10 vols. [Lisbon, 1938–50]); and *MB* (Serafim Leite, ed., *Monumenta Brasiliae*, 4 vols. [Rome, 1956–60]).

[1] Wallerstein, *The Modern World System* (New York, 1974). His bibliography provides an excellent introduction to the historical and theoretical literature. Wallerstein has recently anticipated my criticism and similar observations made by Domenico Sella by stating that "the alternatives available to each unit are constrained by the framework of the whole even while each actor opting for a given alternative in fact alters the framework of the whole." Thus, the difference in our positions may be one of emphasis, although it is clear that, for him, the alternatives are still determined by the system and not by the actors. See his essay, "The Three Stages of African Involvement in the World Economy," in Peter C. W. Gutkind and Immanuel Wallerstein, eds., *The Political Economy of Contemporary Africa* (London, 1976), 30. Also see Domenico Sella, "The World System and Its Dangers," *Peasant Studies*, 6 (1977): 29–32. For an important set of essays on these problems, see Carlos Sempat Assadourian *et al.*, eds., *Modos de producción en América Latina*, vol. 40 of Cuadernos de Pasado y Presente (Buenos Aires, 1973).

New World, not the reverse.² Thus, the nature of the dominant mode of production and the creation and use of a labor force have become issues crucial to an understanding of the New World.

The broad sweep of these theses tends to leave the impression that the growth of capitalism was an inexorable process resulting from conscious decisions and choices made at the center of the world economy. This impression is Eurocentric, not in the old sense of cultural myopia but in the emphasis on Western desires and decisions without much regard for the objective cultural and physical realities in the "colonial" areas. Examining the nature of Indian slavery in the formative period of a plantation economy in northeastern Brazil can be a way of demonstrating how accurate the impression is, how local conditions and the specific cultures of non-Europeans shaped the formation of the various colonial regimes.

Throughout the Americas European powers attempted to make use of the American Indians as a source of labor. With a few major exceptions these attempts proved unsuccessful. Still, from the Carolinas to Santo Domingo and Brazil attempts to enslave native Americans preceded the period of African slavery.³ For plantation agriculture Indian slavery proved transitory, but in frontier regions like northern Mexico and the Amazon it lasted until the nineteenth century. The attempt to use Indians as a coerced labor force, in any case, cannot be simply dismissed as a "false start." The complex interplay of European and Indian perceptions and actions determined the ways in which Indians did—and did not—become integrated into the colonial regimes. In Brazil, Indian slavery had a short history in legal terms (roughly from 1500 to 1570), but various forms of coercion were used well after those dates to acquire indigenous laborers. Even after the large-scale introduction of African slaves, Indians continued to be a major source of labor. An examination of both the general outlines of the history of Indian labor in northeastern Brazil and the nature of life for Indians on the sugar plantations should help explain the role that Indians played in this plantation economy and the reasons that impelled the Portuguese to turn to Africans in the Atlantic slave trade.

A FULL ETHNOGRAPHY OF THE INDIGENOUS PEOPLES of Brazil on the eve of European colonization would be pointless here, but some aspects of the society and economy of the major groups encountered by the Portuguese on the Brazilian coast do help clarify the process of their absorption into a plantation economy. The most numerous and widely dispersed of the Indian peoples who came into contact with Europeans in the first two centuries of Brazilian colonial history spoke the Tupi-guaraní languages⁴ and controlled

---

² Fernando Novais, *Estrutura e dinâmica do antigo sistema colonial (Séculos XVI–XVIII)*, Caderno CEBRAP, no. 17 (São Paulo, 1974). Also see his "O Brasil nos quadros do antigo sistema colonial," in Manuel Nunes Dias et al., eds., *Brasil em Perspectiva* (São Paulo, 1968), 53–71.

³ Winthrop Jordan, *White Over Black* (Chapel Hill, 1968), 89, provides a perceptive discussion of English attitudes toward Indians. Also see note 83, below.

⁴ See the survey presented in Estevão Pinto, *Os indígenas do Nordeste*, 1 (São Paulo, 1935): 168–246. Also see Carlos Ott, *Pre-história da Bahia* (Salvador, 1958), 11–33; Alfred Métraux, "The Tupinambá," in Julian Steward, ed., *Handbook of South American Indians*, 3 (Washington, 1948): 95–135; and Julio Cezar Melatti, *Indios do Brasil* (Brasília, 1960). For the best single sixteenth-century source on the indigenous peoples of

much of the littoral from Maranhão to Santa Catarina. Of these groups the best ethnographic information is available on the Tupinambá, who dominated the coast around the Bay of All Saints. They lived in villages of four to eight hundred individuals organized into large family units which shared some four to eight long houses. Patrilineal kinship was central to their societal organization, but divisions of sex and age also defined responsibility and privilege.[5]

The early colonists adopted the Tupinambá practice of calling all non-Tupi speakers by the general term "Tapuya"—speakers of twisted tongues. For many years anthropologists believed that these peoples belonged to the Gê linguistic family. Many did, but it is now clear that the peoples the Portuguese called Tapuya belonged to a number of linguistic and cultural groups.[6] Since the Tupinambá occupied much of the coast, most of the other peoples had far less contact with the Portuguese in the sixteenth century. Their hunting and gathering economy tended to produce lower population densities, a simple material culture, and a nomadic existence. Migration, intertribal warfare, and Portuguese slaving did, however, bring these peoples within the European orbit. The Portuguese naturally considered the Tapuya particularly barbarous and irrational, in part because their very mobility made groups like the Aimoré effective military opponents.

In Tupinambá society, the acquisition of status, the choice of marriage partners, and progress through the ranks of age largely depended upon the manly activities: the capture of enemies in war and their eventual death as victims in a feast of ritual cannibalism.[7] This need for captives necessitated constant warfare among the various Indian peoples; and ritual cannibalism, naturally abhorrent to the Portuguese, became a principal excuse for the enslavement of the Tupinambá and other peoples. Although warfare and ritual cannibalism perhaps provided the underpinnings of the Tupinambá's view of the universe and social organization, other features of Tupinambá culture are important for understanding Indian relations with the Portuguese. Unlike some of their neighbors, the Tupinambá practiced agriculture; it was well suited to their habitat and needs and it formed an important part of Tupinambá life.[8] Essential to their subsistence was manioc, which they grew as their primary crop. The root flourished in a wide variety of soils, was resistant to the ravages of most insects, provided a high number of calories in relation to the area planted, and needed little care after planting. Made into a flour, manioc became the principal food of European as well as native Brazilians. And, along with maize, manioc was one of the primary Indian contributions to the world's diet.

Bahia, see Gabriel Soares de Sousa, *Tratado descretivo do Brasil em 1587* (4th ed.; São Paulo, 1971), 299-341.
[5] Florestan Fernandes, *Organização social dos Tupinambá* (2d ed.; São Paulo, 1963), 149-309; and Métraux, "The Tupinambá," 119-26. Also see Egon Schaden, *Aspectos fundamentais da cultura guarani* (São Paulo, 1962).
[6] Ott, *Pre-história da Bahia*, 11-33; and Soares de Sousa, *Tratado descretivo do Brasil em 1587*, 78-80.
[7] Florestan Fernandes, *A função social da guerra na sociedade Tupinambá* (2d ed.; São Paulo, 1970).
[8] A. Métraux, *La religion des Tupinambás* (Paris, 1928), 170-71; and Fernandes, *Organização social dos Tupinambá*, 82-98. Métraux calls the Tupinambá an "agricultural people," but it is clear from his work and that of Fernandes that agriculture was not a major ceremonial force in Tupinambá society.

46   *Stuart B. Schwartz*

The Tupinambá employed a form of slash-and-burn agriculture (*coivara*) still in use in parts of Brazil today. The men undertook the heavy labor of opening a field by felling the large trees and then set fires to clear the underbrush; the ashes provided a natural fertilizer. To the women fell the tasks of planting, harvesting, and food preparation. Thus, routine farming was almost exclusively women's work. The men supplemented the Tupinambá diet with game and fish; but only in the heavy communal labor of clearing timber did the men enter the agricultural cycle.[9] The Tupinambá economy was, therefore, basically one of subsistence with a communal or reciprocal attitude toward production. Land and sustenance were distributed according to lineage affiliations; every member of the group through familial ties was assured a minimum part of the whole. "These Indians maintain the time of the Apostles," one Jesuit commented. "They have nothing of their own; everything is common among them."[10] Father Manoel da Nóbrega also suggested the Tupinambá "owned all things communally and that which one has must be divided with the rest, principally if it be foodstuffs." Another Jesuit, Juan de Azpilcueta Navarro, made a classic observation of the more nomadic Indians when he wrote that the Tapuyas of Porto Serguro were "very poor, owning nothing of their own privately; they eat in common that which each day they hunt and fish."[11]

If the Indian economy was essentially communal and subsistent, it was autoconsumptive as well. Each village produced what it needed and depended very little on trade in foodstuffs.[12] It was, therefore, an economy based on production for use rather than for exchange, a system which provided a comfortable livelihood without concern for profit in the Western sense. There was no need to produce to the capacity of time or technology. In such an economy, the tempo of work and of production was intermittent and discontinuous;[13] time was always left for leisure and other "nonproductive" activities like warfare and ceremony. The habitat of coastal Brazil facilitated such attitudes, since an adequate food supply could be obtained without extraordinary effort. This relatively secure food supply also made it quite easy for the Portuguese in Bahia to obtain manioc flour (*farinha*) and other food by trade in the period of early contact; and during the 1550s they purchased large

---

[9] On women in agriculture, see the cross-cultural study by Michael Burton, Lilyan A. Brudner, and Douglas R. White, "A Model of the Sexual Division of Labor," *American Ethnologist*, 4 (1977): 227-50. Also see George P. Murdock, "Factors in the Division of Labor by Sex," *Ethnology*, 12 (1973): 203-25.

[10] Martin da Rocha (September 1576), as quoted in *HCJB*, 2:90. Also see Ambrósio Fernandes Brandão, *Diálogos das Grandezas do Brasil*, ed. José Antônio Gonçalves de Mello (2d complete ed.; Recife, 1966), 199.

[11] Nóbrega, "Informação das terras do Brasil, 153; and *MB*, 2: 249 (1555). Observations like these are probably not quite true. Most Indian peoples recognized individual ownership of goods and objects of production (bows, axes, etc.) but collectively made use of goods of consumption. See Melatti, *Indios do Brasil*, 68-69. On the economy of a Gê group, see Roberto da Matta, "Notas sobre o contato e a extinção dos indios Gaviões do medio Rio Tocantins," *Revista do Museu Paulista*, new ser., 14 (1963): 192-93.

[12] Fernandes, *Organização social dos Tupinambá*, 84-85. Trade was not entirely lacking. For an important theoretical essay, see Claude Levi-Strauss, "Guerra e comércio entre os indios de America do Sul," *Revista do Arquivo Municipal* (São Paulo), 87 (1942): 131-46.

[13] These comments are based on Marshall Sahlins, *Stone Age Economics* (Chicago, 1972), 1-41. Although his analysis deals with Stone Age economics in general, it is directly applicable, contemporaneous observations indicate, to the Amerindian economies under consideration here.

Indian Labor and New World Demands

amounts of farinha from villages along the coast.[14] The Tupinambá, after satisfying their own needs, attached little importance to their surplus and willingly traded it for useful European goods on a limited basis. But, since the limits of the Tupinambá's willingness were finite, they proved to be an undependable source of food and, later, of labor.

A communal or reciprocal attitude toward production and consumption, the domestic mode of production, a society on which status was not derived from economic ability, and the subordination of the economy to other forms of social organization determined Indian responses to European demands. The divergent outlooks of Portuguese and Indians toward the nature and goals of labor and production lie beneath the change in Portuguese-Indian relations in the sixteenth century and help explain the subsequent history of Indian slavery.

Undependability, seeming prodigality, and lack of interest in profit, surplus, and savings grated upon European sensibility; and more than once such attitudes were offered as proof of Indian irrationality and thus evidence of their lack of "humanity." In 1610 Governor Diogo de Meneses wrote, "These Indians, Sir, are a very barbarous people who have no government and are unable to govern themselves; they are so lacking in this regard that even in their sustenance they will not save for tomorrow that which is in excess today. . . ." "Inconstant and changeable," Jesuit Simão de Vasconcellos called them. "That which they struggle to gain today with great labor and sweat tomorrow they disregard."[15] Among the agricultural Tupinambá, the Portuguese were aghast at the "idleness" of the villages, where the men seemed to loll about smoking and preparing for battle.[16] Portuguese attitudes toward the "barbarism" of the agricultural Amerindians were intensified in the face of tribes who lived only by hunting and gathering. It was the confrontation of two peoples whose economic systems and visions of life were worlds apart.

European perceptions and European demands had to be harmonized with the colony's need for food, labor, and defense. The earliest European activity in Brazil was the cutting and export of dyewood logs from the famous brazilwood trees. As logging depleted the supply along the coast, the Europeans increasingly turned to the Indians to provide the wood. Because communal labor, especially the felling of trees, was a characteristic masculine chore in Tupinambá society, this activity was easily integrated into the traditional patterns of indigenous life. Indians seemed quite willing to cut the trees and drag the heavy logs to the coastal region where they could be exchanged for trinkets and other trade goods. Between 1500 and 1535 the

---

[14] Alexander Marchant, *From Barter to Slavery* (Baltimore, 1942), 87–95.

[15] Diogo de Meneses to the crown, September 1, 1610, ANTT, Fragmentos, caixa 1, no. 6; and Simão de Vasconcellos, *Chronica da Companhia de Jesu do Estado do Brasil*, 3 vols. in 2 (Lisbon, 1865), 1: lxxxii.

[16] "They are great friends of leisure (*folgar*)," Os Capitulos de Gabriel Soares de Sousa, *ABNR*, 62 (1940): 373. For an interesting comparison see Edmund Morgan, *American Slavery, American Freedom: The Ordeal of Colonial Virginia* (New York, 1975), esp. chap. 3: "Idle Indian and Lazy Englishman"; and Richard R. Beeman, "Labor Forces and Race Relations: A Comparative View of the Colonization of Brazil and Virginia," *Political Science Quarterly*, 76 (1971): 609–36. Royal administrators in Spanish America used the "natural idleness' of Indians to justify forced labor. See Josefina Cintrón Tiryakian, "La imagen económica del indio," in *Actas del XLI Congreso Internacional de Americanistas*, 3 vols. to date (Mexico, 1976–), 2: 429–35.

Portuguese used barter as the principal means to obtain both brazilwood and, secondarily, manioc flour from the Indians. Barter was also an indirect method for obtaining labor. We do not know whether the Tupinambá and others traded logs, manioc flour, and labor individually or communally during this period, but records of purchases made in the 1540s indicate that the latter was true. If so, the barter system functioned within the parameters of community activity and was all the more adaptable to traditional patterns of life.

With the introduction of the donatary system in the 1530s, the situation in colonial Brazil began to change. The crown granted rights to Portuguese noblemen who, by settling colonists and establishing a secure economic basis, made new demands on the Indian inhabitants. The donataries and the new colonists continued to barter for the brazilwood, food, and even labor needed on a short-term basis for town building—all suitable to the traditions of intermittent effort, sexually defined roles, and communal advantage. But the sustained and continuous agricultural labor of the kind needed to raise and harvest the new crop of the settlement—sugar—could not be obtained by barter. In Bahia, Pernambuco, and elsewhere along the coast, the Portuguese increasingly turned to chattel slavery as a means of securing labor for the canefields and mills. They moved, as Alexander Marchant so ably put it, "from barter to slavery."

In his important study, Marchant has argued that the barter system began to collapse because of a series of economic decisions made by the Portuguese and, to a lesser extent, by the Indians as well. First, there was a glut on the "trinket" market. As Indian demands shifted to more expensive ironware and firearms, the Portuguese costs of supply rose markedly. The increased number of colonists and the presence of royal brazilwood contractors, moreover, created competition for labor among the Portuguese and gave the Indians several markets for their goods and services. Faced with this disadvantageous situation, the Portuguese had to shift increasingly to slavery.

Marchant's interpretation, while correct in its broad outlines, disregards two important aspects of the problem crucial to our understanding of the interplay of cultural and economic forces that created the specific forms of Indian labor in Brazil. First, not only did the value of the goods cause a crisis in the barter relationship but the nature of the goods did as well. Axes and firearms must have had a profound impact on the nature of Indian economy by transforming two of the most difficult and time-consuming tasks, tree-felling and hunting.[17] Iron tools increased the effectiveness of Indian labor and reduced the amount of time needed for certain activities. Yet the change from stone to iron probably did not increase production. By enabling the Indians to satisfy their material needs more quickly, it left them with more free time to

---

[17] Alfred Métraux, "The Revolution of the Ax," *Diogenes*, 25 (1959): 28–40. Indian concern for firearms and ironware is apparent in a curious dialogue in Tupí and French printed as part of Jean de Lery's account of Brazil; see *Viagem a terra do Brasil* (3d ed.; São Paulo, 1960), 251–68, in which the notes of the linguist Plínio Ayrosa are helpful.

A nineteenth-century artist imagines a scene of Portuguese and Indians trading at the beach of Jiquitaya, "where there was always the greatest commerce with the Indians of Bahia." From Anastácio de Santana (O Pardo Velho), "Guia dos Caminhantes (1817)," an unpublished manuscript atlas in the Biblioteca Nacional de Rio de Janeiro. (Courtesy of Dona Lygia da Cunha of the BNRJ.)

engage in ceremonies and make war. Such a hypothesis makes the Indians appear less than "rational" in terms of economic maximization, and this is exactly the problem which underlies Marchant's explanation. Second, Marchant assumes that the Indians were "economic men" involved in a self-regulating labor market, ready to make decisions on the basis of personal or communal economic self-interest. But in many primitive "economies," production and distribution of goods are part of, and usually subordinate to, other considerations of social organization like kinship.[18]

Marshall Sahlins has stated this position with a precision worth quoting: "Even to speak of *the* economy in a primitive society is an exercise in unreality. Structurally, 'the economy' does not exist . . . . Economy is rather a function

[18] Karl Polanyi, *Primitive, Archaic, and Modern Economies*, ed. George Dalton (New York, 1968), 3-37, provides a brief introduction to the vast literature on primitive economics. Also see George Dalton, ed., *Tribal and Peasant Economies* (New York, 1967). I have quite consciously taken a "substantivist" position in the continuing controversy over the nature of primitive economies, because I feel the weight of evidence on precontact Indian cultures in Brazil does not indicate that social forms were produced out of or by the means or modes of production. When faced with a situation such as contact with Europeans, Indian societies were forced to adapt certain institutions to new economic purposes and were sometimes transformed in the process, but the divergence between Indian and European economic concepts remained great. Thus, even if the arguments of the formalists (who see all societies organized around general economic principles) or of the neo-Marxists (who see the forms of social organization in primitive societies as responses to economic needs) have some validity, the continuing disparity between Indian and European economic concepts and forms still served as a major barrier to the integration of Indians into the colonial economy. For an interesting review of the literature on this controversy, see B. Marie Perinbaum, "Homo Africanus: Antiquus or Oeconomicus? Some Interpretations of African Economic History," *Comparative Studies in Society and History*, 19 (1977): 156-78.

of the society than a structure."[19] Barter fitted, quite simply, with traditional patterns of culture, even when what was traded was the communal labor for short-term building. Plantation labor did not. Of course, Indian cultures had the capacity to adapt, but what the Portuguese demanded struck at fundamental aspects of Indian life and thought. To the Indians, agriculture was "women's work." Once a man had enough to eat and a few new tools and weapons, why should he want or work for more? Here was a common colonial situation, noted and commented upon in so many places. The natives—obviously capable of great exertion—were seen as congenitally lazy and undependable.[20] Placed on plantations, they would not work; they were given instead to sulky absenteeism or simply running away. For the Indians refused to respond to the objective conditions of the market created by the Portuguese. Thus, the modes of production established were not simply a matter of European choice but were influenced by the nature of Indian society and the internal dynamics of Indian perceptions and needs.

SLAVERY WAS NOT, IN FACT, THE ONLY FORM OF LABOR the Portuguese were willing to accept. To make Indian workers available as food producers or plantation laborers, the Portuguese attempted a variety of labor systems. The first—that employed by the colonists—was outright coercion in the form of chattel slavery. The second—that tried by the Jesuits and, later, by other religious orders—was the creation of an indigenous "peasantry" by acculturation and detribalization. The third—used by both laymen and ecclesiastics alike and often presented as the ultimate objective of the first two—consisted of the slow integration of the Indians into a capitalistic self-regulating market as individual wage-laborers. In some ways these three modes of production were stages in the history of Portuguese-Indian relations during the colonial era, but the divisions were never clearly marked and the process did not always develop at the same speed in the same direction at all times in all places. During the sixteenth and early seventeenth centuries in the northeast, the Portuguese tried all three techniques simultaneously. And the struggle between the Jesuits and the colonists resembled a clash between two differing strategies with the same goal: the Europeanization of native Americans. In economic terms it was a contest between colonists bent on the imposition of a colonial slave regime and Jesuits in pursuit of a Christianized indigenous peasantry capable of becoming an agricultural proletariat.

The conflict between colonists and Jesuits has already been the subject of intensive historical interest, and there is no need to repeat that story in detail here.[21] Of significance, however, is that this confrontation took place within a

---

[19] Sahlins, *Stone Age Economics*, 76.
[20] *Ibid.*, 86. Also see the observations of S. F. Cook in *The Conflict between the California Indian and White Civilization* (Berkeley and Los Angeles, 1976), 100.
[21] For a review of this struggle in some detail, see my *Sovereignty and Society in Colonial Brazil* (Berkeley and Los Angeles, 1975), chap. 4: "Judges, Jesuits, and Indians," 122-39. Also see Dauril Alden, "Black Robes versus White Settlers: The Struggle for 'Freedom of the Indians' in Colonial Brazil," in Howard Peckham

Indian Labor and New World Demands

specific economic and theological context which placed limitations on colonist and Jesuit alike and on the crown's response to each. The Portuguese monarchs, both Aviz and Habsburg, were morally and doctrinally impelled to recognize the "humanity" of the Indians, to take seriously the royal obligation to convert them to the Roman Catholic faith, and to prohibit, as royal subjects, their illegal enslavement. With the advocacy of the Jesuits, the crown in 1570 began to legislate against the enslavement of Indians; under the Habsburgs further restrictive legislation followed in 1595 and 1609. At each juncture, however, the crown was also faced with the economic realities in Brazil, which imposed a logic of their own. The colony's value lay in sugar production, a point which the colonists never tired of making, and sugar demanded a large labor force. The sugar planters did not yet have the capital or credit necessary to supply their need for labor entirely through the expensive transatlantic slave trade, and thus they remained dependent on indigenous workers. And, left to their own decisions, the Indians would not meet the colony's needs for a variety of cultural reasons. The crown was therefore forced to reconcile its conscience with its treasury receipts.

To solve this problem, loopholes in the legislation permitted the colonists to obtain slaves taken in a "just war." The crown, moreover, supported Jesuit-controlled mission villages (*aldeias*): if the Fathers could convert the Indians and make them available for useful activities such as growing food or working in the canefields while still preserving their freedom, so much the better. What is striking here is the strength of the colonists, especially of the sugar planters, in countering the religious and economic arguments of the Jesuits with pragmatic defenses of their own and in forcing the crown to listen. From the Indian point of view, both the Jesuit and colonist strategies were physically and culturally destructive, albeit in different ways. Colonists and Jesuits agreed on the need to introduce the Indians to "civilized" life and eventually to have them become Christian workers of benefit to the European settlements. The Portuguese, in other words, forced Indian acculturation to Western patterns at a variety of levels. But the Jesuits and the colonists differed in who each believed was better prepared to carry out the directed acculturation process. Both sides did seem to agree that change was desirable and that the ultimate goal was the creation of an indigenous population as similar to Europeans as possible.

Attempts to create an Indian peasantry basically ended in failure, although partially for reasons beyond the Jesuits' control. In Bahia the order created twelve villages (*aldeias*) in the 1550s and 1560s, with a total of about forty thousand Indians. By 1590 flight, disease, and dislocation had reduced the number of *aldeias* to three and the Indian population to only four thousand. By that date Pernambuco, the other major northeastern captaincy, had only two

---

and Charles Gibson, eds., *Attitudes of Colonial Powers toward the American Indian* (Salt Lake City, 1969), 19–46. For the best overall study of Portuguese Indian policy, see Georg Thomas, *Die portugiesische Indianerpolitik in Brasilien, 1500–1640* (Berlin, 1968). Also see Matias Kieman, *The Indian Policy of Portugal in the Amazon Region, 1614–1693* (1954; reprint ed., New York, 1973).

thousand Indians under Jesuit control.[22] Although the level of Jesuit activity varied from captaincy to captaincy, the policy remained the same. The Indians received an education in how to live a Christian life, a concept which included not only European morality but work habits as well. Since the Jesuits basically agreed with the colonists that Indian culture was barbaric, little attempt was made to accommodate or preserve indigenous life: unlike their actions in China and India, the Jesuits showed little cultural relativism in Brazil.[23] As far as possible, they created a full Catholic religious life in the *aldeias*.[24]

At first glance the *aldeias* seem to provide a communal village existence parallel to the pre-European organization of life. The analogy, however, is deceptive, for the Christian communalism of the ecclesiastics bore little resemblance to indigenous patterns, especially when major integrative features of Indian culture were eliminated or transformed. Understandably, the fathers sought to exterminate such fundamental features of Tupinambá life as polygamy, marriages between uncles and nieces, ritual cannibalism, and warfare, but this policy obviously undermined traditional life. For reasons of missionary convenience, moreover, the villages very quickly lost their cultural integrity as peoples of various tribal groups and tongues were mixed together. The Jesuits used *lingua geral,* a simplified form of Tupí, as a *lingua franca* and thereby further reduced Indian culture to a common base that could be controlled by the fathers. Such methods made preaching and conversion easier but increased the pace of detribalization. By their actions the Indians demonstrated how they felt about this assault on their culture. Despite the rosy glow of missionary zeal that pervades much of their comments on the *aldeias,* many Jesuits recognized flight as a chronic problem. In 1566 Father Inácio de Azevedo wrote that "many Indians wish to go with them [colonists] and serve them rather than live in the *aldeias.*"[25] The Indians clearly did not always perceive the protection of the Jesuit *aldeias* as necessarily preferable to the rigors of plantation slavery.

The effect of intentional European interference in traditional practices was compounded by more subtle disruptions in the *aldeias*. Their physical plan, for example, was European. The Jesuit *aldeias* were physically organized according to European norms with a central plaza, a church, and rows of house units flanking the open space. Tupinambá village organization with its layout of four to eight long houses shared by many related families was quite different;

[22] *HCJB*, 2: 58–59. Also see Francisco Soares, *Coisas notaveis do Brasil* (Rio de Janeiro, 1966), 71. Soares' account dates from 1589. Also see Eduardo Hoornaert et al., *História da igreja no Brasil primeira época* (Petropolis, 1977), 128–31.

[23] Thales de Azevedo, "Catequese e aculturação," in his *Ensaios de antropologia social* (Bahia [1957?]), 33–62. For example, the following statement is typical: "Como a gente era rude y sin ninguna policia humana," in "Historia dos collegios do Brasil," *ABNR*, 19 (1897): 75–144. Also see Vasconcellos, *Chronica da Companhia de Jesu do Estado do Brasil*, 1: lxxv.

[24] Cardim, *Tratados de Terra e gente (1583)* (Rio de Janeiro, 1925), 280. Also see Letter of Father António Pires (Pernambuco; June 5, 1552), *Cartas Jesuiticas*, vol. 2: Afranio Peixoto, ed., *Cartas Avulsas, 1550–1568* (Rio de Janeiro, 1931), 124.

[25] (Bahia; November 19, 1566) *MB*, 4: 369–70. Also see *HCJB*, 2: 73; and Hoornaert, *História da igreja no Brasil primeira época*, 131.

The Jesuit Aldeia de Espirito Santo, which became the town of Abrantes, Bahia, in 1758. (The drawing is based on an original in the Arquivo Histórico Ultramarino, Lisbon.)

the complex organization of the Gê-speakers with the village divided into moities and clans, with separate residences for certain age and sex groups was even farther still from the European plan. The traditional village and residence patterns reflected the Indian social and religious cosmos. To change those patterns was to breach the security of the traditional universe and to disorient the Indians in the literal meaning of the word. As Claude Levi-Strauss has pointed out in another context, "All feeling for their traditions would desert them, as if their social and religious systems . . . were so complex that they could not exist without the schema made visible in their ground plans and reaffirmed to them in the daily rhythm of their lives."[26]

[26] Plans of some Jesuit *aldeias* are preserved in the AHU, secção de iconografia. For a reproduction of the plan of the Vila de Abrantes in Bahia, see Nestor Goulart Reis Filho, *Evolução urbana do Brasil (1550–1720)* (São Paulo, 1968), fig. 23. And Levi-Strauss, *Tristes Tropiques* (New York, 1961), 203–05; this discussion focuses on the intentional modification of a Bororo village by Salesian Fathers.

From the Jesuit viewpoint, of course, the success of the *aldeias* was measured by the degree to which Indian culture was modified. The Jesuits argued that the *aldeias* not only protected the Indians from slavery and facilitated their conversion but provided an auxiliary military force for use against hostile tribes, foreign interlopers, and restive slaves. "And work for the Jesuits for nothing," muttered the colonists. The fathers of the Company claimed in response that the *aldeias* also provided labor and sustenance to the estates of the colonists. By 1600 the Jesuits claimed to have fifty thousand Indians in the *aldeias* of Brazil available to both crown and colonists.[27] What disappeared relatively early from the Jesuits' defense of their Indian policy is the creation of an indigenous peasantry.

As Portuguese colonization became increasingly based on export agriculture, conflict with the Indians over land became a central problem. Jesuits directed their efforts toward the protection of village lands and the restoration of property seized illegally. Land grants (*sesmarias*) were given to the Jesuit *aldeias* to insure the Indians against starvation and to provide a basis for Indian production of foodstuffs.[28] Generally, Indian *aldeias* and lands were not located in the regions of the rich *massapé* soils considered ideal for sugar production. Colonists in Bahia claimed in the early seventeenth century that Jesuit *aldeias* could produce over thirty thousand bushels of manioc flour a year. By that time, however, the struggle between Indian subsistence agriculture and the export crops was clearly defined. In 1610, the town council of Paraíba argued that, since sugar was the basis of the Brazilian economy, "the harm caused by their [Indian] farming is much greater than any benefits that might accrue."[29]

To create an indigenous peasantry in Brazil, the Portuguese were forced to start from scratch. In Brazil, unlike Yucatan or the Andean highlands, there was no pre-Colombian tradition of communal agriculture linked to a larger state system. As in Paraguay, the Jesuit *aldeias* of the Brazilian northeast attempted to create peasant communities where none had existed—at least not in ways that would serve the interests of the colony. The Jesuits were not only the defenders of the Indian communities, they were also the creators of them.[30]

With the support of the crown and of sympathetic administrators like Governor General Mem de Sá, the Jesuits were able to secure lands for their Indian charges, but the Jesuit attempt to create a peasantry that was not only

---

[27] AGS, Secretarias provinciales, 1461, ff. 104-08.

[28] Letter of Nóbrega (May 8, 1558) in Serafim Leite, ed., *Cartas do Brasil e mais escritos do P. Manuel da Nóbrega* (Coimbra, 1955), 292. In 1558 Father Nóbrega sought a grant of land to the *aldeia* São Paulo (in present-day Brotas, a suburb of Salvador). He felt that the owner of the land, the count of Castanheira, would cede the property since, lacking water for an engenho, it served for very little. Not even Nóbrega would have tried to secure lands for the Indians which could be used for sugar cane cultivation.

[29] *HCJB*, 5: 21. ANTT, *Corpo Cronológico*, p. 1, maço 115, no. 108.

[30] Regimento of Tomé de Sousa, *DHA*, 1: 53. Also see the discussion in Marchant, *From Barter to Slavery*, 90-91. For an interesting theoretical discussion, see Juan Carlos Garavaglia, "Un modo de producción subsidiario: La organización economica de las comunidades guaranizadas durante los siglos xvii-xviii en la formación regional altoperuana-rioplatense," in *Modos de producción en América Latina* (Cordoba, Argentina, 1973), 161-92.

self-supporting but also a supplier of the colony's needs was never realized. The first royal governor, Tomé de Sousa, arrived in 1549 with specific instructions for establishing a weekly market where Portuguese and Indians could conduct business. This system was designed to supply the food requirements of the Portuguese and, at the same time, protect the Indians from the worst aspects of extortion and fraud by prohibiting the colonists from entering villages at will. But even at the outset concessions were made to the planter class, for only the planters and their people were allowed to trade with the Indians for food whenever it was convenient. Voluntary supply failed, in part because the Indians would not respond to the conditions of the labor market. Aside from the royal instructions (*regimento*) of Tomé de Sousa, there are almost no other references to the weekly fair he was supposed to have established. Lands were given to the Indians primarily to assure their livelihood and therefore their availability to the colony. Mem de Sá stated this position clearly in justifying his grant of a *sesmaria* to the Aldeia de Espirito Santo in Bahia: "seeing how beneficial and necessary they are to this Bahia and that they cannot sustain themselves without lands to farm. . . ."[31]

Neither the sugar planters nor the monarchs were anxious to recognize the failure of an indigenous peasantry. In the ideal mental landscape of sugar producers, the sugar plantations (*engenhos*) would be ringed by canefields as far as possible and at the outer limits by villages of "domesticated Indians," who would keep out the unreduced tribes beyond while growing large quantities of manioc and other foods. Of course, if the Indians worked at the engenho on occasion, that was all the better. Even at the end of the sixteenth century the crown still made reference to the benefits of having Indians farming close by the engenhos of the Portuguese. Certainly, the engenho population sometimes acquired foodstuffs from Indian growers, but this was on an intermittent and haphazard basis. Eventually, what emerged was a geographical division between the export crop, sugar, and the locally consumed food crops. Large-scale production of manioc was pushed into areas that could not profitably support sugar agriculture. The peasantry that developed to work this crop was not Indian, but a predominantly mixed population of *mestiços* and mulattoes.[32]

After failing to create an indigenous peasantry, the Jesuits justified their continued control of the villages by emphasizing the military and labor services which their charges provided. In Bahia during the early 1580s the Jesuit *aldeias* supplied some four to five hundred workers to the colonists under a system of contract labor. The Indians received a meager wage of four hundred réis per month (scarcely a third of a common boatman's salary), but even this sum was often never paid. The Jesuits tried to limit the time of

---

[31] *MB*, 3: 530-31. On Mem de Sá's support of the Jesuits, see Herbert Ewaldo Wetzel, *Mem de Sá: Terceiro Governador geral* (Rio de Janeiro, 1973), 205, 215-17.

[32] *Alvará* (August 21, 1587), in *DHA*, 1: 321-22. Shepard Forman and Joyce Riegelhaupt, "Bodo Was Never Brazilian: Economic Integration and Rural Development among a Contemporary Peasantry," *Comparative Studies in Society and History*, 12 (1970): 188-212; and Francis Jennings, *The Invasion of American Indians, Colonialism, and the Cant of Conquest* (New York, 1975), 58-85.

Indians under European control in Colonial Brazil; an *aldeia* is mobilized to fight for the Dutch. From Joan Bleau, *Le grand atlas ou cosmographie Blaviane* . . ., volume 12· *Amerique* (Amsterdam, 1667), between 263 and 264. (Courtesy of the James Ford Bell Library, University of Minnesota.)

service to three months and usually refused to permit women to accompany the men. Such restrictions made the planters uncomfortable with Jesuit control over their laborers.[33]

Sugar planters were certainly not averse to hiring wage laborers if they could be obtained in sufficient numbers and under conditions favorable to the employers. As early as 1561 settlers in Bahia had sought to hire Indians for wages (*soldada*), and the accounts of Engenho Sergipe indicate that Indians did provide services for pay, although usually at a rate far below that of whites, free blacks, and mulattoes.[34] The royal Indian legislation of 1596 was a clear demonstration of the crown's attempt to integrate the "tame" Indians (*indios mansos*) into the colony as wage laborers. The law also stated unambiguously that Indian laborers were not to remain on the engenhos for more than two months of continual service and that no advance payments were to be made to them. This restriction is a hint that planters may have been turning to a form of debt peonage as an answer to their labor needs. The crown forbade either Jesuits or colonists from using Indians unless they were paid "like free men and treated as such."[35] We will never know whether *aldeia* resi-

[33] Soares, *Coisas notaveis do Brasil*, 77. The wages paid to a boatman in 1574 at Engenho Sergipe were 16$500 réis or an average of 1$375 per month. See Mircea Buescu, *300 anos da inflação* (Rio de Janeiro, 1973), 60–61. *HCJB*, 2: 94–95.

[34] *Moradores* of Bahia asked in 1561 that the probate judge "desse a soldada os moços e moças orfans e outros pedião os casados." Padre Luís de Grã to Padre Miguel de Torres (Bahia; September 22, 1561), in *MB*, 3: 431. On Indian workers at Engenho Sergipe, see Paul Silberstrein, "Wage Earners in a Slave Economy: The Laborers of a Sugar Mill in Colonial Brazil," unpublished paper, University of California, Berkeley (1970).

[35] *DHA*, 1: 404–05.

Indian Labor and New World Demands

dents or other Indians within the radius of Portuguese control could have filled the labor demands of the sugar industry as free laborers. The planters were unwilling to pay wages sufficient to create an adequate wage-labor market so long as an alternate type of cheap labor was available in the form of Indian slaves.[36] That the Jesuits would not permit Indians to neglect their own farming in order to work for the colonists—for fear that starvation would result—indicates that the salaries paid were insufficient to meet necessary minimum requirements of subsistence.[37]

The period of intense Jesuit activity in coastal Brazil roughly coincided with the apogee of Indian slavery. By 1545 the southern captaincy of São Vicente had six engenhos and three thousand slaves, almost all of them Indians. Indian captives could also be found on the plantations of Pernambuco, Bahia, and Porto Seguro. In 1583 Pernambuco had sixty-six engenhos and some two thousand African slaves. Since an engenho probably drew on the labor of one hundred slaves, Indians still accounted for two-thirds of Pernambuco's engenho work force even during a period of transition to African labor.[38]

In Bahia the establishment of royal government in 1549 considerably aided the expansion of the sugar economy. Enslavement of the local tribal groups accompanied this expansion. In the 1550s a number of military campaigns were carried out in the Recôncavo. Both Duarte da Costa and his successor, Mem de Sá, protected established engenhos, secured land for new mills, and obtained captives in a series of punitive expeditions carried out by the Portuguese and their Indian allies.[39] In Pernambuco and Bahia as in other captaincies, the colonists (*moradores*) obtained Indian slaves by "ransoming" (*resgate*) them from other Indians who had already taken them as war captives. More common, however, were Portuguese raids (*saltos*) made for the specific purpose of obtaining slaves. Denounced by the Jesuits and the crown as unlawful and prohibited in the regimento of Tomé de Sousa and his successors, saltos remained the most prevalent form of enslavement. As one Jesuit put it, "only by a miracle does one find here a slave not taken by assault."[40]

The practice of encouraging free Indians (*forros*) to marry or accompany those already enslaved reflects the plantations' increasing demand for Indian workers of any type. This tactic enlarged the supply of labor but was severely denounced by crown and clergy alike. In 1566 the Jesuits were able to obtain legislation against the practice,[41] but engenho inventories of the 1570s and

---

[36] Marvin Harris, *Patterns of Race in the Americas* (New York, 1964), 20–21. Harris follows the argument of John Phelan that Indians "took to earning a living European fashion when adequately compensated." It should be remembered, however, that Phelan is referring to highland peoples already integrated into larger state structures before the arrival of the Europeans. See his "Free versus Compulsory Labor: Mexico and the Philippines, 1540–1648," *Comparative Studies in Society and History*, 1 (1958–59): 189–201.

[37] *HCJB*, 2: 95.

[38] Marchant, *From Barter to Slavery*, 131, based on a number of contemporary sources.

[39] J. F. de Almeida Prado, *A Bahia e as capitanias do centro do Brasil*, 2 vols. (São Paulo, 1945–48), 2: 7–123, contains a survey of these actions. The expedition commanded by Mem de Sá—the War of Paraguassú—was caused to some extent by the fact that Indians in that region refused to return runaway slaves to the Portuguese. Also see Leite, *Cartas Nóbrega* (Bahia; July 5, 1559), 343; and Wetzel, *Mem de Sá*, 59–68.

[40] *HCJB*, 2: 96, 194.

[41] "Resoluções da Junta da Bahia sobre as aldeias dos padres e os Indios," *MB*, 4: 354–57.

seventeenth-century wills from São Paulo indicate that the practice of considering forros as little different than slaves did not disappear easily.⁴²

Jesuit criticism of the enslavement of Indians on the sugar plantations is well known. The Jesuits denounced immorality, concubinage, lack of religious instruction, and brutal treatment, but they did not deny the necessity of Indian labor.⁴³ The question was how best to use them, under whose control, and under what conditions. The colonists generally argued that Indians in contact with Europeans would naturally become Christians and would learn skills of use to the colony, that there was little need for specific Jesuit supervision.⁴⁴ Neither Jesuits nor colonists were prepared for the persistent Indian resistance to a European work regime and for the disastrous demographic decline of the Indian population of the 1560s.

Brought into close contact with Europeans in *aldeias* and engenhos, the Indians suffered the ravages of European diseases. In 1559 or 1560 smallpox killed over six hundred enslaved Indians in Espirito Santo alone; and no one knew the number of dead among free Indians. By 1561 it had spread to the Recôncavo of Bahia. The epidemic reached its height in 1563. Estimates placed the figure at thirty thousand dead among the Indians under Portuguese control in Bahia alone, to say nothing of countless more in the *sertão* to which the disease was spread by Indians fleeing from the coast.⁴⁵ Father Leonardo do Valle, an eyewitness, wrote of children who died at their mother's breast for lack of milk, of people so debilitated that they could not even dig graves for the dead or draw water for the living. One-third of all the Indians in the Jesuit *aldeias* died. Those on the sugar estates suffered similar losses. On some plantations ninety to one hundred slaves perished.⁴⁶ The following year brought no respite. In 1563 a second epidemic, this time measles (*sarampo*), struck an already weakened population. Perhaps another thirty thousand died.⁴⁷ Not surprisingly, measles proved far more lethal to Indians than to Portuguese. Europeans were stunned by its effect: "the population that was in these parts twenty years ago is wasted in this Bahia, and it seems an incredible thing for one never believed that so many people would ever be used up, let alone in such a short time."⁴⁸

The epidemics of 1560–63, not surprisingly, disrupted the social and economic fabric of the colony. Portuguese concentration on the export crop of

---

⁴² *DHA*, 2: 64, 350. Edmundo Zenha, *Mamelucos* (São Paulo, 1970), 130–34; and Alcântara Machado, *Vida e morte de um bandeirante* (São Paulo, 1943).

⁴³ See, for example, the pertinent sections of *HCBJ*, especially vols. 2–3, 5.

⁴⁴ Diogo de Campos Moreno, *Livro que da razão as estado do Brasil* (Rio de Janeiro, 1968). See Engel Sluiter, ed., "Report on the State of Brazil, 1612," *Hispanic American Historical Review*, 29 (1949): 523. The classic statement of the colonist position is presented in Serafim Leite, ed., "Os capitulos de Gabriel Soares de Sousa," *Ethnos*, 2 (1941): 5–36.

⁴⁵ For a brief but accurate account of the plague, see Fernandes, *Organização social Tupinambá*, 40–41. Also see *Cartas Avulsas*, 207–08.

⁴⁶ Leonardo do Valle to Brothers of the Company (Bahia; September 23, 1561), *Cartas Avulsas*, 334. Vasconcellos estimated that three-quarters of all the Jesuit Indians died; see his *Chronica da Companhia de Jesu do Estado do Brasil*, 3: 6. Also see *Cartas Avulsas* (Bahia; May 12, 1563), 378–93.

⁴⁷ Fernandes, *Organização social Tupinambá*, 40. "História dos collegios do Brasil," 84, places the plague in 1563–64.

⁴⁸ Father José de Anchieta, as quoted in Fernandes, *Organização social Tupinambá*, 30.

Indian Labor and New World Demands

sugar and their dependence on indigenous supplies of foodstuffs had, even in the best of times, created an unstable situation. Now with the decimation of the Indians, the main sources of food supply were radically depleted and famine set in. The Portuguese went hungry and the Indians starved to death. Some Indians faced with starvation sold themselves into bondage rather than perish. If those who took this course believed that servitude would be temporary, they soon discovered that such was not the case.[49] Although these vital crises of the 1560s in some ways facilitated the enslavement of Indians still within the range of Portuguese operations, epidemics and famines also made clear the dangers inherent in dependence on Indian labor.[50]

Continuing Indian resistance to Portuguese settlement and to the plantation labor system provided the colonists with further proof of the problems of employing an indigenous labor force under a colonial regime. Indian resistance took a number of forms. Primary resistance of unreduced groups who sought to protect traditional territories characterized the period from 1540 to 1580 as the lands brought under sugar cultivation expanded. Indian opposition was eliminated in some areas by Portuguese military campaigns as the Indians were driven into the interior, enslaved, or killed. Resistance did not cease, however, once the Indians were brought under Portuguese domination. In 1567 a general slave insurrection swept the region of the Bahian Recôncavo. Masters were killed and slaves fled the canefields in large numbers. Only the intervention of Jesuit-led *aldeia* Indians brought the situation under control.[51] More lasting was a large-scale millennarian resistance movement called *Santidade*, which flourished among escaped Indian slaves and former *aldeia* inhabitants in the region of southern Bahia. Combining Roman Catholic and native beliefs, the *Santidade* followers began burning sugar mills and plantations in the 1560s, and despite Portuguese military reprisals their activities continued into the seventeenth century. In 1610 their numbers were reported at twenty thousand and included escaped blacks as well. As late as 1627 their raids continued in the southern Recôncavo and served as a beacon to those still in captivity.[52]

---

[49] Pero de Magalhães de Gandavo, *The Histories of Brazil,* trans. John B. Stetson (New York, 1922), 229. A similar situation occurred in Pernambuco in 1583-84 when a famine in the *sertão* drove three to four thousand Indians onto the coastal plantations.

[50] The susceptibility of Indians to European disease continued. In 1565 so many died in the Jesuit Aldeia de São João in Espirito Santo that the site had to be abandoned; *MB,* 4: 267-68. In 1616-17 Indian and African slaves alike were decimated by smallpox; see Brandão, *Diálogos das Grandezas do Brasil,* 64. The impact of disease in the *aldeias* was not lost on the colonists: in 1610 the Câmara of Paraíba argued against the *aldeias* for exactly this reason; ANTT, *Corpo Cronológico,* p. 1, maço 115, no. 108.

[51] "História dos collegios do Brasil," 89.

[52] For the most complete study to date, see José Calasans, *A santidade de Jaguaripe* (Salvador, 1952). This account and all others have been based on the sixteenth-century materials contained in the Jesuit letters and the Inquisition records. The short account here, drawn from my forthcoming book, presents for the first time evidence of the existence of the movement in the seventeenth century. See Diogo de Meneses to the crown (Bahia; September 1, 1610), ANTT, Fragmentos caixa 1, n. 6; the king to Gaspar de Sousa (Lisbon; January 19, 1613 and May 24, 1613), BI, Correspondência de Gaspar de Sousa, ff. 185-85v., 218-18v.; and Provisão de Diogo Luis de Oliveira, Arquivo da Câmara Municipal do Salvador, Provisões e portarias 1624-42, bk. 155, ff. 245-46.

The colonists were not ready to abandon Indian labor for the growing number of sugar mills in the late sixteenth century, but the uncertainties of Indian resistance and their health and life expectancy made indigenous labor an investment of great risk. This situation helps to explain why Indian slaves were priced far below Africans, why planters were more likely to invest in the training of Africans for the skilled tasks in the sugar mills, and why the colonists did not entirely oppose the development of a wage labor system. The 1570s and 1580s witnessed not only military expeditions but also a number of other schemes designed to bring still unreduced Indians from the interior to meet the labor needs of the engenhos.[53] But, faced with the increasing opposition of the crown to enslavement, the growing demands of the sugar economy, and the disastrous example of the 1560s, the colonists turned to the labor to be found in the Atlantic slave trade. It is no accident that the importation of large numbers of Africans began in the 1570s following the peculiar conjunction of demographic, economic, and political circumstances that demonstrated the risks of an economy based on captive or coerced Indian labor.

The use of Indian labor was, therefore, subject to a number of constraints and limitations. The deadly triad of warfare, disease, and famine that accompanied the conquest and occupation of Brazil limited the nature and availability of an indigenous labor force. The competing strategies of the Jesuits and the colonists over the form and control of this labor system determined much of the history of Portuguese-Indian relations, but this rivalry should not mask the basic agreement shared by planters and missionaries alike that Indian labor was vital to the colony's success. Each side justified its position to the crown by arguing that its control would more rapidly lead the Indian to European standards of religion, morality, and habit, including integration of the Indian into a wage-labor market. But most Indians refused to respond to either: they refused to be shaped at will by alien policies and historical processes no matter how seemingly inexorable. Indian actions and responses varying from armed resistance to accomodation and acculturation limited and defined the nature of the colonial regime. By examining the inner structure of these processes in relation to the establishment of the plantation regime, in relation to the formation and definition of the colony's dominant mode of production, it is possible to define that regime and to suggest the reasons for the abandonment of Indian slavery in the New World's first successful plantation colony and economy.

---

[53] Between 1571 and 1575, about twenty-five thousand Indians were brought under Portuguese control in Bahia. *HCJB*, 2: 82–83. The duke of Aveiro, donatary of Porto Seguro, received permission to "bring down" Indians as did the Count of Linhares but these were exceptions. See ANTT, *Cart. Jesuitas*, maço 16; AGS, Sec. prov. 1487 (October 7, 1603), ff. 33–33v. In 1608–10 Governor Diogo de Meneses favored colonists' desires to bring Indians under plantation control, but the crown prohibited such efforts in 1613. Also see *ABNR*, 57 (1939): 37–40; BI, Correspondência Gaspar de Sousa (March 25, 1613), f. 207. On the famous case of the Potiguares of Pernambuco sent as allies and workers to Ilhéus, see Robert Southey, *History of Brazil*, 2 vols. (London, 1810), 1: 404–05; ANTT, *Cart. Jesuitas*, maço 8, n. 108; and Frei Vicente do Salvador, *História do Brasil* (São Paulo, 1968), bk. 4, chap. 35.

Indian Labor and New World Demands

MOST OF THE DOCUMENTATION and thus the historiography of Indian labor and slavery has, unfortunately, been concerned with the legality and the abuses of the system. Extant but little-used plantation records and parish registers offer another option.[54] The forms, usages, and structures of Indian labor as practiced on the plantations of northeastern Brazil can provide a key to our understanding of the process by which a colonial—Indian supplanted by African—slave regime became essential to a tropical plantation economy.

The terminology of Indian labor is, in itself, revealing of its position within the plans and perceptions of the Portuguese. The categories of social definition and of social structure in Brazil were determined to a great extent by the nature of the agricultural enterprise and by the previous Continental and overseas experience of the Portuguese. Whatever the philosophical and theological problems provoked in Europe by the discovery of a new "race" of men, the Portuguese on the scene in Brazil tended to draw upon familiar models, especially the recent past of African contacts and Atlantic plantations. This tendency is revealed in the widespread use of "*negro da terra*" by Jesuits as well as colonists to describe Indians.[55] "*Negros da terra*" was a parallel phrase to the description of Africans as "*negros de Guiné.*" In medieval Portugal the word "negro" itself had become almost a synonym for slave, and certainly in the sixteenth century "negro" carried implications of "servitude." The phrase therefore reveals Portuguese perceptions of both Africans and Indians, not so much of skin color as of their relative social and cultural position vis-à-vis the Portuguese. Over the course of the sixteenth century, "*negro da terra*" slowly disappeared from common usage as more and more Africans were introduced into the colony. It vanished, in fact, with Indian slavery itself.

For those Indians not enslaved but still under Portuguese control and direction, the colonists applied a variety of terms. Such people were *indios aldeados* (village Indians) or *indios sob a administração* (Indians under the control of . . .) or, most commonly, *forros*. This last term is somewhat confusing

---

[54] One of the reasons why this topic has not been studied in greater depth is the lack of materials. Historians of sixteenth-century Brazil depended to a large extent on Jesuit letters and reports and on government correspondence and legislation. For much of the discussion thus far, I have done the same. Jesuit observers and colonial officials usually wrote of general conditions, not the specifics of slavery in sugar agriculture. For this reason, the existing materials from Engenhos Sergipe and Santana are particularly valuable. Extant inventories from 1572-74, 1591, and 1638 that include lists of slaves enable us to come to some conclusions about the structure of the labor force. Account books of individual harvests from 1607 to 1748 also exist in the Arquivo Nacional da Torre do Tombo, Lisbon. And Engenho Sergipe's chapel served, until the late seventeenth century, as the parish church of its district, and the chapel register has survived, albeit in fragmentary condition; it permits us to examine various social relationships in the period from 1595 to 1626. These sources, limited as they admittedly are, at least provide a glimpse of life on Bahian engenhos in the period of Indian slavery. Whenever possible, supporting materials have been used to confirm the findings of the Sergipe and Santana documents.

[55] The term "*negro da terra*" seems to have persisted in São Paulo well into the seventeenth century. In Bahia, although it was occasionally employed in the period after 1600, it was gradually replaced. See Zenha, *Mamelucos*, 52-72. For an introduction to the question of Indian labor in the sugar industry, see Luis Viana Filho, "O trabalho do engenho e a reacção do indio Estabelicimento de escravatura africana," *Congresso do Mundo Portuges*, 10 (Lisbon, 1940): 10-29. Also see Gilberto Freyre, *The Masters and the Slaves* (New York, 1956), 81-184.

since it was also used to describe a slave who had gained his freedom through manumission (*alforria*), but in sixteenth-century Brazil it was not used exclusively in this way. Instead, *indios forros* were not only freedmen but also Indians who were not enslaved and were under Portuguese control—especially, though not exclusively, that of the Jesuits. The engenhos made use of all three categories of Indians during the sixteenth century and continually sought to increase the pool of available labor. At various times colonists sought the creation of the *encomienda*, the system of personal award of Indian service or tribute as was done in Spanish America. Although never formally instituted, the various arrangements for the use of "free" Indian workers amounted to a system of labor supply parallel to the *repartimiento* of the Spanish colonies.[56]

Aside from the supply of labor from the *aldeias*, the plantations acquired Indians by enslavement, barter, and wages. The law of 1570 had prohibited the legal enslavement of indigenous peoples, but loopholes in the legislation permitted the acquisition of captives by ransoming them through trade with their captors. This trade (*resgate*) was theoretically designed to save those already destined for a cruel death at the hands of their traditional enemies. Rescue was a favor, to be repaid by the captive's labor. Abuses were both obvious and widespread, but, "because of the necessity that the estates have for Indians," few in the 1570s wanted to restrict *resgate*. Individual plantations often arranged for the acquisition of workers by this means. In 1574, for example, Engenho Sergipe had over fifty Indians recently brought in by a *resgate* expedition, and an inventory of the property from about that time contains references to axes, cloth, and knives specifically intended for such trade.[57] *Resgate* enabled the Portuguese settlers to obtain *negros da terra* without having to call them slaves, and it permitted the continuance of slavery long after its apparent abolition.

By the 1580s royal legislation and the increasing effectiveness of the Jesuits had begun to create problems for those who wished to obtain Indians by both *resgate* and "just war." After a visit to the captaincy of Bahia during 1588 and 1589, Jesuit *Visitador* Cristóvão de Gouvea recommended that the Church refuse the sacrament of confession to anyone involved in the *resgate* of Indians. Jesuit effectiveness in overseeing the activities of free Indians within the Portuguese sphere of influence, moreover, began to create difficulties for the colonists. In 1589, Ruy Teixeira, administrator of Engenho Sergipe, complained to his absentee employer, the Count of Linhares, that new legislation had made the Jesuits "masters of the land and of the Indians who with the title of 'forros' serve them, being in reality more captive than the Guiné slaves." He lamented that no Indians remained for *resgate*, but he realized that nothing could be

---

[56] For some data on this problem, see my *Sovereignty and Society in Colonial Brazil*, chap. 4. Also see Thomas, *Die portugiesische Indianerpolitik in Brasilien, 1500–1640*, 90–98.

[57] As quoted in Wetzel, *Mem de Sá*, 216–17. The inventories of Engenhos Sergipe and Santana made between 1572 and 1574 are printed along with the will and testament of Mem de Sá and other relevant materials in *Documentos para a História de Açúcar*, vol. 3: see Inventário Engenho Sergipe (1572), *DHA*, 3: 65; and Livro de Contas do procurador (1574), *DHA*, 3: 406.

Indian Labor and New World Demands

done. "I will speak no more of this for these are matters that have no remedy—may God grant us His mercy."[58]

A variety of labor arrangements for Indians existed simultaneously on the Brazilian plantations, and, although their relative importance varied with time and place, forms of remunerated labor increasingly replaced slavery. The owner of Engenho Sergipe paid the tithe for a nearby *aldeia* that provided workers while using slaves and other free Indians.[59] In addition to the local labor pool, some Indians from the interior were forced by hunger or military pressure to present themselves for work at the plantations.[60] Indians continued to participate as laborers in the plantation economy and life of the colony, (although their presence is largely ignored in traditional historiography, which emphasizes the role of African labor).[61] The planters tended to employ these workers according to their status or manner of acquisition. Slaves were employed in those tasks central to the sugar-making process where returns to investment were highest, regardless of skills required; the contention that slavery cannot be profitable in production that demands skilled labor is, therefore, not supported by the data on Indian plantation slavery in Brazil.[62] Evidence from Engenho Sergipe demonstrates that "free" or *aldeia* Indians were used primarily as an auxiliary work force in tasks of maintenance or service peripheral to the business of sugar production. They cleaned and repaired the water system, worked in the boats, hunted and fished, and cut firewood.

Access to nonslave Indian labor allowed the planters to concentrate the capital invested in slaves in those crucial aspects of production where continuous labor justified the fixed capital the slaves represented. The almost ten-month harvest season (late July through May) and the relatively mild rainy period meant that slaves could work during the entire year in northeastern Brazil.[63] Thus, the conditions of, rather than any imagined lack of complexity

---

[58] "O que pareceo ao Padre Visitador Cristóvão de Gouvea ordernar na visita deste Collegio da Bahia," January 1, 1589, Biblioteca Nacionale di Roma, Fondo Gesuitico 1367; and Teixeira to the count of Linhares (Bahia; March 1, 1589) ANTT, *Cart. Jesuitas*, maço 8, no. 136.

[59] Sebastião Vaz to Diogo Cardim, Provincial of the College of Santo Antão, Bahia, June 5, 1629, ANTT, *Cart. Jesuitas*, maço 69, no. 74. This letter describes the history of Indians in the village near Sergipe do Conde. They had been brought in at great expense by the count of Linhares but, by the time his wife and heir had died, there were few left. When the village was moved elsewhere, however, the engenho Indians went along, a situation that moved Vaz to petition for their return. Also see Safra 1611-12, ANTT, *Cart. Jesuitas*, maço 14, no. 4, f. 24; and *DHA*, 3: 102, 298, 311, 406.

[60] *Feitor* of Engenho Santana to the count of Linhares, August 15, 1599, ANTT, *Cart. Jesuitas*, maço 8, no. 105. The *feitor* called them "gentio do sertão tapuyas do catingua."

[61] For example, see the account of Peter Carder (1578) in Samuel Purchas, *Hakluytus Posthumous: or, His Pilgrims*, 5 vols. (London, 1625), 5: 1190. Although Indians did not come in to make depositions, except in one instance, many who appeared before the Inquisitors spoke of their relations with Indians, of the *aldeias*, of entradas to the *sertão* to bring more Indians down to the coast, and of considerable interaction. A number of mestiços also admitted to practicing Indian customs and of speaking Indian languages. J. Capistrano de Abreu, ed., *Primeira visitação do Santo Oficio as partes do Brasil: Confissões da Bahia, 1591-92* (Rio de Janeiro, 1935); see, for example, 34-37, 64-65, 93-97, 104-05, 123-24, 164-72.

[62] Wallerstein, *The Modern World System*, 88-90. Manuel Moreno Fraginals, *The Sugar Mill*, trans. Adric Beltrage (1964; English ed., New York, 1976), presents a developed analysis based on this classic Marxist premise. But Robert S. Starobin, *Industrial Slavery in the Old South* (New York, 1970), for example, demonstrates that there was no necessary incompatibility between slavery and highly technical operations.

[63] Compare Wallerstein, *The Modern World System*, 88-90, and Ralph Anderson and Robert E. Gallman, "Slaves as Fixed Capital: Slave Labor and Southern Economic Development," *JAH*, 64 (1977): 24-46.

TABLE 1
ETYMOLOGY OF SELECTED TUPI PERSONAL NAMES
ENGENHO SERGIPE, 1572–74

Name	Probable Derivation
Pejuira	*peju* = to blow; *ira* = to detach (interrog.)
Pedro *rari*	*rari* = to be born
Itaoca	*Ita* = stone; *oka* = house
Ocaparana	*oka* = house; *parana* = sea
Mandionaem	*Nhae* = pan; *mandio* = manioc
Antonio Jaguare	*jaguare* = iaguara = jaguar
Francisco tapira	*tapira* = tapiira = ox
Birapipo	*Bira* = ybyra = wood; *pipo* is an interrogative
Cunhamocumarava	*kunhamuku* = a girl of marriageable age; *marava* = marabi = child of an Indian and a stranger
Ubatiba	*Tyba* is a plural ending; *Uba* = port, thighs, fish roe

SOURCE: *DHA*, 3: 89–103.

in, sugar agriculture are more useful in explaining the planters' preference for slaves. But planter calculations and preferences were only part of the equation. There remained the problem of Indian adaptation and adjustment to the engenho regime. The records of Engenhos Sergipe and Santana provide us with a glimpse, albeit fragmentary, of this process and the constraints it placed on actions of the sugar planters.

Indian slave names in the engenho inventories yield some tentative conclusions about the ethnic composition of the Indian slave force of Bahia in the late sixteenth century. As expected, many listed in the Sergipe and Santana inventories of 1572–74 were Tupinambá, native to the coastal area of Bahia. Some carried further references by location—such as *tapariqua* (Itaparica Island), Tamamaripe, *tapecuru* (Itapicuru), and *peroaçu* (Paraguassú River)—or by common Tupi-guaraní descriptive terms—such as "*açu*" (big) or "*merim*" (small). Other names clearly seem to be of Tupinambá origin. Table 1 presents some of the more obvious Tupi names.

A variety of problems, some of which themselves reveal facets of engenho life, beset etymological analysis. Aside from the usual difficulty of parallel words in more than one language, the Portuguese who recorded these names transformed them into the sounds and orthography suitable to a Romance language. What remains is what a Portuguese heard, not what an Indian said. Sometimes, the Portuguese themselves were puzzled by the Indian languages; nor were the planters always certain of their slaves' origins. Phrases like "pela lingua q. não he cristão" (by his language not a Christian) indicate that the Portuguese were unsure of the linguistic stock of some slaves.[64] Despite these

---

[64] *DHA*, 3: 92. On Engenho Sergipe one of the field overseers, Tristão Pacheco, also served as translator (*lingoa do gentio*), a talent of benefit to the engenho. *DHA*, 3: 393–94. Portuguese and *mestiços*, lay and cleric,

Indian Labor and New World Demands

differences, it is evident that the engenhos drew their Indian slaves from a wide range of geographical and cultural backgrounds. Engenho Sergipe counted among its slaves not only local Tupinambás, but peoples brought from Sergipe de El-Rey to the north, Rio das Contas to the south, and the *sertão* of the São Francisco River to the west, as well as a large contingent brought from Pernambuco. The inventories also include Carijos, Tamoios, and Cayetes, all Tupi-speakers and all from regions hundreds of miles from Bahia.

Nor were all the engenho Indians Tupians. The Sergipe and Santana inventories of the 1570s and 1591 often contain Tapuya names. Here and there other ethnic references appear—a few Tapanhuns (a Gê group), for example, and Nãmbipiras. Both inventories of the 1570s also contain many identifications that seem to reflect ethnic or tribal origins although they cannot be positively identified—Tinguas, Tarabes, Taipes, and the like. Engenhos Sergipe and Santana in particular, and probably most of the engenhos that depended on Indian labor, thus employed a heterogeneous Indian labor force. Whether this policy was intentional—that is, designed, as it later was for African slaves, to prevent their close cooperation and forestall rebellion—or was simply a response to the shortage of local laborers and market availability is moot. Northeastern planters were certainly willing to buy Indian slaves from other areas of Brazil and seemed to realize, as did the masters of Indian slaves elsewhere in the New World, that the effectiveness of Indian slavery increased when those enslaved could be moved from their native areas, making flight more difficult. Indians from São Paulo taken in the Jesuit missions of Paraguay reached Bahia and Pernambuco by sea, but their numbers were probably fewer than we have been led to believe. More common seems to have been the capture of Indians from the interior of the northeast.[65] Planters naturally preferred to acquire young male slaves.

---

who spoke Indian languages were proud of this accomplishment and usually pointed it out to the crown or other authorities since it was a necessary and valuable skill in the sixteenth and early seventeenth centuries. *Confissões da Bahia, 1591–92*, 87, 104–05, 167–72. Also see the petition of Luís de Aguiar, AGS, *Guerra antiga*, legajo 906.

[65] Such long-distance movements of Indian captives was common. The survivors of King Philip's War in New England were sent to Barbados, Carolina Indians went to New England and the Caribbean, Indians from northeastern Mexico were shipped to Hispanola, and an active slave trade sent Indians from Nicaragua to Peru, Panama, and the Caribbean. The list could be easily expanded. See Alden T. Vaughan, *New England Frontier: Puritans and Indians, 1620–1675* (Boston, 1965); Verner Crane, *The Southern Frontier, 1670–1732* (Ann Arbor, 1929), 113–15; Silvio Zavala, *Los indios esclavos en Nueva España* (Mexico, 1968); David Radell, "The Indian Slave Trade and Population of Nicaragua during the Sixteenth Century," in William Denevan, ed., *The Native Population of the Americas in 1492* (Madison, 1976), 67–76; and Jerome Handler, "The Amerindian Slave Population of Barbados in the Seventeenth and Early Eighteenth Centuries," *Caribbean Studies*, 8 (1960): 38–64. In their report to the crown of October 10, 1629, Simon Maceta and Justo Mancilla noted the arrival in Bahia of Indians from the south; see Jaime Cortesão, ed., *Jesuitas e bandeirantes no Guaira*, no. 1 of Biblioteca Nacional, Manuscritos da Coleção de Angelis (Rio de Janeiro, 1951), 310–39. The claims of authors such as Alfredo Ellis Junior, of large numbers of Indians being sent from São Paulo to the northeast in the seventeenth century seems greatly inflated given the paucity of documentation on the existence of an intercaptaincy trade. Moreover, the real shortage of African slaves in the northeast came after the Dutch seizure of Luanda in 1645, after the major slaving raids in the south had declined. See Ellis Junior, *Meio Seculo de Bandeirismo* (São Paulo, 1948), *passim*; and "O bandeirismo na economia do século xvii," in *Curso de bandeirologia* (São Paulo, 1956), 55–76. For the opposite interpretation, see Zenha, *Mamelucos*, 193–96.

TABLE 2
SEX DISTRIBUTION, ENGENHOS SERGIPE AND SANTANA, 1572–91

	Married Males	Unmarried Males	Percent Males	Married Females	Unmarried Females	Percent Females
Engenho Sergipe, 1572	51	41	61%	51	8	39%
Engenho Santana, 1572	18	47	60%	18	26	40%
Engenho Sergipe, 1591	17	19	58%	17	9	42%

The sex distribution of the Indian slave force was remarkably similar to that encountered later among black slaves. Usually about 60 percent of the slaves were male, and there was an understandable tendency for the men to be young adults. The marital status of Indian slaves and their ability to maintain family ties is difficult to determine from the Engenho Sergipe and Santana inventories because those records offer conflicting pictures. At Engenho Santana only eighteen married couples were listed in an adult population of one hundred and nine (sixty-five men and forty-four women). At Engenho Sergipe, closer to Salvador and perhaps to ecclesiastical observation, the percentage of married couples was much higher. Of the ninety-one men there, fifty-one were married. Whether this difference is explained by variant procedures for recording slave conjugal units or by a different composition of the slave force is impossible to establish. Perhaps more revealing is that among Engenho Sergipe's regular work force there were only eight unmarried women, and of these three were widows and two were related to other slaves at the mill. These figures, shown in Table 2, underline the expected preference of slave-owners for young males and suggest that little value was placed on female reproductive capacity.

Despite this preference, the nature of Indian slavery resulted in the presence of family units on the engenhos. Wives, children, siblings, or other relatives often accompanied the men into slavery. This pattern placed many people on or near the sugar mills and cane farms whose contribution to the process of sugar-making was marginal at best. In 1548 the great Schetz engenho in São Vicente had some one hundred and thirty slaves, half of whom were children or old people and thus of little use to the owners. Nevertheless, a contemporary observer called this slave force the best in the region.[66] The lists of Engenhos Sergipe and Santana also suggest a high ratio of semi- or unpro-

[66] Eddy Stols, "Um dos primeiros documentos sobre a engenho dos Schetz en São Vicente," *Revista de História*, 37 (1968): 407–20.

Indian Labor and New World Demands

ductive slaves. At Engenho Santana some 25 percent of the total work force was too old, too young, or too sick to contribute very much to the mill's activity. Obviously, this percentage was greatly increased in the years of epidemic disease.

Women made up a significant part of the Indian labor force, but they were not as a rule considered to have skills that contributed to the making of sugar. Female slaves in the sixteenth-century inventories are invariably listed without occupation and their values resulted from combinations of age and health more than anything else. Thus, women appear as an omnipresent but not particularly valued sector of a plantation's operations. Some evidence from Engenho Sergipe suggests a recognition of the traditional role of Indian women in subsistence agriculture: a separate *roça* (farm) was maintained to supply the plantations' food needs, and a group of fifty slaves was assigned there. Two-thirds of these slaves were women—a proportion quite unlike the general sex ratio of the total population of the engenho.[67]

Even when epidemic disease was not important, mortality rates were high. In 1572, a relatively plague-free year, five slaves died at Engenho Santana—a crude mortality rate of forty-three per thousand. In 1606 the chapel of Engenho Sergipe recorded thirty-two Indian deaths and only thirty-five baptisms. Because in this period accurate statistics for compiling general mortality rates are unavailable, there is some advantage in comparing figures with data from other regions of a similar social or economic composition. The crude mortality rate in Pernambuco in 1774 was almost thirty-three per thousand, and it remained at about that level until the late nineteenth century. In Maranhão the crude mortality rate for Indians in 1798 was close to twenty-two per thousand while that for black slaves was just over twenty-seven per thousand. Thus, the figure from Engenho Santana seems high, although it does not approach the rate of seventy per thousand experienced by African slaves in Jamaica and Barbados in the late seventeenth century.[68]

For those Indians who survived and for their Portuguese masters and employers, the remaining major problems included introduction to the regime of large-scale export agriculture and adoption of cultural patterns acceptable to Portuguese religious and social sensitivities. The Portuguese did not permit the Indians to select those aspects of European culture that they found most suited to their needs but instead often forced them to adopt or accommodate those material and mental elements of culture on which the Portuguese placed a high priority. Even so, the process of Indian acculturation to the plantation work regime appears to have been slow and incomplete. Whatever kind of workers the planters hoped to create, elements of Indian culture reinforced at

[67] *DHA*, 3: 348–49.
[68] These figures are my calculations based on the census of Pernambuco in 1774 in *ABNR*, 40 (1918): 21–111; undoubtedly they reflect underregistration of high infant mortality of the eighteenth century. For the nineteenth century, see Peter Eisenberg, *The Sugar Industry of Pernambuco* (Berkeley and Los Angeles, 1974), 148–51; and Bainbridge Cowell, "Cityward Migration in the Nineteenth Century: The Case of Recife, Brasil," *Journal of Interamerican Studies and World Affairs*, 17 (1975): 43–63. For Maranhão, see BNL, Fundo geral, Códice 6936. And for African slaves in the Caribbean, see Michael Craton, *Sinews of Empire: A Short History of British Slavery* (New York, 1974), 194–95.

times by the "free" Indians who also worked at the mills, slowed the acculturation process.

The first superficial sign of acculturation was the adoption of a Portuguese name. The inventories of 1572-74 list many Indians who were still using their indigenous names exclusively, even though the Portuguese tended to assign easily recognizable and, for them, pronounceable names. Of the almost two hundred Indians at Engenho Sergipe, fifty still used only their original names. The inventory of Engenho Santana demonstrates that a period of transition existed during which the Portuguese used one name and the Indians another: entries such as "by her tongue Capea and by ours Domingas" or "Salvador, by his tongue Itacaraiba" frequently appear.[69] A comparison of the Engenho Sergipe inventories of 1572-74 and 1591 suggests gradual acculturation. Whereas the former contains some fifty slaves using only their Indian names, the latter has none. Thus, the Indians apparently began conforming to Portuguese patterns and, quite likely, newly captured peoples were relatively rare on the engenhos by the end of the century.[70] Their places were instead filled by free Indians, by Indians born in captivity and baptized with Christian names at that time, and by increasing numbers of African slaves.

The Indians' eventual recognition and acceptance of their Portuguese names constituted steps toward their integration into the engenho community. Whenever possible, this process culminated in baptism of the formerly pagan Indians. The assumption of a new name was an important feature of Tupi social ascendancy that accompanied various events in life; among the Gê naming also served to recruit individuals for social and ceremonial roles. Thus, Indians could easily comprehend the significance of the baptismal ceremony and the relationship between a new name and a new status.[71] Religion itself, in fact, provided a major avenue of acculturation. The participation of slaves in Church ritual and the sacraments, forced though that involvement may have been, provides a rough measure of the Indians' integration into Portuguese society. For this reason, the chapel register of Engenho Sergipe is a valuable document, despite its incomplete and fragmentary condition. In its record of marriages (1600-26), burials (1598-1627), and especially baptisms (1595-1608) the basic patterns of sexual relationships and ritually defined responsibilities among the three principal racial groups are apparent. The period covered by the record is crucial, because between 1570 and 1630 the plantations crystallized into the distinctive social structure that characterized the area for the next two hundred years.[72]

---

[69] *DHA*, 3: 348-49.

[70] Fernandes, *Organização social Tupinambá*, 178, 269, *passim*. Roberto da Matta presents evidence on naming among the Kraho and Apinaye; see "Notas sobre o contato e a extinção dos indios Gaviões do medio Rio Tocantins," 189.

[71] Inventory of 1591, ANTT, *Cart. Jesuitas*, maço 13, no. 4.

[72] The chapel register is at present housed in the Arquivo da Curia Metropolitana of Salvador. It is bound and titled erroneously as the first register of baptism of Conceição da Praia. I consulted this register in 1968 and realized that it was not what its title indicated, but it was not until 1973 while working at the ACM again that David Smith recalled it to my attention and led me to examine it more thoroughly. Since the chapel, dedicated to Nossa Senhora da Purificação, served as the parish church for the region until a

Indian Labor and New World Demands

The chapel register contains 234 complete adolescent baptisms for the period 1595 to 1608, or about 75 percent of the total baptismal entries. Of the 234 baptized, 171 (74 percent) were the children of slave mothers and, therefore, were slaves. The racial origins of the 134 mothers (for whom a racial designation has been determined) can be used as a rough gauge of the ethnic proportions of the population. It results in the following distribution: whites, 32 percent; Indians, 40 percent; Afro-Brazilians, 28 percent. Given the preference for males in the Atlantic slave trade, the sex ratio of Afro-Brazilians would be distorted and these figures thus underestimate this segment of the population. Still, Indians continued to be an important part of the plantations' population at the beginning of the seventeenth century—equaling, if not outnumbering, Africans and their descendants. The baptismal formula allows us to examine the relationship among four individuals—the father, mother, godfather, and godmother.[73] The patterns of these relationships reveal additional information about social organization and contacts during the period of Indian labor. Table 3 presents a distribution by ethnic/racial group of the four categories.

Baptism was somewhat less formal for slaves than it was for free people—expecially for whites. Whereas a godmother and godfather were always present at the baptism of the child of a white couple, godparents were not always in attendance for slave children. There were some twenty-five instances (13 percent) when the godmother, the godfather, or both were omitted (or unregistered) at the ceremony. On one occasion a group of slaves served as godparents and on another there were two godfathers and no godmother. Such assymetry and irregularity were never found among the Portuguese baptisms at Engenho Sergipe. When adult pagans were baptized, they usually received the sacrament in small groups of three or four. On these occasions one set of godparents might sponsor all of the baptized individuals.

According to Roman Catholic doctrine and practice, the roles of godfather and godmother were vital to the child's guidance. The relationship between child (*afilhado*) and godparents (*padrinhos*) was as binding as that between

---

new parish church was erected in Santo Amaro in 1722, the register contains information about the population of the whole surrounding parish and not solely of Engenho Sergipe. Unfortunately, there are drawbacks that limit the utility of the register for historical analysis. First, its present physical state is poor. Many entries are illegible because pieces of the pages are missing, and, in fact, most of the pages covering the sixteenth century have been lost, as have most of the marriage entries. The term "negro" was used to describe both Indians and Afro-Brazilians; thus, it is impossible to distinguish between them on this basis alone. Whites were never identified as such in the chapel register; therefore, when an individual had both a Christian name and family name and no other ethnic or color designation, I have assumed him to be white. This method probably results in a slight inflation of the white category at the expense of mulattoes and *mestiços*; but, since my major concern here is with Indians and Africans, this distortion is not serious. In addition, many individuals are simply described as "escravo" without more specific identification about color or origin. These problems complicate any analysis and the results presented here are tentative at best. Still, this singular source, an engenho chapel register from the late sixteenth and early seventeenth centuries, presents a rare opportunity to examine some aspects of relations between the various elements of the Recôncavo population.

[73] Each registration follows a formula: for example, "5 August Joana, young daughter of Tomé de Sousa, unmarried and of Luiza, an Indian of Domingos Ribeiro; godparents, Bras Dias and Antonia of the same Domingos Ribeiro." Chapel Register, Engenho Sergipe, 1595–1628, f. 96.

TABLE 3
RACIAL/ETHNIC DESIGNATIONS OF PARENTS AND GODPARENTS
ENGENHO SERGIPE, 1595–1608

	White	Indian	African	Negro/Crioulo	Mulatto	Unknown
Father	61	42	27	6	0	98
Mother	43	54	33	8	3	93
Godfather	132	9	6	7	0	70
Godmother	59	21	8	7	7	114

NOTE: The many unknowns (98) in the case of the fathers result from illegitimate unions and no father present at baptism. In the cases of mothers, godfathers, and godmothers, unknowns are due to failure to report this information or gaps in the documentation.

TABLE 4
INDEX OF COMPADRIO PRESTIGE

	Parents	Godparents	Ratio
Whites	104	191	1.84
Indians	96	30	.31
Afro-Brazilians	64	13	.20

SOURCE: ACMS, Conceição da Praia (Engenho Sergipe) bautismos.

child and parents. The parallel set of ties between the parents and godparents established a set of mutual obligations and dependencies. The *padrinho* was the baptismal sponsor and usually assumed the expenses of the service. More importantly, the role of godfather placed very real obligations on those holding it, for it was not uncommon for godchildren eventually to depend on their padrinhos for economic assistance or protection. The position of the godfather had, therefore, status and prestige. Not surprisingly, a very high percentage of whites were godfathers in the baptisms examined. Not only did white parents always choose a white godfather, but Indians and Africans also sought whites to assume this role. In over 80 percent of the baptisms of the children of Indian mothers, white men served as godfather. Indian men served as godfathers in only nine cases; of the six cases in which the ethnic origins of the parents can be identified, all were Indian couples. African men were in a somewhat similar position and like the Indians continually looked to whites (or had to accept them) as godfathers for their children. Table 4 shows this white predominance.

The most pronounced pattern to emerge from the data concerning godparents is the marked difference between godfathers and godmothers: godmothers, it seems, were selected on a different basis from that of godfathers, for more Indian and Afro-Brazilian women were godparents than were men. White women rarely presented slave children for baptism. Instead, Indian mothers sought Indian godmothers. The godfather could be white for, as protector and benefactor, a white was best equipped to aid the child; the godmother, however, was an auxiliary to the child's upbringing and was a surrogate parent if the biological mother died. Although there were a few instances in which Indian women served as godmothers for children born to African women or in which mulatto women were godmothers to the children of Indian mothers, in the vast majority of the cases Indians, Africans, and whites selected women of the same racial category as *comadres*. Usually the godmother (*madrinha*) was a slave of the same master as either or both of the parents.

A few statistics illustrate this situation. Whereas male slaves comprised less than 12 percent of all godfathers registered, female slaves were over 30 percent of the godmothers. If only slave and forro baptisms are considered, the percentage of slave and forro madrinhas rises to more than 80 percent. The ties among the Indians were strong and in over 60 percent of the baptisms of children born to Indians, Indian women were godmothers. When both parents were Indians, 90 percent of the godmothers were also Indian. On occasion, a slaveowner, his relatives, or his employees stood as sponsor for a child of one of his slaves. But, significantly, planters did not often assume the godparent relationship—that is, it was not frequently used to reinforce the ties between slave and master. In less than 4 percent of the slave baptisms did the owner himself or a close immediate relative stand as sponsor. Patriarchalism may have existed, but it did not express itself very often through the role of *compadre*. Instead, what appears to be the dominant pattern is the selection of a white man who might intercede with the master in case of some future difficulty.

Despite this tendency toward endogamy, sexual relationships across the color lines did take place. White males were most easily able to take advantage of their dominant role in order to select sexual partners from among the slave and free populations. White men fathered over 11 percent of all children born to Indian mothers and 8 percent of those born to African women registered at Engenho Sergipe. If these figures are adjusted to include those cases in which no father was reported, a sign of illegitimacy and unstable or secret reationship, white men fathered 18.5 percent of all Indian children and almost 30 percent of the children of Afro-Brazilian women. Opportunities for association between Africans and Indians also existed in the slave quarters. The Engenho Sergipe inventories list a number of cases in which Africans and Indians formed permanent family units.[74] More impressive, however, is the relative lack of such unions in the chapel register, which contains only two cases of children born to such couples. Both of these were between Indian men

---

[74] Inventory, Engenho Sergipe, 1591, ANTT, *Cart. Jesuitas*, maço 13, no. 4.

TABLE 5
SLAVE MARRIAGES, ENGENHO SERGIPE, 1601–26

Indian/Indian	6
Same "Nation" African	7
Mixed "Nation" African	6
Crioulo/Crioulo	1
Unidentified Origins	9

and African women. In the period of transition from Indian to African slavery, therefore, the majority of Indians married or had sexual relations with other Indians and to a large extent remained sexually separate from those of other cultural backgrounds. What miscegenation that did take place occurred most frequently between whites and Indians or whites and Africans.[75]

Among the three principal racial groups there was a strong tendency toward endogamy, at least in formal, Church-sanctioned unions. Between 1601 and 1626 in thirty slave marriages in which the origins of the partners could be determined, all were between individuals of the same racial group, although not of the same ethnic or linguistic group.

THE TRANSITION FROM A PREDOMINANTLY INDIGENOUS SLAVE FORCE to one composed mainly of Africans occurred gradually over the course of approximately half a century. As individual engenho owners acquired sufficient financial resources, they bought a few African slaves and they added more as capital and credit became available. By the end of the sixteenth century, engenho labor forces were racially mixed and the proportion increasingly changed in favor of imported Africans and their offspring. In the 1550s and 1560s there were virtually no African slaves at the northeastern sugar mills.[76] By the mid-1580s Pernambuco had some sixty-six engenhos and a reported two thousand African slaves. If we estimate an average of one hundred slaves per engenho, then Africans composed one-third of the captaincy's slaves.

In Bahia, the change can be observed in the transformation of a single engenho's population over time. In 1572 Engenho Sergipe had two hundred and eighty adult slaves of which only twenty (7 percent) were African. Twenty years later, in 1591 the engenho had a slave population of one hundred and three, of which thirty-eight (37 percent) came from Africa. When, in 1638, Engeho Sergipe was rented to Pedro Gonçalves de Mattos, it had eighty-one slaves, all of whom were African or Afro-Brazilian.[77] The transition to an

---

[75] This pattern continued throughout the colonial period in Bahia. In the parish of Inhambupe between 1750 and 1800, 80 percent of the 1,294 registered marriages were between couples of the same racial category. See Conseulo Pondé de Sena, "Relações interétnicas atraves de casamentos realizados na freguesia do Inhambupe, na segunda metade do século xviii," unpublished paper, Universidade Federal da Bahia (Salvador, 1974).

[76] Mauro, *Le Portugal et l'Atlantique au XVIIe siècle* (Paris, 1960), 192–94. The Jesuits of Bahia asked for two dozen Africans in 1558, "and these can come together with those the king sent to the [royal] engenho because often he sends ships here loaded with them." Leite, *Cartas Nóbrega* (Bahia; May 8, 1558), 288.

[77] Inventário de Mem de Sá, 1472, in *DHA*, 3: *passim*; Engenho Santana had the same distribution with 7

Indian Labor and New World Demands

African labor force was made in the first two decades of the seventeenth century at a time when the sugar industry experienced rapid expansion and considerable internal growth arising from high international sugar prices, a growing European market, and, perhaps, the peaceful maritime conditions brought about by the twelve-year truce between Spain and the Netherlands (1609–21). A comparison of the positions and roles of Indian and African slaves should help explain why the transition to African labor took place.

The shift to African labor depended in part on Portuguese perceptions of the relative abilities of Africans and Indians. With a long history of black slavery in Iberia which had intensifed during the expansion of the sugar industry in the Atlantic, the Portuguese were well acquainted with Africans and their skills. By the end of the sixteenth century Africans had already impressed the Portuguese with their ability to master the techniques of sugar production on Madeira and São Tomé. The Portuguese in Brazil, long familiar with the use of blacks as servants, urban artisans, and skilled slaves in Portugal and the Atlantic islands, began to look toward Africa as a logical source for these skills. The first black slaves in Brazil came as body servants or skilled laborers, not as field hands. The three extant engenho slave lists from the sixteenth century indicate a high percentage of Africans with various skills, and invariably the most complicated tasks assigned to slaves were given to Africans. In 1548 Engenho São Jorge de Erasmus in São Vicente had about one hundred and thirty slaves "of the land" as well as seven or eight Africans. All of the Africans were *oficiais*, that is skilled at various tasks, and one was sugar master, the most important managerial position on any engenho. The director of Engenho São Jorge's operations proudly wrote to the absentee owners, the Schetz family of Antwerp, that sugar masters on Madeira usually received thirty milréis a year, a sum which their engenho now saved by using this skilled black slave.[78] Three other Africans were employed in positions requiring skilled judgment, one as purger (*purgador*) and two as kettleman (*caldereiro*).

A similar situation is found in the inventory lists of Engenhos Sergipe and Santana of the late sixteenth century. At Engenho Sergipe Indians and Africans were used in different ways during the period of transition. Because the engenho could afford Portuguese technicians and managers, the occupational pyramid of the slaves was truncated. The work force was heavily indigenous, out of 134 male slaves, 115 were Indians. The same proportion of Africans and Indians were listed with a specific occupation; but, when certain jobs such as fishermen, hunters, boatmen, are not included, the proportion of Indians with special occupations drops considerably. Table 6 shows these differences.

Africans in a slave force of 107 or 6.5 percent. Inventory, Engenho Sergipe, 1591, ANTT, *Cart. Jesuitas*, maço 13, no. 4. This listing of slaves was made in a legal dispute between Francisco de Negreiros, representative of the count of Linhares, and Francisco Araújo and his wife, who claimed to have purchased the engenho in 1590. "Treslado do Inventário do Engenho Sergipe," 1638, ANTT, *Cart. Jesuitas*, maço 30, f. 1040.

[78] Stols, "Um dos primeiros documentos sobre a engenho dos Schetz," 418–20.

Table 6
Occupational Structure
Engenho Sergipe 1572, 1591

	1572 Africans	1572 Indians	1591 Africans	1591 Indians
Sugar-making skills				
mestre de açuçar (sugar master)			1	
ajuda do mestre (assistant)			1	
purgador (purger)			2	1
ajuda do purgador (assistant)	1	2	1	
tacherio (small kettleman)	1	2	3	
escumeiro (skimmer)	1			
ajuda do escumeiro (assistant)		3		
caldereiro (kettleman)		6		
moedor (mill tender)		3	2	
premseiro (presser)	1	1		
virador de bagaço (bagasse feeder)		1	1	
caixeiro (crater)		2	1	1
dos melles (molasses)	1	1		
Artisan skills				
carapina (carpenter)		1		
ferreiro (blacksmith)			1	
calafate (boatcaulker)		1		
falleiro (?)	1	1		
Auxilliary skills				
vaqueiro (cowboy)	1	1	2	1
carreiro (carter)	1	1		
boieiro (herdsman)				3
pescador (fisherman)		11		
serrador (sawyer)		7		
lenadeiro (firewood)				1
porqueiro/ovelheiro (pig/sheep tender)		2		
"barcas" (boatman)	1	4		
Management				
feitor (overseer)		1		
Total	9/19	51/115	15/30	7/65

Indian Labor and New World Demands

The inventory of 1572—taken when Indians were still plentiful and relatively inexpensive to obtain, when Africans were still not available in large numbers, and when the legislation against Indian slavery was not yet effectively in force—represents a specific period in the history of Indian slavery. Twenty years later the situation had changed considerably. By 1591 the sugar economy of the northeast was expanding rapidly to supply a growing European demand. The Atlantic slave trade had been regularized to the extent that the supply, while not yet great, was at least dependable. The majority of Engenho Sergipe's slave force was still Indian, but Africans and Afro-Brazilians now filled almost all of the skilled occupations on the estate. Angolan and Guinean men were employed as sugar master, purger, assistant purger, blacksmith, kettleman, and sugar crater. Others were employed in the milling operations of the engenho and a few as cowhands. The Indian occupations were far more rudimentary and, aside from one sugar crater, only three were listed with occupations, one woodcutter and two herdsmen. In other words, when possible, the Portuguese turned to Africans to provide skilled slave labor.

This policy, like the relative price of Indians and Africans, is to some extent explained by demographic and cultural features of both peoples. Many West Africans came from cultures where iron working, cattleherding, and other activities of value to sugar agriculture were practiced.[79] These skills and a familiarity with long-term agriculture made them more valuable to the Portuguese for the specific slavery of sugar. Africans were certainly no more "predisposed" to slavery than were Indians, Portuguese, Englishmen, or any other people taken from their homes and bent to the will of others by force, but the similarity of their cultural heritage to European traditions made them more valuable in European eyes. The Indian susceptibility to European disease at all ages made riskier the investment of time and capital in training them as artisans or managers. Africans, of course, also suffered in the environment of Brazil, but the highest rate of black mortality was always found among the newly arrived (*boçal*) and among infants. Thus, once a slave was "seasoned" and had passed through infancy and childhood, the chances of survival and therefore of safe investment in skill were very good.

African health and skill, as well as lack of resistance, may explain the reluctance of planters to invest in the training of Indian slaves, but it does not respond to the question why, even in cases where Indians were free workers earning a wage, the value of their labor was considered unequal to that of whites, mulattoes, and free blacks. At Engenho Sergipe an Indian carpenter received only 20 percent of the wages paid to whites for the same task. During the seventeenth century Indian workers received only twenty réis a day and

---

[79] The Portuguese slave trade in the sixteenth century was concentrated in the Senegambia. On the cultural and agricultural traditions of the peoples in that region, see Phillip D. Curtin, *Economic Change in Precolonial Africa: Senegambia in the Era of the Slave Trade* (Madison, 1975), 3–58; and Walter Rodney, *A History of the Upper Guinea Coast, 1545–1800* (Oxford, 1970), 1–38.

skilled artisans averaged thirty réis. In the 1630s the municipal council of Salvador paid Indian laborers a daily wage of thirty réis and Paraíba Indians could be paid in manioc and cloth a daily wage of about fifteen réis. Black slaves, by contrast, could earn an average of 240 réis per day.[80]

The wage labor system, therefore, constantly proposed as the ideal way to integrate the Indian into colonial society, was a failure. Indians were often reluctant to participate in the labor market; and the Portuguese, furthermore, did not really allow that market to operate freely: the wages paid to Indians were always below existing rates.[81] The colonists placed Indian wage earners on a scale of reward and labor different from that of other workers. At Engenho Sergipe they were usually paid by the month rather than by the day or, even more commonly by the task. Their work did not usually require completion at a specific time, and often they received payment in kind rather than in cash. Manioc flour, trade cloth (*pano*), and alcohol were the common "wages" for Indians from Maranhão to São Paulo.[82] Obviously, the Portuguese seemed to believe, for whatever reasons, that Indian workers could not be treated like others.

There was, in fact, a remarkable similarity among all of the colonial regimes in the New World in the low value placed on Indian laborers in comparison with Africans. In times and places as widely different as sixteenth-century Mexico, seventeenth-century Brazil, and eighteenth-century Carolina, Spaniards, Portuguese, and Englishmen held similar opinions of Indian and African laborers. The colonists in each situation usually valued Africans three to fives times higher than the Indians.[83] Certainly, market availability, demographic patterns, opportunities for flight or resistance (management costs), and European prejudices entered into these calculations. Still, despite the racist implications of arguments about the relative adaptability of one people over another to tropical labor, the similarity of opinion among all the New World slaveholding regimes suggests that there was a comparative advantage, especially in the formative period of slaveholding, in the use of African rather

---

[80] Paul Silberstein, "Wage Earners in a Slave Economy"; and Biblioteca Nacional de Madrid, Códice 2436, ff. 105–09.

[81] See Antonio Garcia, "Regimenes indigenas de Salariado: El Salariado natural y el Salariado capitalista en la historia de America," *América Indigena*, 8 (1948): 250–87.

[82] Silberstein, "Wage Earners," based on *DHA*, 2: *passim*; and Leite, *HCJB*, 2: 63. Also see Adrien van der Dussen, *Relatório sobre as capitanias conquistadas no Brasil pelos holandeses*, ed. José Antônio Gonçalves de Mello, (Rio de Janeiro, 1947), 88–89.

[83] Almon Wheeler Lauber, *Indian Slavery in Colonial Times within the Present Limits of the United States*, Columbia University Studies in History (New York, 1913), 298–300, presents scattered references to relative prices from New England, New York, and the Carolinas. Crane, *The Southern Frontier, 1670–1732*, 113–15, provides data showing that Indians were valued at one-half to one-third the price of black slaves. Peter H. Wood, *Black Majority* (New York, 1974), 38–40, reviews the literature on Indian slavery in Carolina but is silent on this point. Instead see John Donald Duncan, "Servitude and Slavery in Colonial South Carolina, 1670–1776," 2 vols. (Ph.D. dissertation, Emory University, 1972). On French Canada, see Marcel Trudel, *L'esclavage au Canada Français: Histoire et condition de l'esclavage* (Quebec, 1960); and Guy Fregault, *La civilisation de la Nouvelle-France* (Montreal, 1944), 83–84. Colin Palmer, *Slaves of the White God: Blacks in Mexico, 1570–1650* (Cambridge, Mass., 1976), 34, provides considerable evidence from the 1520s in Mexico as does Zavala, *Los indios esclavos en Nueva España*. Most important is Gonzalo Aguirre Beltrán, "El trabajo del indio comparado con el del negro en Nueva España," *México Agrario*, 4 (1942): 203–07.

than Indian slaves and that this advantage was based on productivity in terms of return on investment. The statement of one observer in the Carolinas in 1740 that "with them [Indians] one cannot accomplish as much as with Negroes" was echoed everywhere in the Americas.[84]

In Brazil, the relative position of Indian and African slaves within the sugar labor force can be seen in its simplest and crudest form in the comparative prices of the two peoples. The average price of an African slave listed with occupation in 1572 was 25$000 while Indians with the same skills averaged only 9$000. The only skilled Indians whose price equaled, or even approached, that of African slaves were those who were truly practicing skilled crafts—carpenters, sugar craters, and boat caulkers, for example, or those engaged in the specialized positions of a sugar mill. The vast majority of Indians listed with some occupation, but not an artisan skill, were priced far below the average value of unskilled Africans. The price difference between skilled and unskilled Indians was greater, moreover, than that among Africans.

There is evidence that these values represent real differences in the productivity of Indian and African labor. Production figures from Bahia at the close of the sixteenth century support this interpretation. Although there is some discrepancy in the reported total of engenhos, a number of accounts list fifty mills operating in the captaincy of Bahia by 1590.[85] For the year 1589 Father Francisco Soares reported that there were fifty engenhos, eighteen thousand slaves, and thirty-six thousand *aldeia* Indians.[86] If we assume that two-thirds of the slave force were involved in sugar agriculture, then the ratio of slaves to engenhos was 240 to 1. This figure—which does not include any of the settled Indians that also provided labor to the engenhos—is extremely high. It represents not only slaves owned directly by the mills, but those owned by tenants, sharecroppers, and others as well. Father Soares estimated an annual production per mill of four thousand *arrobas* or fifty-eight tons. Thus, each slave produced at the time almost seventeen *arrobas* (over five hundred pounds) a year—a very low level of productivity, since the later calculation in Brazil based on black slaves was forty to seventy *arrobas* annually.[87] Even allowing for technological changes and inexact information, the only conclusion that can be drawn from such figures is that Indian labor was characterized by low

---

[84] As quoted in Duncan, "Servitude and Slavery in Colonial South Carolina, 1670–1776," 36. For the traditional racist arguments of the nineteenth century, see Herman Merivale, *Lectures on Colonialism and Colonies [1861]* (London, 1967), 283.

[85] Gabriel Soares de Sousa lists thirty-six engenhos for Bahia, but he also speaks of eight *casas de melles* (molasses-producing units). He gives an annual production of 120,000 *arrobas* for the captaincy or somewhat less than 4,000 *arrobas* per mill. Fernão Cardim also speaks of thirty-six engenhos in Bahia, but José de Anchieta lists forty-six. I have taken Father Soares' figure of fifty because it yields the lowest ratio of slaves to engenhos as a control on my argument that the ratio is extraordinarily high. Using the estimates of Soares de Sousa or Cardim yields over 333 slaves for each mill; see Mauro, *Le Portugal et l'Atlantique*, 193; and Mauricio Goulart, *Escravidão africana no Brasil* (São Paulo, 1950), 100.

[86] Soares, *Coisas notaveis do Brasil*, 11.

[87] See Ward Barrett and Stuart B. Schwartz, "Two Colonial Sugar Economies: Bahia and Morelos, 1600–1800," in Enrique Florescano, ed., *Haciendas Latifundias y Plantaciones* (Mexico, 1975), 550–55. Also see Ward Barrett, *The Sugar Hacienda of the Marqueses del Valle* (Minneapolis, 1970), 98–99.

productivity. As Magahães de Gandavo claimed, "if the Indians were not so fickle and given to flight, the wealth of Brazil would be incomparable."[88]

At the time Father Soares made his estimates, Bahia had between three and four thousand African slaves; thus, three-fourths of its slave force was still Indian in the last decade of the sixteenth century.[89] With the low level of Indian productivity, the price differential between African and Indian slaves becomes readily understandable. In the inventories of 1572–74 African slaves were valued at an average price of 20$000, while adult Indians averaged about 7$000.[90] This ratio of roughly three to one is also the ratio between the estimates of African and Indian productivity in sugar agriculture. It would appear that the Portuguese made a reasonable economic calculation of the comparative profitability of their two alternate work forces. Africans surely cost more to obtain, but in the long run they were a more profitable investment.

WE CAN SPECULATE THAT THE EARLY PRESENCE of large numbers of Indians allowed the mills to begin production with a small original capital outlay in slaves. The expansion of the sugar economy in the 1550s and 1560s depended on the availability of this source of "cheap" labor. During the 1570s, however, resistance, plague, and anti-enslavement legislation reduced the availability—and profitability—of Indians. Plantation owners now found that the cost differential between Indian and African laborers no longer outweighed differences in productivity between the two labor forces. This disparity in overall productivity also helps to explain why the Portuguese preferred imported Africans to coerced but "free" Indians. While there were occasional proponents of *encomienda* or of peonage in Brazil, the colonists believed that, given the high mortality and low productivity of Indians, Africans were a better investment. The Bahian historian and planter, Sebastião da Rocha Pitta, probably summarized majority opinion when he observed that Indians suffered from "working by obligation rather than out of desire, as they had in the state of freedom; and in its loss and in their repugnance to, and thought of, captivity so many die that even at the lowest price they are expensive."[91]

A discussion of profitability in strictly neoclassical economic terms will not suffice as an explanation of the transition of the labor force. There were always cultural and political determinants as well. Not everyone in Brazil was convinced of the wisdom of the shift. Portuguese colonists were generally unwilling to surrender control of Indians, especially when they could be obtained for nothing. The colonists demonstrated their reluctance by political remonstrance and demonstration—most notably in 1609 and 1640. Gaspar da Cunha, overseer of Engenho Sergipe, wrote to the Count of Linhares in 1585 that Africans "cost too much and are prejudicial to the plantation and to the

[88] Gandavo, *Histories of Brazil*, 153.
[89] Goulart, *Escravidão africana no Brasil*, 100.
[90] Buescu, *300 anos da inflação*, 44–45.
[91] Sebastião da Rocha Pitta, *História da America portugueza* (1730; 2d ed., Lisbon, 1880), 196–97.

Indian Labor and New World Demands

neighborhood; they are neither as necessary nor as beneficial as the Indians of this land."[92] He then petitioned for more free Indians to be brought to the engenho. By the early seventeenth century such requests and sentiments were far less frequent. The shift to African labor was well on its way, especially in the northeastern sugar region where capital had accumulated and the patterns of international commerce were securely established. Colonial slavery had emerged as the dominant mode of production, and the process of its emergence was not dictated by the market so much as by the organization of production. The system of labor and nature of the labor force was determined not only in the court at Lisbon or in the counting houses of Amsterdam and London but in the forests and canefields of America.

[92] Gaspar da Cunha to the count of Linhares (Bahia; August 28, 1585), ANTT, *Cart. Jesuitas*, maço 8, no. 9. For a similar opinion see Martim Leitão's *Parecer*, in which he valued one Indian equal to four Guiné slaves. BA, 44-XIV-6, ff. 185-93v.

3
Slave Families on a Rural Estate in Colonial Brazil

Richard Graham

The family structure of the lower classes once received relatively little attention from historians and that of slaves perhaps even less. It was long assumed, for instance, that nuclear families were created by the industrial revolution rather than having been the typical family structure among the lower classes long before that. Slaves were believed to have had almost no family life at all, but to have been housed in virtual barracks, promiscuously procreating children who grew up with absent fathers and domineering mothers. In recent years social history has made long strides to cast aside the old certainties, although new ones have not yet emerged to take their place. What is already established is that past family life was both more varied and more complex than was once thought.[2] Before new generalizations can be made bits and pieces of information must be gathered, sometimes painfully, from diverse cultural and societal contexts.

The present article examines some information on 1347 slaves counted in 1791 on one estate in Brazil. Their age, sex, and fertility, the relative ages of husbands and wives, types of family groups, and workforce characteristics can be established from some recently uncovered sources. Although, as I will note, there are several reasons to be cautious about drawing conclusions from this data for Brazil as a whole, it is striking that for this estate — with what was surely one of the largest holdings of slaves at any time anywhere in the Americas — many of the accepted generalizations about slave life in Brazil and elsewhere do not hold true. For the most part, these slaves lived in separate family units built around husbands and wives. Our entire view of what it meant to grow up as a slave may thus be altered. The typical married woman bore two children and began bearing them just before she turned 20. Men married mostly after age 25, but the spread in ages between husband and wife tended to be five years or less. Separation (or "divorce") occurred, but was not the norm. The proportion of skilled workers in this group was very high for a rural estate, reaching 30% for middle-aged males and including at least one literate pharmacist. Less than 2% were handicapped. In all, the characteristics of this group are so different from what we would have imagined that its examination may also suggest important questions for social historians of other places and other times.

When the Jesuits were expelled from the Portuguese empire in 1759, their extensive properties were taken over by the crown. Among these estates was the immense Fazenda Santa Cruz about 35 miles west of the city of Rio de Janeiro. The Jesuits had early devoted attention to its development and disbursed large sums in draining its low-lying terrain to avoid flooding and open up grazing land. By 1729 they possessed 11,000 head of cattle and had installed a pottery, a carpentry, a manioc-flour mill, a limekiln, and a fishery, not to mention a warehouse, an inn, and the church. Later they added a tannery and a pharmacy.

SLAVE FAMILIES ON A RURAL ESTATE

Instead of sugar cane, as in northeast Brazil, the major focus of their activity was cattle raising and the production of staples for the urban market. Many of the estate's products (such as hides, manioc-flour, pots, and bricks) were processed in semimanufactories. Presumably other Jesuit establishments in the region depended upon it for necessary supplies, although the full history of the estate in Jesuit days has never been studied with much attention to these matters.[3]

Under royal control after 1759, the *fazenda* continued to render a large income despite repeated reports of mismanagement.[4] In addition to cash income, it provided meat for the crews of the royal navy and charcoal for public facilities.[5] A new administrator who took over in 1783 devoted most of his energy to developing and using the pasture lands for fattening beef bought in Sorocaba, a trading center to the south, and sold in Rio de Janeiro at a butcher shop owned by the *fazenda*. He also planted some land in rice and corn and built two new large manioc-flour mills. By the end of 1790 the viceroy was considering the recommendation of one of his advisors that a sugar mill be installed on the property.[6] Although the mill was built and some prosperity ensued, the inconstancy of government policy and the vagaries of bureaucratic appointments soon led to renewed decline.[7] Nevertheless, the land was not sold off and was still listed as a source of some public income at the very end of the nineteenth century.[8]

In 1791 a complete inventory was made of the Fazenda Santa Cruz at a time when responsibility for its direction passed from one bureaucrat to another. Among the property inventoried were 1347 slaves listed by name, age, sex, and family group.[9] It is from this inventory that data on family life have been derived. The slaves were apparently summoned for enumeration to their huts which are known to have stretched in two rows perpendicular to the church and convent around a large square.[10] The inventory-taker began by listing the male he considered head of the household, if there was one, whom he then described as married, widowed, or single. He then must have asked the slave how old he was or made an estimate himself. This figure was listed with great precision, a precision that can hardly be justified even by the *fazenda's* ex-status as a Jesuit property in which more than customary attention was presumably paid to baptism. The man's wife and children were listed next, along with their respective ages. When there was no male head of the household, the woman was usually listed as widowed, although occasionally simply as "married." In many households there were unmarried daughters with children of their own. Finally, some men and women were listed as single and a few of these women also had children. What determined the differentiation between married-women-without-husbands and single mothers was probably not the record of a marriage contract but the answers rendered by the blacks or the vagaries of the annotator. Contrary to the conventional view we have had of slave relationships, this latter had the fixed idea of a relatively stable family life among the slaves and reported on it without surprise.[11]

Of the 1347 slaves on this plantation, 363 were men, 448 women, and 536 children 14 years of age or younger. Of the adult males, 115 were classified as single and, of the females, 145 (including 19 single mothers). The median age of all these slaves was 23.8 years, with the women being slightly younger (23.6) and the men slightly older (24.0). If only adults 15 years of age and older are counted, the difference in age between men and women somewhat increases, being 35.2 for the men and 33.8 for the women. The typical single male slave was 20.9 years old and his potential mate 20.3. Figure 1 and Table 1 provide a profile of the entire slave population of the *fazenda* by age and sex.

Figure 1. Sex-Age Pyramid

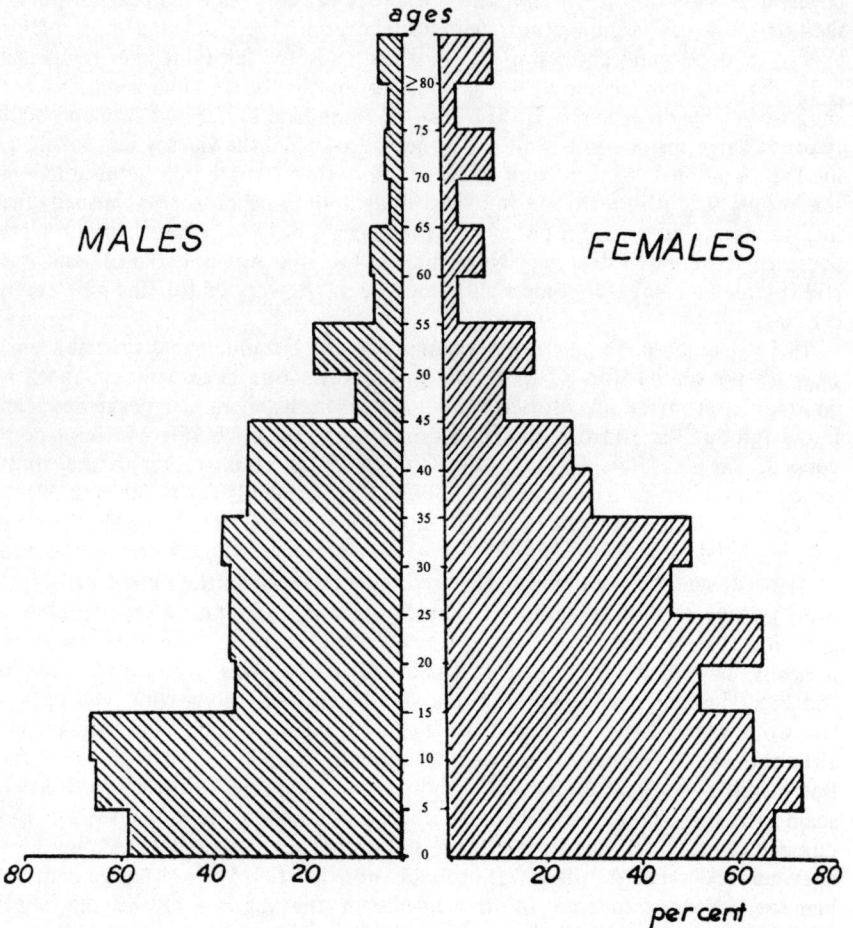

SLAVE FAMILIES ON A RURAL ESTATE

Table 1. Males And Females By Age Group

Age Group	Males		Females		Males per 100 Females	
0-4	80		92		88	
5-9	89		100		89	94
10-14	90		85		106	
15-19	48		70		68	
20-24	50	201	88	288	56	69
25-29	50		62		80	
30-34	53		68		77	
35-39	44		40		110	
40-44	44	126	35	115	125	109
45-49	13		16		81	
50-54	25		24		104	
55-59	8		3		262	
60-64	10	26	11	30	90	86
65-69	3		3		100	
70-74	5		13		38	
75-79	3	10	2	15	150	66
80 and over	7		13		53	
TOTALS	622		725		82	

The disproportionately large number of women overall contrasts sharply with what we know about the Brazilian slave population in general, at least for the first half of the nineteenth century. As long as the slave trade continued and for a long time after its end in 1852, slave men in Brazil outnumbered the women. There is no indication in this inventory as to whether any slaves were born in Africa. None of their names reflects African tribal origins, and I assume they were all creoles.[12] Of course, even for creoles, any given plantation would be expected to deviate from the populational pattern as a whole.

The traditional view of slave life in the United States immediately leads to the suspicion, given such a large number of women, that this Brazilian estate was used for slave breeding. Although there was apparently no need to purchase Africans, an administrator had purchased 80 to 100 new slaves between 1783 and 1789.[13] Were they women? We do not know. But the relative ages of men and women can shed some light on the question of breeding. Table 1 shows that after the age of 35 their numbers were roughly balanced, except for the aged. But in the age group 15-34 the women outnumbered the men by almost half again as many.[14] On the other hand, peasant communities often reveal a low number of men of working age who are out of the village in search of work elsewhere. The analogous situation here would be that slaves were rented or loaned to other enterprises; there is evidence of this practice but no indication of its precise dimension. The structural parallelism is striking. Table 1 also reveals a puzzling imbalance among children. Not only are there significantly fewer boys than girls below age 10, but the ratio of males to females is lower in the 5-9 age group than in the 10-14 group, that is, precisely when some might be entering the labor force and be of marketable age. The figures for the 0-4 age group are similarly mystifying, but it is to be expected that the age-sex pyramid will be distorted since it is based on a single observation at a precise moment in time for a total population that, although impressive in terms of slaves on one estate, is very small in a statistical sense.

There is no way of knowing the number of children for every family in 1791, since the parentage of those offspring who no longer lived in the same household with their parents was not indicated. Also, and for the same reason, the inventory cannot indicate with any certainty the average age of parents at the birth of their first child. To suggest the lower age limits of fertility and estimate family size we may isolate married fathers in the 25-29 age group (N = 35) since even their oldest surviving child would still tend to be listed at home. For them the median age of children was 4.7 years and the number of children per couple was 1.23. It is probably more valid to isolate married mothers in the same age group (N = 48). The median age of their children was 5.0 years and the average number of children 2.04. In general, slave women on this plantation began to bear children in their late 'teens and early twenties and bore two live children apiece. The fertility of single women this age is low (0.64 children per woman), but presumably this is partly because it was those who did not bear children who were considered still single. The median number of children per woman this age (whether single or married) was 1.73. Of the "married" women this age 6 or 12.5% were childless; another 7 single women this age were childless, making the percentage of childless women in the 25-29 age group a surprisingly high 21%. Little in these figures lends strength to the belief that the slaves were used as breeders.[15]

A consideration of families among this slave population shows that almost half (47.75%) of them were formed by mates and children and that almost a full 55% were built around man and wife (see Table 2). Married sons and daughters at Santa Cruz left or were forced to leave their parents and a married daughter rejoined them only if she were widowed, abandoned, or in effect divorced, and not often even then. In contrast, a large number of unmarried single men and women remained in their parents' household into their thirties.[16] The high percentage of married or widowed slaves represented by this and other tables below contrasts sharply with figures reported for Brazil as a whole a century later.[17]

Table 2. Households By Type

	No.	%
1. Male and female	27	7.16
2. Male, female, and children	171	45.36
3. Male, female, children, and grandchildren	8	2.12
4. Male, female, and grandchildren	1	0.27
Subtotal	207	(54.91)
5. Female and children	50	13.26
6. Female, children, and grandchildren	3	0.80
7. Female and grandchildren	1	0.27
Subtotal	54	(14.32)
8. Male and children	17	4.51
9. Male, children, and grandchildren	3	0.80
10. Male and grandchildren	2	0.53
Subtotal	22	(5.84)
11. Brother(s) and sister(s)[a]	8	2.12

SLAVE FAMILIES ON A RURAL ESTATE

12. Male alone	35	9.28
13. Female alone	50	3.26
14. Groups of males	1	0.27
Subtotal	94	(24.93)
TOTAL	377	100.02

[a]In only one of these households were there members of both sexes over 15.

The enumerator assumed that the head of the household was the male whenever there was one of suitable age. The children were listed as his children. One of the most surprising features discovered here is the number (22) of men without wives who headed families. A comparable study for Jamaica seems to have found no instance of this type of family,[18] whereas at Santa Cruz there were two families headed by a father alone for every five headed only by a mother.

Later on, at the height of nineteenth-century coffee prosperity in Brazil, it is believed to have been common practice to house single slaves in large undivided barrack-like quarters, separating males from females.[19] This was not the case on the Fazenda Santa Cruz. Seventy-seven of the single men aged 15-24 were listed with their parents as against only 14 listed separately. Among the same age group of childless single women, 102 lived with their parents and 16 alone. Furthermore, those males and females who did not live with their parents were listed in scattered fashion, either alone or in groups of 2 or 3, and not all together as would probably have been the case if they shared common quarters.[20]

Men tended to marry in their late twenties: as can be seen from Table 3, column 3, 88 percent of the men in their early twenties were unmarried, whereas this figure for men in their later twenties falls to 30%. It could be argued that this drop was an indication that the saleability of younger slaves encouraged the managers to regard them as single regardless of their relationships with women; one could speculate that, as slaves grew older and less saleable, the unions which they had established were considered more or less permanent. But larger societal factors must also be considered. It probably reflected, at least to a point, the accepted norm regarding marriageable age.

In Table 3, columns 2 and 4 show the clustering of ages of the male and female partners. Men were older than women, as might have been expected, and has been noted. For the age groups over 35, men characteristically outnumbered the women, whereas the situation was reversed for groups under 35. Columns 6 and 7 list available marriage partners. There were 128 marriageable women under the age of 25, but only 92 men in the same age group. Even if the ages considered marriageable are extended to under 35 for the women (N = 163) and under 55 for the men (N - 137), the disproportion remains noticeable. The sharp decline in the number of women after age 35 is not matched by the men for another 10 years. Possibly late childbirth accounts for the decline of the female population at that age.

It is conventional to think of the age spread between husband and wife in colonial Brazil to have been very large. Probably this was the case only among the upper levels of society where marriages were arranged for the sake of acquiring or preserving property. Slaves had no such concerns. As can be seen from Figure 2, the men on this estate were generally older than their wives: in the case of 178 couples the man was older than the woman whereas she was older than he in only 20 cases (in 13 they were the same age). But in the great majority of the cases (3 out of 4) the difference in age was only 5 years or less. Among couples in which the man was significantly older than the woman, the evidence suggests it was frequently his second marriage. In one case, for instance, his 25-year-old daughter lived under the same roof with his 25-year-old wife.

Table 3. Marital Status of Men and Women by Age Group

Age Group	Men with wives	Percentage of married men to total men	Women with husbands	Percentage of married women to total women	Men w/o wives	Women w/o husbands	TOTAL MEN	TOTAL WOMEN
1	2	3	4	5	6	7	8	9
15-19	0	0.0	0	0.0	48	70	48	70
20-24	6	12.0	30	34.1	44	58	50	88
25-29	35	70.0	43	69.4	15	19	50	62
30-34	43	81.1	52	76.5	10	16	53	68
35-39	38	86.3	27	67.5	6	13	44	40
40-44	36	81.8	30	85.7	8	5	44	35
45-49	11	84.6	7	43.8	2	9	13	16
50-54	21	84.0	12	50.0	4	12	25	24
55-59	5	62.5	2	66.6	3	1	8	3
60-64	7	70.0	5	45.5	3	6	10	11
65-69	1	33.3	1	33.3	2	2	3	3
70-74	3	60.0	1	7.7	2	12	5	13
75-79	2	66.6	0	0.0	1	2	3	2
80 and over	4	57.1	2	15.4	3	11	7	13
TOTALS OR AVERAGES	212	58.7	212	47.3	151	236	363	448

Appendix A presents information on the slave women on this estate. Of the 448 women, 212 or 47.3% lived with their husbands, 126 or 28.1 percent lived with their parents, and the remaining 110 or 24.5% lived independently. Whenever the enumerator dealt with married women without husbands, he revealed a marked preference for the category "widow" over "married-without-husband." But we may assume these two categories (columns 3 and 4) are almost synonymous, at least for the younger women. One-quarter of the widows were, in fact, under the age of 40. Whereas only 36 men were listed as widowed or single heads of households, this number among women reached 91 (columns 3 plus 4). The difference between the ages of men and women — already noted — is further evidenced by the sharp decline in the number of women past age 44.

SLAVE FAMILIES ON A RURAL ESTATE

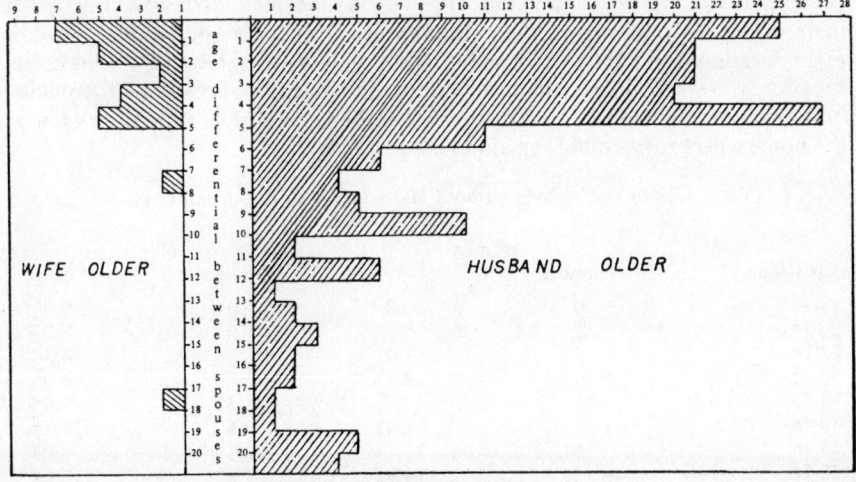

Figure 2. Number Of Occurrences

NOTE: In 13 cases spouses were of the same age

Thirteen women, however, were over 80 years of age. There was only one childless woman over age 30 listed as single in the entire population. Could this suggest an androcentric society in which there was no room for the "old maid"? Of single mothers, almost all (17 out of 19) lived with their parents; but there were 10 married women, all of them under age 40, who lived without a husband, in addition to the 81 widows referred to above. The fact that no single mothers past age 35 (or 40 if we include wives without husbands) are listed probably indicates that children born to young single mothers did not remain with them to age 20, while older mothers were no longer considered single but allowed to live with a man.

In the case of slave societies elsewhere it has been alleged that the male field hand may have been relatively divorced from the role of father and head of household, while the woman's location closer to the administrative headquarters, her use in domestic service, and her occasional role as the master's sexual partner gave her access to information about the master's world and thus, in a sense, to power denied her husband.[21] The Fazenda Santa Cruz, of course, was not the type of agricultural unit reflected in this view. But, as we can see from Appendix B (column 6), only 64 or 17.6% of the men lived alone, relying on neither a woman or a parent, in contrast to the 24.5% (110) of the women who lived independently (Appendix A, column 13). These numbers tend to suggest that as a matter of fact more women than men were forced to make daily family decisions alone.

Another way to get at the question of female independence within this slave society is to return to the question of the number of men and women without mates but with dependent children by age. Table 4 presents this information. The women are more than twice as numerous as the men and the disproportion is especially marked (5.6::1) for ages 20-40. Even allowing for those cases where children may have been born out of wedlock, the men were evidently much more readily separated from their children, sold off the estate, or dispatched to other locations, and their role within the family must surely have been correspondingly weakened. We should remember, however, that many of the men — both married and widowed — were not field hands and thus worked in or near the home where they would remain in control.

Table 4. Men and Women Without Mates and With Dependent Children by Age of Parent

Age Group	Women		Men	
	Number	%	Number	%
15-19	0	0.0	0	0.0
20-24	3	5.56	0	0.0
25-29	4	7.41	1	4.55
30-34	10	18.52	1	4.55
35-39	11	20.37	3	13.64
40-44	4	7.41	6	27.27
45-49	6	11.11	2	9.09
50-54	8	14.81	3	13.64
55-59	1	1.85	2	9.09
60-64	2	3.70	2	9.09
65-69	0	0.0	0	0.0
70-74	5	9.26	1	4.55
75-79	0	0.0	0	0.0
80 and over	0	0.0	1	4.55
TOTALS	54	100	22	100

The corollary to the disproportionate number of solitary women with dependents is the large number of children who were being brought up by their mother only. Table 5 shows that of those children under the care of only one parent, 77 were with a mother while only 23 were with a father. Forty-six boys would thus grow up without a father-image and 43 girls would mature having seen a mother coping with life alone. The opposite figures would be, for boys with only fathers, 9, girls without mothers, 23.

Once again, however, one is struck by the apparent stability of family life. Over three-quarters of the children were growing up with both a father and a mother and for the younger children this figure reaches four-fifths.[22] The number of orphans is surprisingly low.[23]

Table 5. Children Under 15 With and Without Parents by Sex and Age

	Boys							
	0-4	%	5-9	%	10-14	%	Total	%
With neither parent	0	0.0	5	5.62	7	7.78	12	4.63
With mother only	11	13.75	12	13.48	11	12.22	34	13.12

SLAVE FAMILIES ON A RURAL ESTATE 391

With father only	1	1.25	2	2.25	6	6.67	9	3.47
With both parents	68	85.0	70	78.65	66	73.33	204	78.76
TOTAL BOYS	80	100.0	89	100.0	90	100.0	259	100.0
Girls								
With neither parent	0	0.0	3	3.0	6	7.06	9	3.25
With mother only	20	21.74	11	11.0	12	14.12	43	15.53
With father only	2	2.17	4	4.0	8	9.41	14	5.05
With both parents	70	76.09	82	82.0	59	69.41	211	76.17
TOTAL GIRLS	92	100.0	100	100.0	85	100.0	277	100.0
All Children								
With neither parent	0	0.0	8	4.23	13	7.43	21	3.92
With mother only	31	18.02	23	12.16	23	13.14	77	14.37
With father only	3	1.74	6	3.17	14	8.0	23	4.29
With both parents	138	80.23	152	80.42	125	71.43	415	77.43
TOTAL ALL CHILDREN	172	100.0	189	100.0	175	100.0	536	100.0

What changes occurred over time in the composition of this slave population? There is another inventory, made in 1768, that sheds some light on this question.[24] Unfortunately, it fails to indicate the ages of the slaves and is therefore not susceptible to the kinds of analysis so far undertaken. But the totals in each category are suggestive. As can be seen in Table 6, the total population had increased by 109 over the 23-year period. Interestingly, 17 freedmen, married to slave women, were included in the earlier count and this fact leads one to believe that perhaps the same group, undifferentiated, was included in the 1791 figures cited heretofore. The number of women rose by 118 and that of men by 58 whereas the number of children decreased by 67. The most significant increases arise among single men (up 82) and women (up 88). However, many of those whom the 1768 inventory cited as children may have been over 15 and therefore included in what I have considered the single population in 1791. If one were to assume that the difference in the number of children was all accounted for in this way, dividing this difference equally into male and female, then the additional single men would be 49 and single women 54. Such an estimate would almost precisely account for the increase in the total population. In any case, the increase of single men and women in almost equal proportion could be accounted for either by new acquisition or by the delaying of the marriage age. Married couples decreased by 43 while there were 49 more widows and 19 more widowers.

Table 6. Changes in the Slave Population over a 23-Year Period

Category	1768[a]	%	1791	%
Men with women	255[b]	20.6	212	15.7
Widowers	17	1.4	36	2.7
Single men	33[c]	2.7	115[d]	8.5
TOTAL MEN	305	(24.63)	363	(26.94)
Wives with husbands	255	20.6	212	15.7
Wives without husbands	6	.5	10	.7
Widows	32	2.6	81	6.0
Single mothers	—[e]	0	19	1.4
Single childless women	37[f]	2.9	126[g]	9.3
TOTAL WOMEN	330	(26.65)	448	(33.25)
Children	603[h]	48.7	536[i]	39.8
TOTAL SLAVES	1238[b]	100.0	1347	99.8

[a] Figures are drawn from the original summary which has not been checked for accuracy or consistency.

[b] Including 17 freedmen married to slave women.

[c] "Rapazes cazadouros" of unspecified age.

[d] Over 14 years of age.

[e] Not distinguished in 1768; no doubt included in other categories.

[f] "Raparigas cazadouras" of unspecified age; no other single women are listed except widows.

[g] Over 14 years of age.

[h] "Filhos" of unspecified age.

[i] Under 15 years of age.

More significant is the changing proportion of men and women. If one does not count young singles, there were 24 fewer men but 30 more women in 1791 than in 1768. These figures would indeed suggest that during this period a systematic policy had been pursued of selling off the men and keeping the women, thus significantly increasing the number of "widows," wives without husbands, and single mothers. It could be argued that the increase in the number of women cannot be attributed to a breeding intent since the number of children failed to increase. But the acquisition of new slaves may have been made only at the very end of the period in question. We know that a significant expenditure was made for slave purchases sometime after 1783 and before the beginning of 1790.[25] Furthermore, a lot more than 67 "children" of 1768 may have been over 15, in which case the number of children would actually have increased during the period and the number of young singles remained relatively constant or even declined.

Two types of information relating to the workforce — on skills and handicaps — were provided by the enumerator in 1791. Sixty-five of the slaves, including one woman, were listed as having a professional skill. Table 7 shows that the most common trade was that of carpenter, followed by musician.[26] Construction skills, including carpenter, mason, sawyer, and woodturner, accounted for 32 or almost half the number of skilled workers. Some of the potters may also have been brick and tilemakers (or kiln operators), so it is clear that construction

SLAVE FAMILIES ON A RURAL ESTATE

trades predominated. There was no butcher or gunsmith despite other evidence that the sale of meat and the rental of slaves to a gun factory had earlier represented sources of income for the estate.[27] It is possible that additional slaves belonging to the estate, but off the premises at the time, were not listed in this inventory. At the time, the erection of sugar mills was said to be indicated because "almost all the trades and all the shops and installations needed" were present at Santa Cruz.[28] Literacy is not indicated, although in another part of the inventory, one of these skilled slaves, a *boticario,* attested to the accuracy of the list of the pharmacy's paraphernalia and medicines by affixing his signature.[29] Forty of these skilled workers were between the ages of 35 and 54 and in no case does there seem to have been any danger that they would die off before being able to teach younger men their trade. Musicians were among the youngest of this group, but several men in their forties were also so listed. The oldest worker was a weaver and the single woman was a potter.

Table 7. Skilled Occupations by Age Group

	Carpenter	Musician	Potter	Mason	Blacksmith	Tanner	Weaver	Pharmacist	Sawyer	Woodturner	Cooper	Shoemaker	TOTALS
10-14		3											3
15-19		1										1	2
20-24	1	2									1		4
25-29	3					1	1						4
30-34	1	1	2	2									6
35-39	6	1			3		1	2	1				14
40-44	1	4	1	1	3	2							12
45-49	3					1	1						5
50-54	3	3[a]				1			1	1			9
55-59					1		1						2
60-64	1		1										2
65-69													0
70-74	1				1								2
75-79													0
80 and over							1					1	1
TOTALS	20	12	7	7	5	5	3	2	2	1	1	1	66

[a]One is a woman.

The proportion of skilled men to total male population was astoundingly high, reaching 30.16% in the middle-aged (35-54) group. Table 8 indicates either that the longer one lived, the more likely it was that one would have had the chance or be required to learn a trade, perhaps after having been worn out in the fields, or that skilled workers, perhaps by avoiding harsher forms of labor, tended to live longer.[30] If we assume that there may have been an additional male population assigned to other establishments and that these required skilled hands, the number of skilled workers would be still higher. The *fazenda's* position as a publicly owned estate probably lessened the maximization of profit and the concentration on export agriculture that may have characterized other money-making estates, and one must not forget the 60 or 70% of Santa Cruz slaves who were occupied tending cattle, stooping in rice fields, or hoeing beans and corn. Still, the number of skilled slaves implies, for a relatively large number of slaves, some sense of their own creative ability and thus of their selfhood.

Table 8. Proportion of Skilled Males to Total Males by Age Group

Age Group	Male Skilled Workers	Total Male Population	Percentage of Skilled
15-19	2	48	4.16
20-24	4	50	8.00
25-29	3	50	6.00
30-34	6	53	11.32
35-39	14	44	31.82
40-44	12	44	27.27
45-49	5	13	38.46
50-54	7	25	28.00
55-59	2	8	25.00
60-64	2	10	20.00
65-69	0	3	0.0
70-74	2	5	40.00
75-79	0	3	0.0
80 or over	1	7	14.29
TOTALS	60	363	16.52

Table 9. Handicapped

Age Group	Aleijado (Crippled)		Coxo (Lame)		Quebrado das costas (Hunchback)	
	M	F	M	F	M	F
5-9	1					
10-14		1			1	
15-19						
20-24		1				
25-29		1	1			
30-34						
35-39		1				
40-44	1					
45-49						
50-54		1				
55-59						
60-64		1				
65-69						
70-74						
75-79						
80 and over						
TOTALS	2	6	1	0	1	0
	8		1		1	

SLAVE FAMILIES ON A RURAL ESTATE

One final bit of information provided by the inventory dealt with the handicapped. Presumably only those so badly injured or deformed as to be unable to work would have been listed here. More women were handicapped than men, although the total number is too small for any significance to be attached to this fact. The first four categories in Table 9 could all be lumped under the category "crippled," bringing this total to 11 and making the balance between men and women roughly equal for this group. The enumerator was evidently at greater pains to note the specific nature of the handicap for men than for women. Why there should have been more than twice the number of blind females than males is unknown. Aside from mere chance, it is possible that a blind male more easily suffered fatal accidents. A blind woman may have been considered more useful (even for childbearing!) and kept on the estate while blind men were released as beggars. It is impossible to know how many of these unfortunates may have been the victim of their foreman's rage. *Quebrado das costas* and *estropiado* are two categories that imply noncongenital deformities. There are no significant variations by age group. In contrast to what one might have expected from the growing literature on the harshness of slave life in Brazil, only 1.9 percent of the slaves were handicapped, and, despite corrections suggested above, this figure seems low.[31]

The data presented in this essay must be used with caution if one's purpose is to arrive at larger generalizations. To begin with, the number of slaves was exceptionally high for one estate. Why it was so high remains a puzzle. Medium-sized *fazendas* in Brazil at this time would typically have had only about 50

"Esteporado"[a] (maimed)		Cego (Blind)		Gotacoral (Epileptic)		TOTALS		GRAND TOTAL
M	F	M	F	M	F	M	F	
						1		1
					1	1	2	3
							1	1
			1			2	2	4
1		2	1			2	1	3
							1	1
			1				2	2
			1					
			2			1	2	3
								0
			1				2	2
								0
		1				1	1	2
								0
			1				1	1
								0
		1	1			1	1	2
1	0	4	9	0	1	9	16	25
1		13		1		25		25

[a] Misspelled for *estropiado* or *estuporado*, both of which mean crippled, paralysed, or maimed, by implication after birth.

slaves, and one observer, writing from sugar-rich Bahia in 1781, commented on a "formidable" and "immense" ex-Jesuit plantation with "good lands" which had what he considered the large sum of 150 slaves.[32] It was the Jesuits who initially bought the bulk of the slaves for the Fazenda Santa Cruz, possibly for their extensive earthmoving projects. Nine years after confiscation of the estate and before serious efforts had been made to expand the royal income from it, there already were 1237 slaves on the *fazenda.*[33] To be sure, sometime between 1783 and 1789 the administrator spent 4.7 *contos* on the purchase of slaves, but this probably could account for no more than 80 or 100 slaves.[34] Why he should have felt the need for any additional ones is unclear, but evidently he was continuing a Jesuit tradition of maintaining what now appears to have been a labor surplus. One of the best arguments advanced for entering sugar-production in early 1791 was the fact that there were "one thousand, two hundred slaves, more than six hundred of them able to work."[35] It was thus obvious that not all these slaves were needed for cattle raising or for cultivating rice, corn, and manioc. Twenty years later the English traveler John Mawe, who was offered the job of manager, reported there were still approximately 1500 slaves on the estate.[36] The number of slaves clearly distinguishes the Fazenda Santa Cruz from the ordinary estate.

It is not within the scope of this essay to explore in what ways this estate's unrepresentative quality may have been heightened by its Jesuit origin. Whether the Jesuits were more likely to have been humane masters or more rapacious in their economic pursuits than the average landowner is a heated topic within certain historiographical circles in Latin America and the less unfounded speculation added to the debate, the better. As for royal administration, the argument is less heated but surely as moot. Mawe provided evidence for both sides on this point when he wrote about "the negroes on this estate," saying that

> Great pains have been taken to enlighten them. They are regularly instructed in the principles of the Christian faith, and have prayers read to them morning and evening. Plots of ground, of their own choice, are assigned to each and two days in the week... are allowed them to raise and cultivate produce for their own subsistence.... The system of management, however, is so bad that they are half-starved, almost destitute of clothing, and most miserably lodged.[37]

Another factor that limits the utility of this data is our sketchy knowledge of contemporary Brazilian society.[38] Until more is known about it, there is no comparative matrix into which these figures can be inserted and comparing them with data extracted from other societies may obscure much of their real meaning. Historians will want to consider, for instance, What was the age differential in nonslave lower-class Brazilian marriages? What was the life expectancy of all Brazilians at this time? Were free children in the same proportion being brought up by their mothers alone? What was the typical marriage age in Brazil? What was the number of children per couple in the country at large? Were there significant differences between urban and rural populations in regard to these questions? How do the slave populations of privately owned estates compare with those on the Fazenda Santa Cruz? It is doubtful that reliable figures for the entire population can be ascertained, but a sensitivity to such questions and the possibility of comparing the results for other small groups with the group surveyed here should foster our understanding of the nature of family life in Brazil in the late eighteenth century.

SLAVE FAMILIES ON A RURAL ESTATE

The information on family groups presented in the 1791 inventory must be used with special care for still another reason. It represents the vision of reality as seen by a white administrator and may tell us more about him than about the slaves. For this very reason, of course, it is significant and in this sense may be considered indicative of a general societal view of slave family life. No other contemporary document even took note of it, thus implying that their familial arrangements did not escape the normal. The large number of skilled workers was considered a favorable indication of economic potential, but no surprise was expressed at their existence. The estate's management was criticized for general inefficiency,[39] but not for overleniency toward slaves or for any aberrance in the family life of enslaved workers.

So the larger meaning of this inventory must be found in the mind-set of the inventory-taker. The fact that the slaves were so carefully listed by family group suggests a particular image of family life. There is no evidence that breeding a "better stock" explains this predilection, for records were not kept on the physical or mental abilities of the slaves. Possibly all this effort merely reflected an attempt to identify more easily subsequent defalcation of the slave property: "Where is the son of José?" One is left, however, with the impression that it bears a greater significance. It would surely have been much easier to have had a "round-up" of slaves and to have quickly counted off the males, the females, and the children.[40] Instead, each one was listed by name, age, and family relationship. Although slaves might be sold off separately, they were enumerated in families. Clearly, and not really surprisingly, the human quality of this form of property was thus recognized. This manner of enumeration revealed, although not necessarily to contemporaries, the basic contradiction between slaves as property and slaves as human beings, between their will-lessness and will-fullness.

Insight into the history of society is thus gained from even a limited source. Although inarticulate themselves, some aspects of the slaves' life can be perceived through such an inventory. In this particular estate over half the family units included husband and wife. Some families were even headed by men without women. By far the largest number of children grew up with both father and mother. A large proportion of skilled workers enhanced the probability of a father's presence in the home. Marriage age, age at childbirth, and fertility suggest that the proper context within which these data may be placed is not the slave societies of the following century, but contemporary family life in Europe or perhaps in Africa. In any case, the family life of these slaves reflect dimensions of a reality that has only begun to be investigated for Latin America.

FOOTNOTES

1. The author gratefully acknowledges the assistance of Dauril Alden, Robert Conrad, Herbert Klein, Maria Luiza Marcílio, Dudley Poston, Donald Ramos, Susan Soeiro, and Sônia Bayão Rodrigues Viana, as well as financial aid from the American Philosophical Society and the Institute of Latin American Studies of the University of Texas at Austin. Secretarial and computational assistance was provided by the Department of History and the Population Research Center of the same university.

2. Two examples of recent interest in slave family life are Robert William Fogel and Stanley L. Engerman's *Time on the Cross: the Economics of American Negro Slavery,* 2 vols. (Boston, 1974), esp. 126-44, and an extended review of this book by Herbert G. Gutman, "The World Two Cliometricians Made: A Review Essay of F + E = T/C," *Journal of Negro History* 60 (1975): 53-227, esp. 138-227. See also the older view presented by E. Franklin Frazier, "The Negro Slave Family," *Journal of Negro History* 15 (1930): 198-259, and corrections to that view put forward in Gutman, "Le phenomene invisible: la composition de la famille et du foyer noir après la Guerre de Secession," *Annales: Economies, societés, civilizations* 27 (1972): 1197-1218. Also see John W. Blassingame, *The Slave Community: Plantation Life in the Antebellum South* (New York, 1972), pp. 77-103. I have not consulted Bobby Frank Jones, "A Cultural Middle Passage: Slave Marriage and Family in the Antebellum South" (Ph.D. diss., Univ. of North Carolina, 1965).

3. José Saldanha da Gama, "História da imperial Fazenda de Santa Cruz: primeira parte," *Revista do Instituto Histórico e Geográfico Brasileiro* 38 (1875): 165-230; a wealth of information on the Jesuit period of the estate's history can be found in Serafim Leite, *História da Companhia de Jesus no Brasil,* 10 vols. (Rio de Janeiro, 1938-1950), by consulting its excellent index. Also see [Manoel Martins do Couto Reis], "Memórias de Santa Cruz, seu sucessos mais notaveis, continuados do tempo da extinção dos denominados Jesuitas, seus fundadores, ate o ano corrente de mil, setecentos, noventa e nove [enclosed in José Caetano de Lima Rodrigo de Sousa Coutinho, Rio de Janeiro, 27 Sept. 1799]," *Revista do Instituto Histórico e Geográfico Brasileiro* 65 (1902): 201-321. The ruins of Jesuit waterworks can still be seen outside today's bustling city of Santa Cruz.

4. Sônia Bayão Rodrigues Viana, "A Fazenda de Santa Cruz e a crise do sistema colonial (1790-1815)," *Revista de História* No. 99 (July-Sept., 1974): 61-96, also found in Richard Graham, ed., *Ensaios sobre a politica e a economia da província fluminense no seculo XIX* (Rio de Janeiro, 1974), pp. 9-63. Also see Viana, "A Fazenda de Santa Cruz e a política real e imperial em relação ao desenvolvimento brasileiro, 1790-1850" (M.A. thesis, Universidade Federal Fluminense, 1974).

5. In 1771 it was said to yield an annual revenue of 3.6 to 4 *contos de reis* at a time when an entire sugar mill with stock pen was valued at only 0.6 *contos* and ten prime slaves aged 15 to 25, unskilled, but without physical defects, were estimated to be worth something less than 0.5 contos (Dauril Alden, *Royal Government in Colonial Brazil, with Special Reference to the Administration of the Marquis of Lavradio, Viceroy, 1769-1779* [Berkeley and Los Angeles, 1968], pp. 346-7, 509-10).

6. Report of José Joaquim de Silva Castro, Engenho Novo, 18 June 1790, Arquivo Nacional, Rio de Janeiro (hereafter AN), Cx. 507, Pac. 3, Doc. 5; Conta do inspector da Real Fazenda de Santa Cruz do gado que comprou e do rendimento do açougue, 4 May 1790, AN, Cx. 507, Pac. 3, Doc. 1. Report of José Feliciano da Rocha Gameiro, Rio de Janeiro, 19 Feb. 1791, AN. Cx. 507, Pac. 5, Doc. 2. Also see Various reports, 1793, AN, Cx. 507, Pac. 6.

7. John Mawe, *Travels in the Interior of Brazil. Particularly in the Gold and Diamond Districts of that Country...* (London, 1812), pp. 106-7.

8. Brazil. Contadoria Geral da União, *Balanço da receita e despeza da República no exercicio de 1899* (Rio de Janeiro, 1904), pp. 19, 293.

9. Inventario dos escravos pertencentes a Real Fazenda de Santa Crus *[sic]* q o Sarg^{to} M^{yr} Manoel Joaq^m da S^a Castro entregou e ficão em carga a Joaquim Henriques Guerra, Cabo da

SLAVE FAMILIES ON A RURAL ESTATE 399

Esquadra do Regim^to de Artilharia em 12 de junho de 1791, AN, Cod. 808, 4: 165-84. Unfortunately, this inventory tells us nothing about color, place of birth, or value. No parish registry has yet turned up, so it has been impossible to compare these data with those of Nicholas P. Cushner, "Slave Mortality and Reproduction on Jesuit Haciendas in Colonial Peru," *Hispanic American Historical Review* 55 (1975): 177-199.

10. Gama, "Historia," p. 213.

11. Cf. B.W. Higman, "Household Structure and Fertility on Jamaican Slave Plantations: A Nineteenth-Century Example," *Population Studies* 27 (1973): 543. No study of family life among the contemporary lower-class free population has yet been made, but with considerable historical license, one may imagine it to have been highly disorganized as portrayed for the period after 1808 in Manoel Antonio de Almeida's mid-nineteenth-century picaresque novel, *Memórias de um sargento de milícias*, 4th ed. (São Paulo, 1962), or as described in the temporally more removed but more reliable scholarly picture of the late nineteenth century studied by Maria Sylvia de Carvalho Franco, *Homens livres na ordem escravocrata*, Publicações do Instituto de Estudos Brasileiros 13 (São Paulo, 1969), pp. 40-48.

12. Indian slavery had been outlawed and although some may have been kept in bondage, this inventory makes no mention of Indians (the Jesuits had long opposed Indian slavery).

13. He spent 4.7 *contos* on the purchase of slaves (Mapa das despezas que tem feito o Inspector e Administrador da Fazenda de Santa Cruz o Sargento Mor Manoel Joaquim da Silva e Castro desde que entrou no dito emprego em o principio de abril de 1783 ate o fim de 1789, AN, Cx. 507, Pac. 4, Doc. 1). See note 4 on the value of slaves.

14. Cf. the contrasting data for the population as a whole at a later period presented by Eduardo E. Arriaga, *New Life Tables for Latin American Populations in the Nineteenth and Twentieth Centuries* (Berkeley, 1968), pp. 25-42.

15. On the other hand, these slaves were evidently more successful at maintaining their population size than the overall slave population of Brazil as we know it operated in the nineteenth century. Still, the low number of children per woman aged 25 to 29 may indicate a high infant mortality, the sale of children, or the practice of birth control.

16. Appendix B (col. 7) and Appendix A (col. 14) show the number of single men and childless single women living with parents. The fact that the bulk of single mothers also lived with their parents suggests the strength of family ties, ties that were relied upon in time of trouble.

17. Robert Conrad, *The Destruction of Brazilian Slavery, 1850-1888* (Berkeley, 1972), p. 298; see also pp. 32-3.

18. Higman, "Household Structure," pp. 527-50.

19. Stanley J. Stein, *Vassouras: a Brazilian Coffee County, 1850-1900*, Harvard Historical Studies 49 (Cambridge, Mass., 1957), pp. 43-4.

20. Among the families that included both husband and wife there were 7 in which a daughter was less than 15 years younger than the wife. This suggests the existence of a previous wife although a disguised polygamy cannot be absolutely ruled out. There was one case in which the "father" was less than 15 years older than the oldest child and another in which a maleless household included a child less than 15 years younger than the "mother."

21. E.g., Stanley M. Elkins, *Slavery: A Problem in American Institutional and Intellectual Life*, 2d ed. (Chicago, 1968), p. 130. In considering the role of women it is important to remember that at the Fazenda Santa Cruz there was an unusually large proportion of females.

22. Since the inventory was taken only once, it may be too much to speak of "stability." Yet nothing in the records of the estate would lend credence to a belief that partners or children were shifted around and only caught here momentarily in this relationship as in a daguerreotype.

23. Cf. Fogel and Engerman, *Time on the Cross*, 1: 50. It is possible that orphaned children were enumerated with adopted parents to whom they may have been assigned in keeping with familial traditions possibly carried on from Africa.

24. [Alexandre José Mello Moraes Filho, ed.], "Treslado do autto de inventario da Real Fazenda de Santa Cruz e benz que nella se acham que fes o Desembargador dos Aggravos e Juis do Sequestro geral feito aos denominados Jezuitas o Doutor Manoel Francisco da Silva e Veiga [6 May 1768]," in Distrito Federal, Archivo, *Revista de documentos para a historia da Cidade do Rio de Janeiro* 1 (1894): 189-92, 217-24, 333-9, 418-19.

25. See note 13. Brusque fluctuations are presumably more characteristic of slave than of general populations. These figures could reflect momentary characteristics.

26. On the later extensive use of slaves as musicians, see Daniel Parish Kidder and James Cooley Fletcher, *Brazil and the Brazilians Portrayed in Historical and Descriptive Sketches* (Boston: Little, Brown, 1857), p. 441.

27. Alden, *Royal Government*, p. 347.

28. Report of José Feliciano da Rocha Gameiro, Rio de Janeiro, 19 Feb. 1791, AN, Cx. 507, Pac. 5, Doc. 2.

29. Relação do Inventario do q. pertence a Botica da Real Fazenda de Santa Cruz q. o Sarg^{to} M^{l} Manoel Joaq^{m} da S^{a} Castro entregou e fica em carga ao pardo Cirurg^{m} José Alves escravo da mesma faz^{da} em 7 de junho de 1791, AN, Cod. 808, vol. 4.

30. Cf. Fogel and Engerman, *Time on the Cross*, 1: 149-53 and Gutman, "The World," pp. 126-32.

31. Cf. Stein, *Vassouras*, p. 185.

Appendix A. Females by Age and Marital Status

1	2	3	4	5	6	7	8	9
Age Group	Wives w/ husbands	Wives w/o husbands	Widows	Total Wives	Single mothers living independently	Single mothers living with parents	Total single mothers	Single mothers plus wives w/o husbands
15-19	0	0	0	0	1	1	2	2
20-24	30	2	0	32	1	5	6	8
25-29	43	1	4	48	0	7	7	8
30-34	52	4	8	64	0	4	4	8
35-39	27	3	10	40	0	0	0	3
40-44	30	0	5	35	0	0	0	0
45-49	7	0	9	16	0	0	0	0
50-54	12	0	12	24	0	0	0	0
55-59	2	0	1	3	0	0	0	0
60-64	5	0	5	10	0	0	0	0
65-69	1	0	2	3	0	0	0	0
70-74	1	0	12	13	0	0	0	0
75-79	0	0	2	2	0	0	0	0
80 and over	2	0	11	13	0	0	0	0
TOTALS	212	10	81	303	2	17	19	29

SLAVE FAMILIES ON A RURAL ESTATE

32. José da Silva Lisboa to Domingos Vandelli, Bahia, 18 Oct. 1781. In Eduardo de Castro e Almeida, ed., "Inventário dos documentos relativos ao Brasil existentes no Archivo de Marinha e Ultramar," *Anais da Biblioteca Nacional do Rio de Janeiro* 32 (1910): 501.

33. Moraes Filho, ed., "Treslado," p. 419.

34. Mapa das despezas (note 13).

35. Report of José Feliciano da Rocha Gameiro, Rio de Janeiro, 19 Feb. 1791, AN, Cx. 507, Pac. 5, Doc. 2.

36. Mawe, *Travels*, p. 107; a traveler later reported 2500 slaves there (Conrad, *Destruction of Brazilian Slavery*, p. 73n.

37. Mawe, *Travels*, pp. 107-8. The reference to the assignment of plots is another indication that either this case was atypical or that the structure of this slave society was not as simple or as uniform as we might have thought from the previous literature on the subject.

38. Pioneering work is now being done on some aspects of these problems. See, for instance, Maria Luiza Marcílio, *La Ville de São Paulo: peublement et population, 1750-1850* (Rouen, 1972), esp. pp. 127-60, and Donald Ramos, "Marriage and the Family in Colonial Vila Rica," *Hispanic American Historical Review* 55 (1975): 200-225. Demographers may also estimate some general characteristics of the population simply from conditions known to exist for other preindustrial societies.

39. Vianna, "A Fazenda de Santa Cruz," pp. 93-5.

40. Inventories of the latter type were made with regularity. See Mapa dos escravos pertencentes a Real Fazenda de Sta. Cruz... [28 Aug. 1790]; Mapa geral da escravatura da Real Fazenda de Santa Cruz... [26 July 1799]; Relacao dos escravos pertencentes a Real Fazenda de Sta. Cruz... [22 Aug. 1814]; Mapa dos escravos da R. Fda. S. Cruz... [2 Oct. 1815], AN, Cód. 808, 4: 191-2, 193, 195, 199. The point is that in the 1791 inventory a very different procedure was used. It was also used in 1768 when slaves were "checked off by married couples as in the old inventory," apparently a reference to Jesuit precedent. A patient investigator could attempt to identify the slaves listed by both inventories and see the changes taking place in family structure over a 23-year period; but the use of first names only and the lack of data on ages in the earlier inventory would make the results of dubious reliability.

10	11	12	13	14	15	16	17	18
Single mothers wives w/o husbands, widows (cols. 3, 4, & 8)	Idem minus those living w/ parents (cols. 1-7)	Childless single women living independently	Single mothers, wives w/o husbands, widows, and childless single women living independently	Childless single women living w/ parents	Total childless single women (cols. 12 & 14)	Total women living w/ parents (cols. 7 & 14)	Women w/o husbands (cols. 10 & 15)	Total women (cols. 5, 8, & 15)
2	1	10	11	58	68	59	70	70
8	3	6	9	44	50	49	58	88
12	5	0	5	7	7	14	19	62
16	12	0	12	0	0	0	16	68
13	13	0	13	0	0	0	13	40
5	5	0	5	0	0	0	5	35
9	9	0	9	0	0	0	9	16
12	12	0	12	0	0	0	12	24
1	1	0	1	0	0	0	1	3
5	5	1	6	0	1	0	6	11
2	2	0	2	0	0	0	2	3
12	12	0	12	0	0	0	12	13
2	2	0	2	0	0	0	2	2
11	11	0	11	0	0	0	11	13
110	93	17	110	109	126	126	236	448

Appendix B. Males by Age and Marital Status

1 Age Group	2 Widowed or single heads of households	3 Heads of households with wives	4 Total heads of households (columns 2 & 3)	5 Single, over 15, living independently	6 Total without wives living independently (columns 2 & 5)	7 Single, men over 15, living with parents	8 Total single men (columns 5 & 7)	9 Total without wives (columns 6 & 7)	10 GRAND TOTAL
15-19	0	0	0	5	5	43	48	48	48
20-24	1	6	7	9	10	34	43	44	50
25-29	1	35	36	6	7	8	14	15	50
30-34	2	43	45	6	8	2	8	10	53
35-39	4	38	42	2	6	0	2	6	44
40-44	8	36	44	0	8	0	0	8	44
45-49	2	11	13	0	2	0	0	2	13
50-54	4	21	25	0	4	0	0	4	25
55-59	3	5	8	0	3	0	0	3	8
60-64	3	7	10	0	3	0	0	3	10
65-69	2	1	3	0	2	0	0	2	3
70-74	2	3	5	0	2	0	0	2	5
75-79	1	2	3	0	1	0	0	1	3
80 and over	3	4	7	0	3	0	0	3	7
TOTALS	36	212	248	28	64	87	115	151	363

4
African Slave Trade and Economic Development in Amazonia, 1700–1800
Colin M. MacLachlan

Brazil was a sleepy colony of secondary interest to Portugal until sugar turned the northeastern coastal strip into a rich plantation region in the second half of the sixteenth century. With significant development of a new commercial crop, Brazil's developers increasingly talked about ways to find reliable labor for expanding estates in the area of Pernambuco and Bahia. Settlers tried to put Amerindians to work, but these plans failed because of the sparse Indian population and the natives' lack of experience with organized, collective labor in agriculture. Since Europeans did not come to Brazil in sufficient numbers, and since many of those who did preferred to give orders to others rather than work themselves, a new source of labor had to be found. Africa quickly became a popular source of manpower. Settlers in Brazil found that Africans brought over by Portuguese traders made good workers and could be purchased as slaves for very low prices.

Since African labor came to represent an essential element in Brazil's sugar boom, the Portuguese regarded a plentiful supply of Africans as necessary for successful economic development. Gazing at their maps of the huge northern regions of the Amazon basin, bureaucrats in Lisbon deluded themselves into thinking that a second Pernambuco could arise in the jungle by mere manipulation of the slave trade. As MacLachlan observes, where these thoughts were accompanied by an awareness of the economic potentialities and limitations of the area (as in Maranhão, which became an important exporter of rice and cotton), success followed. Where they were accompanied by nothing more than a burning desire to emulate the success of a region with totally different prospects (as in the case of Pará), the importation of African labor was worse than useless. Development of the vast Amazon region would require more than just masses of laborers, as even modern-day economic planners in Brazil have discovered. To the present time, the Amazon remains an area of great potential but a region still far from delivering its long-celebrated promise. Now, several centuries after the grand experiment of Belém do Pará, man is making another attempt to conquer the Amazon, hoping that technology and social organization will help to overcome the long history of grandiose but largely abortive efforts.

R. B. T.

Successful establishment of a plantation economy in the sugar-producing regions of colonial Brazil created a pattern for colonial development. The combination of black slaves and land appears to have been responsible for the transformation of Brazil from a possession of limited utility into a profitable agricultural colony. Understandably, Portugal sought to duplicate the process in other parts of its American empire. A plantation system, supported by African slaves, became the "ideal" model that ensured wealth and prosperity.[1] When the Portuguese extended their authority northward, anchoring themselves at the mouth of the Amazon with the founding of Belém do Pará in 1616, colonial development was firmly identified with black labor and plantation crops.

Faith in the economic benefits of African slavery blinded the Portuguese to the limited agricultural potential of the Amazon. The wild profusion of jungle vegetation resembled a tropical garden that lacked only labor to bring it into a controlled and profitable flowering. When Portugal ousted the foreign interlop-

ers, it fell heir to the myth that Amazonia was an incredibly rich and fertile area.² Giving full range to their imagination, the Portuguese envisioned a vast plantation stretching from Maranhão to Pará, based on what almost had become a fixation—black slaves.

The existence of forest Indians in the Amazon basin strengthened Lisbon's faith in the benefits of introducing African labor, since previous experimentation with Indian labor in the south had not been successful. Brazil's seminomadic forest Indian had only a rudimentary grasp of agriculture and could not be adapted to the needs of a plantation economy. Lisbon was content to organize the Indian within a mission system to protect Portugal's claim to the region until European settlers with their black slaves arrived. The vast northern territory, embracing the captaincies of Maranhão, Piaui, and Pará, which then included the area of the future captaincy of Rio Negro (Amazonas), became the state of Maranhão. The huge state, formed in 1621, remained intact until 1772.³ Inevitably, the political union of the Amazon basin obscured the fact that the economic potential of each captaincy differed in spite of its inclusion within the state of Maranhão. The crown assumed that the entire state could effectively utilize black plantation labor.

Initial Attempts to Introduce African Slaves

The Portuguese did not immediately begin the economic development of their newly secured northern territories. Seizure of Recife by the Dutch in 1630, and the struggle to evict them that lasted until 1654, diverted the attention of Portugal. In addition, the reestablishment of an independent Portuguese throne in 1640, after sixty years of union with the Spanish crown, required considerable diplomatic skill and energy; moreover, the limited resources of Portugal were habitually overcommitted. Such marginal areas of the empire as the Amazon basin understandably received little attention, and consequently the first serious attempt to establish a plantation economy was not made until 1682, when the crown instructed the Companhia de Commercio do Maranhão to import ten thousand African slaves over a twenty-year period.[4]

The company, however, suffering from mismanagement and the greed of its principals, failed to initiate the trade. Unmet labor demands resulted in constant petitions by the European settlers for a larger share of mission Indians. Acute labor shortages caused unrest, which eventually culminated in a sharp but ill-fated revolt in 1684 that forced the government to compromise and liberalize the settlers' access to indigenous labor.[5] This concession appeared in the new *Regimento das Missoes do Estado do Maranhão e Grão Pará* (1689), but in this same document Lisbon reaffirmed its determination to replace Amerindian labor with African slaves. One of its principal provisions called for the formation of a company to import black slaves; however, the lack of private capital made the formation of the projected company impossible.[6] The failure of private investors to initiate the slave trade forced the government to accept the responsibility, and in 1690 the crown financed the formation of the Companhia de Cacheu e Cabo Verde, setting a minimum goal of 145 slaves per year at a price to be officially determined. Additional government financing became available in 1692, when the crown ordered the governor of the state of Maranhão to reallocate 20,000 cruzados originally intended to buy forest products to the purchasing of slaves from the African company of Cacheu e Cabo Verde.[7]

The limited number and high price of black slaves introduced caused constant complaint. Both the municipal councils of São Luis do Maranhão and Belém do Pará believed that the crown set an excessive price. Such complaints appear justified by the rapid increase in the cost of slaves from 55$000 réis in March of 1695 to 130$000 réis in December of that year, reaching 300$000 réis by 1718.[8] Lisbon, however, noting that owners of sugar *engenhos* (sugar mills and surrounding plantation) continued to petition for even more slaves, impatiently dismissed such complaints. The small number of Africans entering the region became an object of competition between São Luis and Belém, and rivalry between the two provincial capitals continued throughout the eighteenth century. Since the port of São Luis do Maranhão was geographically closer to Africa than Belém, slave ships generally landed their cargoes in that settlement first. Quite naturally, the settlers of Maranhão selected the best physical specimens, leaving only the less desirable and the sick for sale in Belém. The inhabitants of Belém understandably believed that their labor needs were slighted. In response to their bitter protests, the crown ordered that each shipment of slaves be divided equally between the two settlements.[9] The unending criticism concerning the number and price of slaves must have angered the authorities. In 1708 the royal government unsuccessfully attempted to extricate itself from its uncomfortable position by urging wealthy settlers, in Maranhão and Pará, to send a ship directly to the slave coasts of Mina and Angola; however, as in the past, the limited capital of the region could not float such a venture.[10]

Unable to shift the burden of the trade to private initiative, Lisbon endeavored at least to control the use of servile labor. In order to obtain the most economically advantageous employment of African slaves, the crown decreed that all slaves must be devoted to sugar cultivation, a directive that ignored the fact that the Amazon basin was only marginally suitable for the cultivation of cane.[11] The poor quality of the region's sugar, coupled with transportation difficulties, had already prompted a switch from sugar production to the distilling of cane brandy.[12] *Cachaça* became the *bebida divina* (divine beverage) of the Amazon, serving as a medium of exchange as well as a stimulant.

In spite of seemingly obvious problems, the government continued to press for the development of the sugar industry. Since sugar depended on slaves, Lisbon continued its efforts to encourage the trade, but with little success. In addition, development of an intercoastal slave trade, linking the north with the sugar regions of the south, failed to materialize. The slave markets of Pernambuco and Bahia absorbed most of the slaves offered; consequently, traders did not turn to the more distant markets in the north.

118   *Colin M. MacLachlan*

Besides supply, the ability of the settlers to finance the regular purchase of slaves posed a problem. In 1749 the governor requested that the crown not only introduce black slaves, but that they also be sold to settlers on liberal credit terms; Lisbon apparently did not respond favorably to the proposal.[13] Subsequently, such suggestions would be reviewed by the future Marquês de Pombal and undoubtedly were taken into consideration when the crown decided to establish a monopolistic trading company. The number of African slaves introduced up to 1750 could not have exceeded a few thousand.[14]

African Slavery and Forced Development

The continued poverty of the Amazon state as well as the necessity of regular financial subsidies, which made the state of Maranhão an economic burden rather than an asset to the mother country, caused concern in Lisbon. The marked contrast between the north and the more favored areas of the south appeared to be explained by the lack of labor. It was observed that the sugar regions of Pernambuco and Bahia had only begun to flourish after the massive introduction of black slaves.[15] In response the crown urged increased efforts to stimulate interest in the slave trade. Subsequently, in 1751, several businessmen of Belém and São Luis petitioned for the privilege of forming a company to import two to four hundred slaves a year and suggested that the government grant them a ten-year monopoly as well as exemption from import taxes.[16] Lisbon rejected the monopoly proposal, but responded favorably to the idea of a tax-free slave trade, extending the exemption to include all slaves introduced into the state.[17] Although the proposed company never materialized, a number of private individuals took advantage of the new regulations to import black slaves.

The slave trade took on greater impetus under the influence of a more active government in Portugal. With the rise to power in 1750 of Sebastião José de Carvalho e Melo, the future Conde de Oeiras and Marquês de Pombal, the home government vigorously attempted to force the economic development of Amazonia. Carvalho e Melo viewed Belém do Pará as the northern anchor of imperial defenses, and he considered Mato Grosso, on the western frontier, as the keystone linking Pará with Rio de Janeiro in the south. A viable Paraense economy appeared vital to the grand design of empire that would be elaborated after 1750.[18] Since Lisbon did not question the basic value and potential of the land, attention focused on the means of production. When Francisco Xavier de Mendonça Furtado, the Marquês de Pombal's brother, assumed the direction of the state of Maranhão, one of his most important tasks was to facilitate the introduction of African slaves. The crown instructed the new governor to meet with the principal settlers to ascertain how many slaves could be effectively utilized and how the trade could be set up and financed.[19]

Governor Mendonça Furtado envisioned a Companhia Geral de Commercio Nacional dedicated to the African slave trade. Under official pressure, local capitalists contributed 32$000 réis, "in truth an insignificant sum" in the words of the governor.[20] Mendonça Furtado voiced the opinion, which by now had become an article of faith with the government, that the countryside would soon be dotted with sugar *engenhos* and plantations if black slaves could be introduced.

Responding to his brother's advice, Pombal concluded that only a well-capitalized company was capable of meeting the needs of such an undeveloped area. Quite obviously the limited capital of Belém or São Luis could not finance the African slave trade; consequently, the Marquês instituted the Companhia Geral do Grão Pará e Maranhão, capitalized with 1.2 million cruzados. Officially established in 1755, the company's charter specifically charged it with the introduction of African slaves as an essential part of its program to stimulate trade and general prosperity.[21] Recognizing the poverty of the inhabitants, the company's charter provided for the sale of slaves on credit, but to protect its interest, such slaves could not be seized to satisfy other creditors or any previous debts. The crown ordered the company to make every attempt to see that those engaged in economically productive tasks received preference. Only in the event of a surplus could slaves be sold to speculators, and such sales had to be strictly on a cash basis.[22] In order to head off the competition for servile labor, the government urged the company to introduce a sufficient number of slaves to meet the demand.[23] As originally envisioned, the company proposed to devote its efforts to stimulating the prosperity of both Maranhão and Pará without favoring one over the other.

Lisbon viewed the company's slaving operations as its most important obligation, assuming that once sufficient blacks had been imported the resultant surge in production would automatically increase profits. Essentially the crown considered the slave trade as pump-priming, the first step necessary to transform the Amazon basin into an economically viable part of the Portuguese empire. Because of the importance of the trade, the government gave the company broad administrative privileges in three African ports: Cacheu and Bissau in Guinea and Benguela on the Angolan coast. The Guinea privileges were in part illusory, Bissau and Cacheu were frequented as much by the English as by the Portuguese, and foreign slave traders frequently ignored the Portuguese presence, directly engaging in trade with African suppliers.[24] In order to take the traffic under firm control, as well as to eliminate foreign price competition, the crown authorized the repair and erection of fortifications and navigational aids. One of the company's major efforts, the fortress of Bissau, was completed in 1773 at a total cost of 147,690$763 réis.[25]

The African terminals, located in the deltas of extensive river networks with easy access to the interior of the continent, were ideally situated for collecting their unwilling cargoes. Their river location permitted the concentration of captives at spots convenient for both the African dealers and the European purchasers.[26] Tobacco, *cachaça* (cane brandy), cloth, hats, muskets, mats, and many other items of European manufacture served as the medium of exchange. To service the company's African complex, a fleet of at least eighteen ships crisscrossed the

Atlantic, bringing trade goods from Portugal and the Cape Verdes Islands in exchange for slaves to meet the needs of the Amazon's slave markets.[27] During the existence of the Companhia Geral do Grão Pará e Maranhão, approximately 75 percent of the slaves imported came from Bissau and Cacheu, and Angola accounted for the remainder.[28] The company thus linked the Amazon region directly with Africa's slave ports.

Although the labor problem had partially been solved, the other part of the economic equation—how to employ the black slaves imported by the new company—remained. Lisbon ordered a study of the region's products and their economic possibilities. Governor Mendonça Furtado investigated a total of thirty-nine different items of which sugar, cotton, rice, coffee, and cacao were reported to be the principal cultivated crops; however, most of the region's production was simply collected rather than cultivated.[29] Extension of the land under cultivation, as well as the adaptation of natural forest products to a plantation system, appeared possible, provided the necessary labor could be imported. Governor Mendonça Furtado recognized that certain areas seemed more suitable for certain types of crops, but failed to understand the agricultural limitation of the Amazon basin.[30]

The ability of the land to respond profitably to the application of African labor depended on its suitability for plantation crops. In spite of its size, however, the state possessed few areas capable of supporting intensive agriculture. In Pará only a small area around the city of Belém and across the delta in Macapá, given the limited technology then available, could support a plantation culture. In Maranhão, on the other hand, the delta formed by the Mearim and Itapecuru and the flood plains between the rivers provided a level and relatively fertile area suitable for plantations.[31]

Prior to the establishment of the company, rice and cotton had entered into the export market only in insignificant quantities. Although the native variety of rice grew profusely, it yielded little and its gains were small and brittle.[32] With the replacement of the *arroz da terra* by Carolina rice, introduced by the company in 1776, the crop became an important export commodity both in Maranhão and Pará. Low production, not poor quality, accounted for the failure of cotton to enter the export market; consequently, the importation of African slaves had a dramatic effect on production in the areas where cotton could be extensively cultivated. The degree of success of each of the major areas, Pará and Maranhão, in developing a plantation economy is revealed in their export statistics. Figures for the last five years of the company (1773-1777) indicate the trend that continued into the nineteenth century. During that period Maranhão exported 153,747 *arrobas* of cotton, compared to the relatively insignificant total of 3,013 *arrobas* shipped from Pará. Rice exports in the same period totaled 437,983 *arrobas* from São Luis and 95,796 from Belém.[33] Whereas rice and cotton became the principal Maranhense export crops, Pará's major export continued to be cacao, a natural forest product, dependent on Indian collectors.

Owing to the marginal economy of Pará, most complaints of high slave prices originated from that area.[34] In 1773, in response to the many pleas for relief, the crown ordered the company to sell slaves at cost, the price to be determined by the original cost in Africa plus the cost of transportation.[35] The regulation appears to have had little effect on the price of slaves; in 1773 the highest price paid for a prime slave was 129$000 réis; the following year, presumably sold at cost, the highest price obtained was 120$000 réis.[36] An embittered settler accused the company of padding the actual price paid for the slave on the coast of Africa to allow for a good profit margin. Allegedly the company purchased trade goods used in bartering for slaves on credit, immediately increasing the cost, and then overvalued the goods as much as 20 percent. Although the company actually paid 100$000 réis, their books indicated an inflated price of 120$000 réis. A steady decline from the high profit levels of the early 1760s may well have forced some shady accounting practices.[37] The company was also accused of indulging in the practice of selling second- and even third-grade slaves as prime specimens.[38] Complaints against the company probably owed more to the marginal financial resources of the settlers than to any other factor. Had the settlers enjoyed the profits of a successful plantation culture, their complaints would have been minimal.

Theoretically the company's charter obliged it to import black slaves into both regions, but economic realities modified the plan. Forest collecting, the economic standby of Pará, did not lend itself to the use of African slaves. A collecting expedition, dispatched into the interior following the extensive river system of the Amazon basin, lasted from six to eight months.[39] The impossibility of proper security to prevent escape was an obvious difficulty. Even the Indians who supplied the labor for the collecting expeditions often deserted even though they received a small salary. Desertion of an Indian, however, involved only a small loss for the settler who financed the expedition, whereas the escape of an African slave meant the loss of a sizable investment. In addition, forest collecting was one task where the native Indian proved superior to the African. Collecting to sustain the Amerindians' subsistence economy had been traditional long before the arrival of the Portuguese. Indians almost instinctively understood the complexities and dangers of river navigation, and settlers soon learned to cherish native river pilots.[40] The limited utility of African labor in the forest collecting industry, which unquestionably became the backbone of the Paraense economy, affected demand. Only those engaged in tasks other than collecting employed blacks; however, the marginal economy of the region barely financed the investment.

Black slaves introduced into Pará could only be purchased by individuals who were sufficiently affluent to pay for them outright or who were able to offer excellent security to underwrite the extension of credit.[41] The company understandably hesitated to tie up capital in such a profitless fashion, and as a result, avoided credit sales whenever possible. An official in Belém even proposed that slave imports be limited to a number sufficient to meet cash demand, in order to avoid the credit problem entirely.[42] Although the company would have preferred to follow such advice, it could not have escaped the wrath of labor-hungry

124   *Colin M. MacLachlan*

settlers. Consequently, the company attempted to restrict credit sales to individuals obviously able to carry the debt; hence the settlers in Macapá, the prime rice-growing area of Pará, received two hundred slaves on credit.⁴³ The owner of two sugar *engenhos* situated on six or seven leagues of land, an excellent credit risk, purchased forty slaves on credit.⁴⁴ Less fortunate settlers, however, found it virtually impossible to conclude such transactions. Bishop João de São José Queiróz, during the course of his pastoral visit in the backlands, reported that many farms had been abandoned, allegedly because "the company won't sell slaves on credit."⁴⁵

Reluctance to extend credit in Pará caused considerable bitterness. As previously noted, the company's administrator in Belém had advised restricting the number of slaves introduced, reasoning that the region already bore a heavy debt because of previous purchases. Such suggestions infuriated Governor Manoel Bernardo de Mello e Castro, who insisted that in order to satisfy its debts the region had to produce and in order to produce it needed more slave labor, not less. He complained that where a settler needed twenty or thirty slaves, he usually had to be satisfied with two or three—not enough to make a difference in production but just enough to establish a debt. Mello e Castro had absolutely no sympathy for the company's profits, and in fact accused the monopoly of making an excessive profit on cacao, paying 2$000 réis per *arroba* in Belém and selling for three times that in Lisbon.⁴⁶

The government, both in Pará and in Portugal, refusing to accept the dominance of the forest collecting industry and the apparent willingness of the company to be satisfied with that line of trade, ignored the economic realities and continued to push the development of agriculture in Pará. Encouraged by initial success with rice culture in Macapá, officials urged the introduction of two thousand blacks for use in the rice fields.⁴⁷ Although the settlers of Macapá had no difficulty paying off their debts on the two hundred slaves previously purchased, the company showed no interest in the scheme.

Although the government in Belém constantly bemoaned the reluctance to remit slaves, implying that the company was responsible for the backward state of the economy, the slaves actually imported were often badly utilized.[48] The municipal council itself absorbed 207 slaves purchased on credit. Acquired ostensibly for municipal services, these slaves apparently were not indispensable, and the council found it profitable to rent them to the royal government for construction work on the fortress of Macapá.[49] The company itself employed a number of slaves in its own warehouses as well as in shipbuilding and logging operations and as a result was accused of hoarding slaves for its own use.[50] Perhaps a more serious drain on the supply of African slaves resulted from Pará's geographical position. Besides meeting the minimal needs of the subordinate state of Rio Negro (Amazonas), slaves landed at Belém do Pará became an item of trade with Mato Grosso. Traders came up to Belém through the river system, purchased European products as well as the number of black slaves necessary to conduct their heavily laden canoes back to Mato Grosso, and then profitably disposed of both goods and slaves.[51]

Ironically the very economic base of Pará encouraged the nonproductive use of African labor. The forest collecting industry obviously could not utilize high-cost servile labor. Slaves not employed as field laborers to meet the needs of the limited agricultural production of the region could be utilized as servants or artisans, thus providing services but hardly adding to production. An interested observer would later note that the ladies of Belém went to church in hammocks carried by Indians and black slaves, with several more slaves trailing along for added effect, whereas the less fortunate attended early morning mass to avoid embarrassment.[52] Like the municipal council, settlers rented slaves to the government, which employed them in fortifying the Amazon basin in the face of menacing neighbors.[53] Government demands for labor apparently provided a constant market for rented slaves, which prompted the governor to propose that the state import black slaves for all royal needs in order to relieve the

Indian to concentrate on the forest collecting industry.⁵⁴ This extremely pragmatic proposal would have required an expansion of the slave trade, as well as a considerable royal investment; consequently, the crown showed little interest.

By 1770 economically viable Maranhão, in sharp contrast to Pará, could rely on its relatively strong economy to finance the importation of African slaves without assistance. The fall from power of the Marquês de Pombal (1777), coupled with other internal and international factors, led to the official dissolution of the company on 25 February 1778, ending the era of planned economic development in Amazonia.⁵⁵

Success and Failure

The suppression of the company of Grão Pará e Maranhão caused initial uncertainty over the future of the slave trade. Governor João Pereira Caldas, a strong supporter of the monopoly, had observed the backward state of both Maranhão and Pará immediately before the creation of the company and was convinced that the continued economic activity of the region depended on the introduction of black slaves. Although Pereira Caldas willingly acknowledged that the results had been more successful in Maranhão, he ventured the opinion that Pará simply needed renewed emphasis on the importation of Africans and dismissed complaints against the monopoly, as well as demands for its abolition, as misguided self-interest on the part of those who hoped to gain from its suppression.⁵⁶ Scornfully he noted that when disgruntled settlers called for the importation of slaves on the same basis as in other parts of Portuguese America, they overlooked the fact that elsewhere slaves were purchased for cash, not credit. To seal the argument, the governor posed the rhetorical question, "Without the company who would introduce slaves?"⁵⁷ Pereira Caldas hastened to remind Lisbon that the reduced number of Indians that still existed in the region could not be expected to fill the gap should the slave trade come to a halt.⁵⁸

The problem of the continued importation of black slaves concerned Pará more than Maranhão. As a result of increased demand for cotton, aggravated by the outbreak of armed rebellion in English North America (1776-1783), Maranhense cotton enjoyed a steady market and the resultant prosperity enabled plantation owners to finance the purchase of black slaves. Settlers in Pará, on the other hand, unable to participate in the cotton boom, had only limited financial resources to invest in slaves. Moreover, private slave traders, unlike the company, could not be forced to sell slaves on credit.

In an effort to compensate for the abolition of the monopoly, an imaginative plan was proposed that combined European with involuntary African immigration. The proposal advocated combining the initial capital investment of the company of Grão Pará e Maranhão with that of the company of Pernambuco e Paraiba to form a colonization company.[59] The capital of the proposed organization would then be applied to the purchase of carefully selected black slaves in the optimum age group of fifteen to twenty-five. In an effort to achieve a stable slave population, the plan envisioned the introduction of an equal number of both sexes. Concurrent with the importation of black slaves, the scheme called for the transporting of European emigrants, in an optimum age bracket between eighteen and thirty, from Portugal and the Azores. The colonization company would then supply each family with six slaves, three of each sex, while a bachelor would be entitled to an allotment of only three slaves, and a credit of 240$000 réis per family. In the unfortunate event more than one-third of the settler's slaves died before he had paid off his debt, they would be replaced on the same liberal terms. Rather than dilute the economic impact of the new arrivals by dispersing them throughout the Amazon basin, the proposal suggested they be settled around the city of Belém. This plan appears to have been the only attempt to institute a balanced immigration into Pará involving both Europeans and Africans.[60]

The crown, however, having acted against a monopolistic trading company, was not inclined to create a new one, preferring instead to trust free commerce to meet the demand for black labor. The switch to free trade resulted in little change in the number of slaves introduced into the Amazon region. Instead of a dramatic drop, the trade remained at approximately the same level as previously, a sign that the company had only met the market demand, especially in the last decade of its existence, without attempting to stimulate the economy by actively introducing a large number of surplus slaves. In Belém the company introduced an average of 581 slaves a year. From 1779 to 1790 the average number declined less than 10 percent, to 547. In Maranhão the upward trend begun by the company continued. From 1779 to 1790 the annual number was 1,605, a difference of three to one over the rate of importation into Pará, reflecting the continuing ability of Maranhão to finance African slaves.[61]

The suppression of the company of Grão Pará e Maranhão did not affect the operations of the African terminus of the slave trade. Nevertheless, the direct, one might say artificial, connection between the African slave ports and the Amazon was severed. Instead of a guaranteed and steady importation of slaves, settlers had to rely on private initiative. The company had necessarily accepted local products in exchange; however, with its abolition the settlers faced the problem of a market for their products in order to finance African slaves among other necessities. In Maranhão a steady demand for its crops presented little difficulty, but Pará, still dependent on the forest collecting industry, had no such advantage.

In spite of past failures Lisbon continued to encourage schemes to introduce black labor into the area, refusing to accept the relentless economic logic that explained the backward state of Pará. Aware of the government's interest, a group of Paraense businessmen proposed the formation of a privileged local slaving company to import slaves.[62] In exchange for a six-year tax exemption, they proposed to place four ships in the African trade. To facilitate purchases the government would guarantee that any slaves sold on credit could not be seized to satisfy previous creditors. As Governor Francisco de Sousa Coutinho observed, the government had very little to lose since theoretically any drop in customs revenue would be offset by the increased productivity of the area. The governor optimistically noted that if the company was exempted from taxes it would be more inclined to sell on credit; moreover, local merchants could judge more effectively the credit standing of their fellow countrymen. Although Sousa Coutinho backed the idea, he favored extending these privileges to any individual or group that agreed to bring slaves directly from Africa.[63]

Although the crown granted the requested privileges, the company failed to materialize; however, individual members of the group did engage in random voyages. In 1797 an obviously frustrated Sousa Coutinho analyzed the history of the slave trade and concluded that the lack of local capital posed the major obstacle. He suggested that the government lend individual businessmen capital, repayable within three or four years, to finance the African slave trade. The idea that Pará would eventually attract a substantial intercoastal slave trade with the sugar regions of the south was dismissed on the ground that the general poverty of the captaincy precluded cash sales. In fact, many settlers still owed the long-suppressed company of Grão Pará e Maranhão for slaves purchased as credit. The inability of the inhabitants to finance African slaves could only be overcome by government guarantees. Governor Sousa Countinho suggested that any slaves purchased as credit be protected from previous debts to eliminate uncertainty. More importantly, the governor urged the crown to authorize the government to buy slaves from traders for cash and to sell them to the settlers on credit.[64]

Lisbon, responding to the governor's pleas in a reasonably imaginative manner declared the reexportation of slaves, from Rio de Janeiro and Pernambuco to Belém do Pará, duty-free.[65] At the same time the crown dispatched a circular to the governors of Luanda and Benguela urging them to do everything in their power to encourage slave exports to Belém.[66] These tentative steps were soon followed by a more comprehensive tax incentive plan. A royal decree of 27 October 1798 exempted the loading of slaves in Africa and their introduction in Pará from all customs duties and in addition ordered that all products carried out of Pará up to the value of the slaves introduced were to be exempt from export taxes on leaving Belém as well as import taxes at Lisbon.[67] Such a measure was calculated to stimulate local production as well as to finance the trade.

In spite of these bold measures, the crown once again failed to consider the ability of the region's products to support the cost of African slaves. It remained for the governor of Luanda, from a position of objectivity on the coast of Africa, to point out the fatal flaw. The governor noted that although, on first consideration, the tax exemptions appeared to be all-embracing, in actual fact the only major change involved cotton exports since other products had previously been exempted. Moreover, Paraense cotton could not compare with the superior quality exported from Maranhão and Pernambuco; consequently, it was only in the total absence of better-quality cotton that Pará's product could be sold on the European market.[68] The governor observed that even intercoastal trading vessels refused to deal in local products, demanding payment in specie or coin so that they could purchase a load of hides or cotton in Ceará to carry back to their home ports. The desirability of exotic forest products was questioned with the observation that many, such as sarsaparilla, met an uncertain demand in Europe. Quite naturally these factors had an adverse effect in the slave traffic. A slave trader had only two alternatives if he chose to trade in the area: either sell on credit and tie up his working capital or accept a cargo of poor quality products for possible sale in Europe.[69] The governor of Luanda's analysis of the situation was one of the first official observations that recognized that the marginal economy of Pará could not sustain or finance African labor.

Besides the inability of the Paraense economy to sustain a regular triangular trade, its very geographical position discouraged links with Africa. The voyage from the Portuguese slaving ports in Africa to Belém was both hazardous and long. Many vessels survived the ocean crossing only to come to grief while attempting to enter the delta.[70]

Although Pará could not have utilized a massive influx of Africans, even the limited demand for slaves that existed could not be satisfied; as a result Belém made a good market for marginal slaves from the south. Each year three of four ships from Bahia, Pernambuco, and even São Luis do Maranhão arrived with small numbers of sick or recalcitrant slaves that could not be profitably disposed of in the more selective slave markets of Brazil.[71] Selling on a cash basis, but at attractive prices, these coastal traders quickly disposed of their cargoes. Such slaves proved a bad bargain at any price, since the sick soon perished and the "troublemakers" did little except sow dissension and violence among the more resigned slaves. The unwary buyer often found himself deprived not only of his new slave, but also of several others who had joined him in flight.[72] Although São Luis also fell prey to unscrupulous traders, Belém in particular made an easy target.

In addition to the individual loss suffered by the purchaser, the marginal intercoastal slave trade had a negative effect on commerce in both Pará and Maranhão. Such traders preferred cash and thus caused a drain of specie from the north. The governor of Maranhão estimated that each vessel removed from 20,000 to 25,000 cruzados from local circulation.[73]

Governor Francisco de Sousa Coutinho of Pará, in spite of the obvious negative effects of the trade on the economy, found the effect on public order to be of more concern. In an effort to control the introduction of "vicious slaves," the governor ordered customs officials to investigate such vessels entering port to ascertain the true origin of the slaves.[74] The governor's concern reflected the traditional problem of security and control in a mixed society of free and slave labor. The surrounding dense jungle offered the runaways the hope of concealment, and the myriad waterways that sluggishly penetrated the backlands permitted access to distant sanctuaries. Unfortunately, once the slave had successfully slipped into the jungle, his sanctuary offered little except loneliness and slow starvation. Without the Indian's knowledge of the jungle, survival became an uneven contest. A fugitive could not rely on making contact with forest Indians, since the Indians themselves frequently turned captured slaves over to the authorities and collected a reward for their diligence.[75] Unsympathetic Indians were not the only danger encountered by the runaway slave. Organized bounty hunters, under the leadership of a *capitão do mato* and composed of blacks, mulattoes, and Indians, scoured the jungles in search of fugitives.[76] Given the hazards and difficulties of sustaining themselves in the jungle for any extended period, many runaways attempted to integrate themselves into the free black population in an area where they were unknown. Fugitives from Maranhão often hoped to find a haven in Pará or, still better, to reach the French colony of Cayenne five hundred miles northwest of Belém. In Pará, Cayenne proved to be the biggest attraction, but runaways also headed for Maranhão and elsewhere.[77] The ease with which a slave could slip into the jungle proved a constant problem, especially in interior areas; one slaveholder in Macapá reportedly lost more than twenty slaves.[78]

Runaway slaves, who preferred not to risk integrating themselves into the free black population, banded together in *mucambos* or jungle encampments. These illegal settlements, by providing a definite destination that eliminated the uncertainty of attempting to flee to a vague sanctuary in an unknown area, attracted dissatisfied slaves from the surrounding areas. In Maranhão many small interior settlements such as Viana, Pinheiro, Alacantera, Guimaraes, and Maraacassume had a *mucambo* nearby that drained off discontented slaves in the district.[79] Such settlements, complete with abducted female slaves or Indian women, became shadowy societies mirroring the settlements of the Portuguese and often engaged in substance farming. Their desire for European goods could be met by raiding isolated settlements. Inevitably, success brought on a punitive Portuguese expedition, and few survived for long periods of time. One daring group terrorized the area around Belém, raising fears that they intended to attack the city and were allying themselves with another fugitive group for that purpose.[80] Soldiers dispatched to thwart these plans recaptured twelve blacks. Another African settlement, situated near the town of Santarém, three hundred miles from the mouth of the Amazon, held off a Portuguese expedition, which had to withdraw after killing one runaway and recapturing two. This expedition brought back word of several other groups of slaves, one rumored to be governed by a fugitive Jesuit priest who allegedly maintained contact with the French and the Dutch on the Caribbean coast.[81]

The existence of the French colony of Cayenne complicated the normal problems of controlling servile labor. Flight across international borders resulted in the convention of 1732, in which the French and the Portuguese agreed to exchange runaways.[82] It is more than probable that during the first half of the century most of the slaves exchanged were Indians, not Africans. In spite of the convention of 1732, Portuguese authorities proved less than zealous in returning runaways. For example, in 1756 the crown approved the distribution of fugitives to local settlers with the proviso that their disposal be officially recorded in case of a French inquiry.[83]

Portugal constantly suspected the French of plotting to seize its Amazon empire and viewed the flight of Indians and black slaves into French territory as a plot to weaken Portuguese control. The fear of suspected French attempts to encourage the desertion of slaves, present before 1789, became almost paranoid after the French Revolution. In 1795 the governor dispatched an expedition to the northern delta region to arrest all unauthorized individuals and put a stop to the loss of "our slaves" to the French.[84] When news reached the Portuguese that the French were freeing their slaves, the Portuguese almost despaired of being able to contain their own blacks.[85] The governor reported in horror that he had received word that the emancipated slaves of Cayenne refused to work; the result was extreme hunger in the colony.[86] The patrol boats posted in the mouth of the Amazon and along the northern coast were urged to make every effort to intercept runaways. War with France increased the fear of a foreign-instigated slave revolt, especially in Pará, whose proximity to the French colony was continually a source of regret. Suspicious gatherings of free blacks and slaves caused alarm; however, such apprehension proved needless since the resistance of the African to his imposed slave status continued to be the traditional one of desertion rather than rebellion.[87]

Although the importation of Africans into the Amazon basin was intended to increase the labor supply of the region, and hence its prosperity, it also had important effects on the native population. The success of the slave trade in Maranhão relieved the Indian of much of the burden of labor. Ironically, the availability of African labor made the constant subjugation of newly contacted Indian tribes unnecessary, and as a consequence, Maranhão would be plagued by hostile Indians into the nineteenth century.[88] In Pará the failure of the slave trade forced the Indian into the labor pool, and as a consequence by the early nineteenth century the depleted Indian population posed absolutely no threat to the state's security.[89] But if African slavery proved a blessing to the aborigines, it was a mixed one; epidemic diseases and slaves often arrived on the same ship.

After the creation of the company of Grão Pará e Maranhão and the start of systematic importation of African slaves, the frequency of smallpox epidemics increased markedly.[90] Scarcely a year passed without an outbreak of the dread disease, and often such outbreaks reached crisis proportions. The slaving operations of the company broke down the relative biological isolation of the Amazon basin. Before the arrival of regular shipments of slaves, the region had achieved a delicate balance. In the early years of the Portuguese presence in the basin, both measles and smallpox decimated the population. After initial havoc, these diseases seldom reached major epidemic proportion except in the late 1720s, early 1730s, and the middle 1740s, when thousands of Indians died from the smallpox.[91] Limited immigration and contact with the outside world offered some protection from fresh contamination; however, the company of Grão Pará e Maranhão, by breaking the isolation of Amazonia, exposed it to the horrors of frequent epidemics. The fact that the company generally sold part of each cargo of slaves in Maranhão and Pará almost guaranteed that disease would simultaneously be introduced into both areas. Once an epidemic had broken out, the authorities could do little to control it and primitive methods of fumigation had scarcely any effect. The sulphurous fumes of black powder fired by cannons in the districts most affected did more to disturb the ill than retard the spread of disease.[92]

The government recognized the importance of isolating infected slaves. São Luis' hospital of Bomfim served as a refuge for many unfortunate black victims.[93] In 1788 the municipal council of Belém proposed the establishment of a quarantine station on an island in the bay; subsequently, the island of Arapiranga and a location outside the city (Val de Caens) served such a purpose.[94] The municipal council of São Luis, noting that Lisbon routinely practiced quarantine measures, requested the governor to establish a quarantine station on the island of Mido.[95]

Unfortunately, quarantine regulations proved hard to enforce, especially since slave traders, after an outbreak of disease, did everything in their power to dispose of their property before more were stricken. When two infected ships arrived in Belém, one inappropriately named *Boa Fortuna* on which two hundred slaves died of smallpox during the voyage, authorities quickly isolated their cargoes; nevertheless, two slaves were surreptitiously sold, reportedly causing twenty cases of smallpox in the city.[96] In another case a smack from Pernambuco carrying diseased slaves arrived in Belém after a stop in São Luis. The captain, aware of the imminent loss of his human capital, evaded quarantine regulations, landed his slaves, and sold many of them before being apprehended. The remaining slaves were placed under three-month quarantine while those already sold were dispersed throughout the region.[97] Although all races and classes endured the ravages of these attacks, the Indian in particular suffered. Governor José Narciso de Magalhaes Menezes observed that even a slight case, in fact "one pox on an Indian," was almost a sure death sentence.[98]

Amazonia in 1800

At the end of the century the contrast between the two principal captaincies was reflected in the number of slaves. Maranhão had 36,880 black or mulatto slaves, who made up 46 percent of the population, as well as 13,613 free blacks, out of a total population of 78,860. Blacks and mulattoes constituted 64 percent of the population; the Amerindians, excluding unsubdued tribes, made up only 5 percent; and the remaining 31 percent were classified as white.[99] Pará, with a population placed at 80,000 in 1801, slightly more than Maranhão, presented an entirely different situation, with only 18,944 slaves or approximately 23 percent. Indians made up another 20 percent; the remaining 57 percent was made up of whites, mulattoes, and other mixtures as well as free blacks.[100] The figures demonstrate to what extent the cotton and rice economy of Maranhão depended on servile labor while Pará, territorially dwarfing the province of Maranhão, continued to depend on forest products and Indian labor to sustain its marginal economy. In 1800 Pará exported 127,181 *arrobas* of cacao, its traditional forest export, compared to 90,836 *arrobas* of rice and only 15,930 *arrobas* of cotton.[101]

Government attempts, especially after 1750, to encourage economic development and the introduction of African labor obviously succeeded in Maranhão, where the combination of suitable land and an international market for cotton provided the basis for a viable economy. In Pará, however, the effort failed. The crown's determination in the face of adverse economic factors stemmed from imperial considerations that viewed the north as vital to the future of Portuguese America. Political considerations motivated the drive to overcome the obstacles to development in Amazonia. In many respects Lisbon's approach is mirrored in modern Brazilian policy, which has recently been directed toward "national integration" of Amazonia. The artificial channeling of investment funds, the cutting of roads to the outer limits of the Brazilian Amazon, essentially are political actions designed to preserve an area viewed as vital to Brazil's future. The Marquês de Pombal certainly would have approved.

APPENDIX A
Number of African Slaves Landed at Belém do Pará, 1757-1800

Year	Number	Year	Number	Year	Number
1757	371	1772	341	1787	710
1758	1,103	1773	817	1788	631
1759	534	1774	307	1789	687
1760	209	1775	696	1790	473
1761	315	1776	870	1791	279
1762	1,637	1777	517	1792	204
1763	147	1778	765	1793	263
1764	885	1779	318	1794	522
1765	832	1780	845	1795	1,096
1766	138	1781	471	1796	—
1767	441	1782	329	1797	176
1768	268	1783	681	1798	400
1769	180	1784	470	1799	—
1770	704	1785	269	1800	514
1771	895			Total	23,884

Such figures must be used with caution. At best they indicate the trend rather than the exact number of slaves imported. The major difficulty lies in determining which figures represent the actual number of individual slaves and which indicate the number of *peça de India* (a standard unit measurement of slaves that could include one to three individual slaves). In addition, official statistics often are contradictory.

The above figures were collected from many different sources in the Arquivo Historico Ultramarino, Lisbon, and in the Arquivo

138   *Colin M. MacLachlan*

Nacional, Rio de Janeiro. However, one key document should be noted: "Mapa dos Escravos que a Companhia Geral do Grão Pará e Maranhão importou neste Estado do Pará . . . 1757 ate . . . 1772," AHU, caixa 32 (Pará). The figures listed here for Pará were erroneously used by Manuel Nunes Dias in his *Fomento e Mercantilismo* (see previous citation) to represent the total figure imported by the company in both Pará and Maranhão. That this is not the case is supported by the document cited above and the statistics contained in "Certidão do Juiz da Alfandiga (Pará) 1 March 1792," ANRJ, cod. 99, Vol. 13, f. 127, and "Os Administradores da Extincta Companhia, Reverdo os livros della . . . Numero de Escravos Introduzidos neste Porto . . . desde 1755 ate . . . 1766," ANRJ, cod. 99, Vol. 13, f. 125. Numbers for the years 1792, 1794, 1795, 1798, and 1800 have been taken from Davidson, Appendix V, pp. 477-481.

APPENDIX B
Number of African Slaves Landed at São Luís do Maranhão, 1757-1800

Year	Number	Year	Number
1757		1788	2,894
	10,616	1789	2,107
		1790	1,411
		1791	1,166
		1792	1,187
1778	–	1793	2,361
1779	1,474	1794	2,186
1780	926	1795	1,740
1781	944	1796	1,854
1782	752	1797	1,536
1783	1,602	1798	–
1784	1,375	1799	–
1785	1,345	1800	637[a]
1786	662		
1787	2,160	Total	40,935

These figures are also more indicative of a trend than exact statistics. The above figures were collected from AHU (Maranhão), caixas 47-49, 52, 55-68, and 70. Statistics for the period 1757-1777 were taken from "Recapitulação dos escravos introduzidos pela Companhia do Grão Pará e Maranhão ... de 1757 ate 1777," AHU, caixa 37 (Pará).

a. The small number of slaves introduced in 1800 probably resulted from the hostile action of French corsairs stationed at Cayenne. The following year (1801) the number was a more normal 1,328.

Notes

[1] The extent to which the Portuguese depended on the African and identified him with plantation labor, especially in the prime Brazilian sugar regions of Bahia and Pernambuco, was succinctly expressed by Antonio Vieira S.J.: "Without blacks there would not be a Pernambuco." Quoted in Alfonso de E. Taunay, "Subsidios para a historia do trafico Africano no Brasil," *Anais do Museu Paulista* 10 (1941): 8. See Vicente Salles, *O Negro no Pará sob o regime da escravidão* (Rio de Janeiro, 1971), for a general view of African slavery in the region.

[2] Ernesto Cruz, *Historia do Pará* (Belém, 1963), 1: 86. Some of the fragmentary details of foreign interest in the region have been drawn together in James A. Williamson, *English Colonies in Guiana and on the Amazon, 1604-1668* (Oxford, 1923).

[3] The state remained intact except for a brief two-year period from 1652 to 1654. Cruz, 1: 55.

[4] Roberto C. Simonsen, *Historia economica do Brasil*, 5th ed. (São Paulo, 1967), p. 318. In 1680 an earlier but less systematic attempt to meet the region's labor needs was undertaken when the crown financed the introduction of 350 slaves in Pará and 250 in Maranhão. See António Carreira, "As Companhias Pombalinas de navegação, comercio e trafico de escravos entre a costa africana e o nordeste brasileiro" (Part 1), *Boletim Cultural da Guiné Portuguesa*, January-April 1968, p. 20. This well-documented work appears in four parts, the other three being in May-August 1968, January 1969, and April 1969. The information supplied by Carreira deals mainly with the African end of the trade.

[5] Simonsen, p. 319.

[6] C. R. Boxer, *The Golden Age of Brazil, 1695-1750: Growing Pains of a Colonial Society* (Berkeley and Los Angeles, Cal., 1962), p. 270.

[7] Annaes da Biblioteca e Arquivo Publico do Para, 1: 104 (hereafter cited as ABAP).

[8] ABAP, 1: 106, and "Carta regia ao Senado da Camara," 6 February 1703. Biblioteca National, Rio de Janeiro (hereafter cited as BNRJ), 15, 4, 8 (Maranhão), and Carreira, Part 1, p. 29.

[9] ABAP, 1: 119. It is interesting to note that after official encouragement of the Madeira River trade with Mato Grosso, merchants from the far west in turn received the "pickings" from Pará. David M. Davidson, "Rivers and Empire: The Madeira Route and the Incorporation of the

Brazilian Far West, 1737-1808" (Ph.D. diss., Yale University, 1970), p. 150.

[10] ABAP, 1: 129.

[11] ABAP, 1: 114.

[12] The refusal of the crown to accept the manufacture of cane brandy in spite of many convincing arguments in its favor approached stubbornness. See Sue A. Gross, "Agricultural Promotion in the Amazon Basin, 1700-1750," *Agricultural History*, April 1969, pp. 269-276.

[13] Francisco Pedro de Mendonça Gorjao to the court, 26 April 1749. Arquivo Historico Ultramarino, Lisbon (hereafter cited as AHU), caixa 32 (Maranhão).

[14] Unfortunately no exact figures are known. The scattered references to African slavery only permit one to state with certainty that a number of blacks were imported into Amazonia before 1700 and during the first half of the eighteenth century. See ABAP, 1: 104, 106, 114, 119, 129; and BNRJ, 15, 4, 8 (Maranhão).

[15] Peracer, Conselho Ultramarino, 15 May 1750, AHU, caixa 32 (Maranhão).

[16] Consulta, 7 November 1752, AHU, caixa 35 (Maranhão).

[17] Carta régia, 22 November 1752, Biblioteca e Arquivo Publico, Belém (hereafter cited as BAP), cod. 884, doc. 29.

[18] Davidson, p. 157. The overall plan also envisioned transforming the Indian into a productive, profit-motivated "Portuguese" settler. See Colin M. MacLachlan, "The Indian Directorate: Forced Acculturation in Portuguese America (1757-1799)," *The Americas*, April 1972, pp. 357-387.

[19] Marcos Carneiro de Mendonça, *A Amazonia na era pombalina: Correspondencia inedita do . . . Francisco Xavier Mendonça Furtado* (São Paulo, 1963), 1: 29.

[20] Francisco Xavier Mendonça Furtado to Diogo de Mendonça Corte Real, 18 January 1754, BNRJ, 11, 2, 43 (Pará).

[21] *Instituição do Companhia Geral do Grão Pará e Maranhão* (Lisbon, 1755), art. 30.

[22] Carta régia, 13 June 1760, BAP, cod. 668, doc. 27.

[23] Carneiro de Mendonça, 3: 1052.

[24] José Mendes da Cunha Saraiva, *Companhias gerais de comercio e navegação para o Brasil* (Lisbon, 1938), pp. 32-34.

[25] José Mendes da Cunha Saraiva, *A Fortaleza de Bissau e a Companhia do Grão Pará e Maranhão* (Lisbon, 1947), p. 38.

²⁶Manuel Nunes Dias, *Fomento e Mercantilismo: A Companhia Geral do Grão Pará e Maranhão, 1755-1778* (Belém, 1970), 1: 471.

²⁷Nunes Dias, 1: 495. A substantial part of the company's trade was in Cape Verde textiles, which were highly prized on the coast of Africa. T. Bentley Duncan, *Atlantic Islands: Madeira, the Azores and Cape Verdes in Seventeenth-Century Commerce and Navigation* (Chicago, 1972), p. 221.

²⁸"Recapitulação dos escravos introduzidos pela Companhia do Grão Pará e Maranhão . . . de 1757 ate 1777," AHU, caixa 37 (Pará).

²⁹Carneiro de Mendonça, 1: 199.

³⁰Carneiro de Mendonça, 1: 200.

³¹See Nunes Dias, 1: 159-162, for a concise geographical description.

³²Jeronimo de Viveiros, *Historia do comercio do Maranhão, 1612-1895* (São Luis, 1954), 1: 76.

³³Nunes Dias, 1: 430, and Manuel Barata, *A Antiga produccão e exportação do Pará* (Belém, 1915), p. 3.

³⁴Such complaints were often petulant and almost invariably included a statement to the effect that Maranhão was favored over Pará. João Pereira Caldas to Martinho de Mello e Castro, 18 June 1777, AHU, caixa 36 (Pará); see also BNRJ, 11, 2, 43, docs. 121 and 189 (Pará).

³⁵Edital, 1 December 1773, BAP, cod. 595, doc. 148.

³⁶Mapa, AHU, caixa 34 (Pará) and caixa 43 (Maranhão).

³⁷H. E. S. Fisher, *The Portugal Trade: A Study of Anglo Portuguese Commerce 1700-1770* (London, 1971), p. 46.

³⁸"Minuta de um papel sobre a escravatura," AHU, caixa 35 (Pará).

³⁹AHU, caixa 42 (Pará).

⁴⁰João de Alburqueque to Francisco de Sousa Coutinho, 2 May 1791, Arquivo Nacional, Rio de Janeiro (hereafter cited as ANRJ), cod. 99, Vol. 13, f. 147.

⁴¹Nunes Dias, 1: 272.

⁴²Manoel Bernardo de Mello e Castro to Francisco Xavier de Mendonça Furtado, 1 August 1759, BNRJ, 11, 2, 43 N 100 (Pará).

⁴³João Pereira Caldas to Martinho de Mello e Castro, 18 June 1777, AHU, caixa 36 (Pará).

⁴⁴João de São José Queiróz, *Visitas Pastorais: Memorias (1761 e 1762-1763)* (Rio de Janeiro, 1961), p. 176.

⁴⁵*Ibid.*, p. 172.

⁴⁶Manoel Bernardo de Mello e Castro to Francisco Xavier de Mendonça Furtado, 1 August 1759, BNRJ, 11, 2, 43 N 100 (Pará).

⁴⁷João Pereira Caldas to Martinho de Mello e Castro, 18 June 1777, AHU, caixa 36 (Pará).

⁴⁸João Pereira Caldas to Marquês de Pombal, 8 March 1774, BNRJ, 11, 2, 43 doc. 189.

⁴⁹Portaria, 14 November 1771, BAP, cod. 592, doc. 914, and Carta régia, 11 May 1779.

⁵⁰João Pereira Caldas to court, 8 November 1774, BAP, cod. 778, doc. 123.

⁵¹The trade between Pará and Mato Grosso, including goods and slaves, was estimated at 300,000 cruzados in 1775. João Pereira Caldas to Marquês de Pombal, 1 April 1775, BNRJ, 11, 2, 42, doc. 216.

⁵²Alexandre Rodriques Ferreira, "Miscellanea historia para servir de explicação ao prospecto da cidade do Pará," ms. dated 19 September 1789, BNRJ, 21, 1, 1, N 2.

⁵³Edital, 21 August 1771, BAP, cod. 858, n.p.

⁵⁴João Pereira Caldas to Martinho de Mello de Castro, 7 April 1773, BNRJ, 11, 2, 43 doc. 176.

⁵⁵Jeronimo de Viveiros, 1: 74.

⁵⁶João Pereira Caldas to Martinho de Mello e Castro, 11 September 1777, BNRJ, 11, 2, 43 doc. 247.

⁵⁷*Ibid*.

⁵⁸João Pereira Caldas to Martinho de Mello e Castro, 1778, AHU, caixa 38 (Pará).

⁵⁹António de Abreu Guimares, "Methodo facil, proprio e util para se animar e prover a importante povoação do Grao Para, suas lavouras, fabricas, navegação e comercio em ulitidade geral e do proprio estado," 1778(?), ms. in BNRJ, 1-28, 28, 32 A N-2 (Pará).

⁶⁰*Ibid*.

⁶¹See Appendix A.

⁶²Francisco de Sousa Coutinho to court, 25 April 1792, ANRJ, cod. 99, Vol. 13, f. 118.

⁶³*Ibid*.

⁶⁴Francisco de Sousa Coutinho to court, 21 August 1797, BAP, cod. 702, doc. 102.

⁶⁵"Circular para Rio de Janeiro, Bahia, e Pernambuco," 16 April 1798, BAP, cod. 683, doc. 42.

⁶⁶"Circular para Angola e Benguela," 16 April 1798, BAP, cod. 683, doc. 41.

[67] Edital, 19 October 1798, BAP, cod. 689, doc. S.
[68] São Paulo da Assumpção de Luanda to court, 12 October 1800, BAP, cod. 764, doc. 55.
[69] *Ibid*.
[70] To help avert maritime disasters, pilots were stationed at Salinas, close to the actual entrance to the river, with crews of Indians to board and direct ships to the port of Belém. José de Napoles Telo de Menezes to Conselho Ultramarino, 17 January 1781, BAP, cod. 714, doc. 81.
[71] Francisco de Sousa Coutinho to court, July 1790, ANRJ, cod. 99, Vol. 11, f. 72.
[72] Portaria, 22 June 1790, ANRJ, cod. 99, Vol. 11, f. 74.
[73] José Telles da Silva to Martinho de Mello e Castro, 31 December 1785, AHU, caixa 57 (Maranhão).
[74] Portaria, 22 June 1790, ANRJ, cod. 99, Vol. 11, f. 74.
[75] João Pereira Caldas to governor of Maranhão, 26 May 1774, BAP, cod. 589, doc. 81.
[76] Jeronimo de Viveiros, 1: 88. For more details, see C. R. Boxer, *Golden Age of Brazil*, p. 170.
[77] Governor José de Napoles Telo de Menezes noted that the Tocantin River was a virtual *"porta franca"* for fugitive slaves and others. José de Napoles Telo de Menezes to court 9 June 1780, BAP, cod. 714, doc. 71.
[78] João Vasco Braun to Francisco de Sousa Coutinho, 8 November 1790, *Instituto Historico e Geografico Brasileiro* (hereafter cited as IHGB), lata 278, ms. 14748 N 2.
[79] Jeronimo de Viveiros, 1: 88.
[80] Governor to court, 1771 or 1772, BNRJ, 1, 28, 27, 5 N 1-10.
[81] Francisco de Sousa Coutinho to court, 1 March 1793, ANRJ, cod. 99, Vol. 14, f. 96. A description of such a settlement in another region of Portuguese America is provided by Stuart B. Schwartz, "Buraco de Tatu: The Destruction of a Bahian Quilombo," *International Congress of Americanists* 3 (1968): 429-438.
[82] João Pereira Caldas to Martinho de Mello e Castro, 29 November 1773, AHU, caixa 34 (Pará).
[83] Carta régia, 26 May 1756, BAP, cod. 667, doc. 19.
[84] "Portaria para Alferes José . . . Rodgrigues Camello," 2 February 1795, BAP, cod. 567, doc. 15.
[85] Francisco de Sousa Coutinho to court 10 September 1795, BAP, cod. 682, doc. 54.
[86] *Ibid*.

[87] Francisco de Sousa Coutinho to court, 21 June 1795, BAP, cod. 682, doc. 41, and Luis Pinto de Sousa to governor, 23 March 1796, BAP, cod. 682, doc. 87.

[88] The Alto-Mearim was effectively blocked to advancing civilization until 1835. Jeronimo de Viveiros, 1: 291. Further references to hostile Indians in early-nineteenth-century Maranhão may be found in BNRJ, 7, 4, 74, docs. 4, 7, 19 (Maranhão) and 3, 4, 24 (Maranhão).

[89] The dependence of Pará on forest products collected by Indian laborers helped deplete the Indian population. In 1800 the governor aptly characterized the state as one of great territorial extension (*extenção*) and small intensity (*intenção*). Francisco de Sousa Coutinho to Rodigo de Sousa Coutinho, 25 September 1800, IHBG, lata 281, ms. 14769.

[90] Arthur Vianna, *As Epidemias no Pará* (Belém, 1906), p. 11. Archival research documents four major smallpox epidemics before 1755. After that date, smallpox outbreaks occurred on at least thirteen different occasions.

[91] *ABAP*, 5: 29.

[92] Vianna, p. 15.

[93] Juiz de Fora to Manoel e Pinto de Almdo e Lima, 17 August 1792, AHU, caixa 64 (Maranhão).

[94] José Narciso de Magalhaes Menezes to court, 29 August 1806, BAP, cod. 706, doc. 41.

[95] Senado da Câmara to governor, 23 April 1788, ANRJ, cod. 99, Vol. 9, f. 72.

[96] José Narciso de Magalhaes Menezes to court, 29 August 1806, BAP, cod. 706, doc. 41.

[97] Francisco de Sousa Coutinho to court, 22 July 1793, ANRJ, cod. 99, Vol. 14, f. 212.

[98] José Narciso de Magalhaes Menezes to court, 29 August 1806, BAP, cod. 706, doc. 41.

[99] "Conta das habitantes, novidades annuaes... de toda a Capitania do Maranhão anno de 1798," AHU, caixa 37 (Maranhão).

[100] Vianna, p. 11.

[101] Manuel Barata, p. 7.

5
Encomienda, African Slavery, and Agriculture in Seventeenth-Century Caracas
Robert J. Ferry

FOUNDED in 1567, Caracas, as was soon evident, had neither the human nor the mineral resources to sustain a prosperous Hispanic settlement. The earliest descriptions suggest a sense of general stagnation and disappointment. In 1578, forty of the sixty vecinos in the rustic, thatched-roofed village held encomiendas, evidence that in itself an Indian grant was not enough to distinguish an elite from among the colonists. A fair measure of the lackluster appeal of early Caracas is the fact that included among the forty encomenderos of 1578 were just eighteen members of the band of 136 men who had overcome the last Indian resistance to found the town such a short time before. By contemporary estimate, these forty enjoyed the benefit of the labor of 4,000 native tributaries, but this was only one-third the number of Indians who had inhabited the region ten years earlier. For an ambitious, adventuresome generation, these small and shrinking encomiendas did not measure up to the promise of the Indies, and the majority of Caracas's conquistadors had moved on to seek their fortunes elsewhere.[1]

A report to the king by the royal treasurer don Diego de Villanueva indicates that by 1607 Caracas had developed a minor, but regular, trade in agricultural products. Tobacco was the principal export, followed by wheat flour and small amounts of sugar, sarsaparilla, and cotton cloth. Hides, from extensive herds of privately owned cattle, were also traded

---

* The author, an Andrew W. Mellon postdoctorate fellow at Tulane University, wishes to thank Stuart Schwartz, John Lombardi, James Lockhart, and Eric Van Young for their helpful criticisms.

1. "Descripción de Santiago de León de Caracas" in Antonio Arellano Moreno, ed., *Relaciones geográficas de Venezuela* (Caracas, 1964), p. 120. The 136 conquistadors are listed in Hermano Nectario María, *Historia de la conquista y fundación de Caracas*, 2d ed. (Madrid, 1966), pp. 63–67. The estimate of Indian population comes from the testimony of Francisco Infante and Garci González de Silva, Jan. 3, 1589, in Archivo General de la Nación, Caracas (hereinafter AGN), *Encomiendas*, 5 vols. (Caracas, 1945–58), I, 230–232.

TABLE I: Caracas Exports, 1607.

Product	Quantity	Unit value*	Total value*	% Total value
Tobacco	1,362 arrobas	25	34,050	42.9
Flour	7,127 arrobas	4	28,508	35.9
Hides	651	8	5,208	6.6
Sugar	139 arrobas	30	4,170	5.2
Sarsaparilla	75 quintales	50	3,750	4.7
Cotton cloth	800 varas	3	2,400	3.0
Biscuit	45 arrobas	16	720	.9
Cacao	4.5 fanegas	96	432	.5
Cheese	25 arrobas	9	225	.3
			79,463	100.0

* Values are given in reales.
SOURCE: AGN, Real Hacienda, legs. 3, 5, 6, passim, *almojarifazgo* tax. A similar table, with slightly different values, is to be found in Eduardo Arcila Farías, *Economía colonial de Venezuela* (Mexico City, 1946), p. 68. This is the last year for which we can be sure that all noncontraband exports were recorded; see n.6.

(see Table I). The hides went with the tobacco and sarsaparilla to Spain, while wheat and the other items found markets in Cartagena and other Caribbean ports. Caracas's imports, on the other hand, consisted principally of Spanish dry goods, Canarian and Andalusian wine, and an occasional African slave. Some few luxury items, a bolt of Chinese silk, silver jewelry and tableware, or several slaves bought in a lot from a supposed "forced entry" ship, were purchased by those few Caraqueños who exported tobacco, wheat, or hides in quantity.[2] By 1607 the more common of these goods had reached—in small amounts, to be sure—as far as the San Sebastián cattle district, thirty leagues south of Caracas. They were paid for there with hides, tobacco, cotton, and "some cacao," the only reference in the Villanueva report to the tropical plant that would shortly revolutionize the Caracas economy.[3]

---

2. "Relación de Diego de Villanueva y Gibaja, [1607]" in Arellano Moreno, ed., *Relaciones geográficas*, pp. 287–301. Few Caracas residents received goods on consignment from merchants in Seville. For a list, see AGN, Real Hacienda, leg. 11, fols. 23–25. The illegal entry of slaves is discussed at length in Miguel Acosta Saignes, *Vida de los esclavos negros en Venezuela* (Caracas, 1967), chap. 3. Examples of slave ships allegedly blown off course on their way from Angola to the Canary Islands can be found in AGN, Real Hacienda, leg. 10, Sept. 22, 1613, and June 25, 1618.

3. "Relación de Villanueva" in Arellano Moreno, ed., *Relaciones geográficas*, p. 280. Thirty leagues was Villanueva's statement of the distance from Caracas to San Sebastián; however, the San Sebastián site was changed five times during the seventeenth century, according to Lucas Guillermo Castillo Lara, *Materiales para la historia provincial de Aragua* (Caracas, 1977), pp. 273–278. The Spanish league used in Caracas was most likely the *legua común* of 5.57 kilometers, judging from the measure of 24 leagues given in the 1578 *relación* as the distance from Caracas to Valencia—a distance of approximately 125 kilo-

ENCOMIENDA, AFRICAN SLAVERY, AND AGRICULTURE 611

The origins of cacao cultivation in this part of the New World are obscure. The plant does not appear in the sixteenth-century *relaciones geográficas* or in any other official documents from the Caracas area before 1607. Most likely for this reason it is commonly assumed that the first cacao trees on the Venezuelan coast were planted there by native and African laborers at the command of Europeans. And yet, from only 0.5 percent of the total value of Caracas's exports in 1607, cacao became the region's foremost item of trade before 1650. It seems unlikely that this boom could have been created on a basis of newly planted trees, especially given the difficult labor situation of the colony. In fact, the earliest testamentary inventories of coastal cacao groves indicate that the first trees to be harvested were native to the region. Labeled *árboles viejos de la tierra*, or simply *de la tierra*, in the documents, these were almost certainly indigenous plants, both because the phrase itself, "of the land," virtually means as much, and because the estimated ages of the inventoried trees shows that many of them were already standing when Hispanic commercialization of cacao began. The Basque Liendo family was among the first to ship Caracas beans to Mexico; their first lot was sent to Veracruz in 1628. A quarter of a century later, in 1653, an inventory of the Liendo estate in the coastal valley of Cepi showed a total of 12,382 trees. Only 1,165 had been planted during the previous twenty-five years, and the ages of these trees, labeled *árboles de Trujillo*, were known exactly. The remainder, 11,217 *árboles de la tierra*, were dated only as "more than twenty-five years old," which means that the Liendo family found them standing in the valley when they arrived at Cepi in the late 1620s. The best proof that cacao was native to the Caracas region is a firsthand observation to that effect. In 1618, Juan de Ibarra, a Basque merchant, sent Portuguese encomendero Diego de Ovalle cloth, wheat, wine, pepper, cinnamon, and copper worth 17,099 reales to be paid for with cacao gathered from the trees that grew wild on Ovalle's Choroní estate. Two years later, in a legal suit brought before the bishop, Ibarra complained that since no one was then interested in the cacao trade, he had been doing Ovalle a favor by offering to take his cacao in exchange for merchandise. Ovalle had refused to pay to have Ibarra's cacao shipped to La Guaira, however, and while the two men argued over who should transport it, the market value of the merchandise fell by half. In his deposition, Ibarra gave evidence that leaves no doubt about the origin of cacao cultivation and commerce:

---

meters. Thus the San Sebastián referred to by Villanueva was located about 160 or 170 kilometers south of Caracas. See Roland Chardon, "The Elusive Spanish League: A Problem of Measurement in Sixteenth-Century New Spain." *HAHR*, 60 (May 1980), 294–302.

Since I am a merchant, there will be no question [when I say] that in this city there was no cacao trade when the said Capitán Diego de Ovalle bought my goods and sold his cacao, and that he had never traded in cacao. [Only] because there came news that cacao was valuable did the vecinos get together and plant it; in other words, when the [Ovalle] sale was made, all there was was wild cacao.[4]

Whether the first large quantities of cacao to be exported from Caracas were harvested from indigenous stands or from groves planted by laborers under the direction of Europeans is a question of more than academic importance. Murdo MacLeod is certain that in the 1620s Caracas and Guayaquil cacao replaced beans from the traditional Central American suppliers, Soconusco and Izalcos, in the New Spain market.[5] Evidently the South American beans were preferred for their flavor, but even with the matter of taste in their favor, it is doubtful that in such a short time Caracas growers, located so much farther from Mexico than their Central American counterparts and with no important labor advantage, could have planted enough trees to capture a significant share of the New Spain cacao trade. A related problem has to do with the very success of this commerce. Once begun, the export of cacao from Caracas expanded at a remarkable rate. The aggregate data compiled by Eduardo Arcila Farías, even though seriously deficient because they do not include the exports of tax-exempt Caracas vecinos, indicate a threefold increase in cacao shipped between the 1630s and the 1650s.[6] This expansion can be

---

4. Archivo del Registro Principal del Distrito Federal, Caracas (hereinafter ARPC), Testamentarías, 1653–55 CL; AGN, Real Hacienda, leg. 14, July 29, 1628. The Ibarra-Ovalle dispute is in Archivo Arquidiocesano de Caracas (hereinafter AAC), Episcopales, Obispo Gonzalo de Angulo. Venezuelan historian Eduardo Arcila Farías suggested that cacao grown on the coast originated in the Andes because cultivation of the plant was recorded in Mérida in 1579; Eduardo Arcila Farías, *Economía colonial de Venezuela* (Mexico City, 1946), p. 88. The designation *árboles de Trujillo* may refer to transplanted Andean cacao, but these trees appear after the initial boom was well under way.

5. Murdo J. MacLeod, *Spanish Central America: A Socio-Economic History, 1520–1720* (Berkeley, 1973), pp. 117, 241, 378.

6. Arcila Farías first published his data on Caracas commerce in 1946 in *Economía colonial*, pp. 96–101. He made no clear reference to the source of his information, but evidently he used the series entitled *Libro común y general de la tesorería*, Real Hacienda, AGN. The tax recorded in these volumes is the *almojarifazgo*, and therein lies a problem: Caracas vecinos, who shipped most of the region's cacao, enjoyed royal exemption from this duty during most of the first half of the seventeenth century. The original cédula granting this favor, dated Apr. 16, 1608, was copied into the cabildo record in 1619; *Actas del Cabildo de Caracas* (hereinafter ACC), 12 vols. (Caracas, 1943–75), IV, 127–128. Arcila Farías was aware of this exemption (*Economía colonial*, pp. 89, 463), but when he compiled his statistics on commerce, he ignored the fact that his source did not include the greater part of all cacao shipped, that which belonged to Caracas vecinos. Due to this error, Arcila Farías's often cited data give the impression that cacao cultivation and trade developed

accounted for in part by new groves planted during these years; we know that much of the work of clearing and irrigating the land was done by African slaves who, in all probability, had been purchased with profits made in the 1620s and 1630s, thanks to encomienda labor at work in virgin cacao *arboledas*. That Caracas growers were able both to sell their beans at a competitive price in the Mexican market and purchase hundreds of slaves with the income from these sales would seem to depend on the existence of substantial groves of indigenous cacao trees.

One clue to this process, which merits much closer examination than is possible here, comes from the record of the Mexican Inquisition. Many of the prosperous Portuguese traders who were accused of being Jews by the Inquisition in the late 1630s and early 1640s had been active in the slave trade and the Caracas-to-Mexico cacao trade.[7] This suggests the following hypothesis: Caracas vecinos exchanged cacao for Africans at a rate that allowed the Portuguese slave traders–cacao merchants to undersell the Central American cacao dealers in New Spain. Eager to sell their fragile human cargoes as soon after the Atlantic crossing as possible, the slavers needed no large margin of profit in Caracas because they stood to make substantial gains in Mexico on the cacao that they took in return for slaves. In this way, Caraqueños both obtained slaves at a moderate price and found an outlet for their cacao.[8]

---

gradually and steadily over the course of the seventeenth century. In fact, exports were more considerable during the thirty years before 1650 than during the thirty years thereafter. The same data are given in Eduardo Arcila Farías, *Comercio entre Venezuela y México en los siglos xvii y xviii* (Mexico City, 1950), pp. 71–73.

7. The close involvement of Portuguese in the commercial life of Caracas may be the reason why both Real Hacienda and cabildo records are missing for most of the decade of the 1640s. The most thorough account of Portuguese Jews who traded Venezuelan cacao in New Spain is by Stanley Mark Hordes, "The Crypto-Jewish Community of New Spain, 1620–1649: A Collective Biography" (Ph.D. Diss., Tulane University, 1980), esp. pp. 81–84, 92, 107–109, 131–132. Also see, J. I. Israel, *Race, Class, and Politics in Colonial Mexico, 1610–1670* (Oxford, 1975), pp. 124–130.

8. In 1638 a Portuguese agent in Angola reported that slave traders were packing ships with 700 and 800 Africans rather than the customary 400, with the result that "at sea it causes the death of many hundreds of them because of the excessive crowding and lack of water." Quoted in Herbert S. Klein, *The Middle Passage: Comparative Studies in the Atlantic Slave Trade* (Princeton, 1978), p. 200 n.47. In these circumstances, Caracas's proximity to Africa and its situation as the first port after the Atlantic crossing where slave cargoes might be absorbed on a regular basis made it welcome sighting for ships' captains and crews. The hypothesis that slaves were used as a medium of exchange to acquire readily sold, highly profitable cacao, and that they were taken in Caracas by individuals who had only limited immediate labor needs (many were encomenderos), but who needed a market for their cacao beans, supports the argument of Brazilian historian Fernando Novais, who would have it that the slave trade created African slavery in the New World, and not the reverse. Fernando Novais, *Estrutura e Dinâmica do Antigo Sistema Colonial (Séculos XVI–XVIII)*, Caderno CEBRAP, no. 17 (São Paulo, 1974). My thanks to Stuart Schwartz for bringing Novais's views to my attention.

Concrete evidence shows that during the infant years of the cacao trade the initial advantage went to a small number of encomenderos with grants located along the Caribbean Coast. Seventeenth-century Caracas encomiendas were of the archaic, *servicio personal* type. Rather than collect a fixed sum, or *tributo,* from their Indians, encomenderos expropriated their labor directly: that is, Indians were put to work gathering cacao beans, planting new groves, or carrying on any other activity considered useful or profitable by the encomendero. Caracas encomenderos insisted that their Indians could not produce anything of value to be used as tribute, and by claiming abject poverty they resisted until the end of the century every attempt by the crown to eliminate *servicio personal.*[9] A royal order demanding conversion to tribute collection was received in Caracas in 1621, but it was not acted upon by Governor Juan de Tribiño Guillamas. When his successor, Gil de la Sierpe, attempted to make law of the king's will, he was arrested by order of the cabildo and sent under guard to Spain.[10] In 1633, in response to this act of lese majesty, the crown called for a new study of the Caracas encomiendas. The data compiled in compliance with this decree provide several clues to the relationship between encomienda labor and cacao cultivation at the very beginning of the trade with Mexico.

This document, dated about 1635, lists four series of facts: the names of the encomenderos and the titles of social distinction (don or doña, *maestre de campo, capitán,* and so forth) of those who were so identified; the geographical location of the encomiendas; the number of Indian tributaries in each grant; and a specific value in pesos defined only as the *renta* of each encomienda.[11] That this value represents the annual income of each grant is confirmed by an existing fragment of the church tithe record. *Diezmos* collected for the region described in treasury documents as *"costa del mar y caraballeda"* totaled 2,600 pesos and 2,979 pesos for the years 1632 and 1634 respectively.[12] These amounts correspond very closely to what would have been the Church's 10 percent share of the

---

9. Eduardo Arcila Farías, *El régimen de la encomienda en Venezuela* (Seville, 1957), chaps. 8, 9.

10. Luis Alberto Sucre, *Gobernadores y capitanes generales de Venezuela,* 2d ed. (Caracas, 1964), pp. 115–118.

11. The ca. 1635 document is in AGN, Fundación de Trujillo. leg. 10, fols. 335–346. The sources listed in notes 49 and 52 *infra* were used to fix a date for this document. It was published, dated tentatively but erroneously as sixteenth century, by Guillermo Morón, *Historia de Venezuela,* 5 vols. (Caracas, 1971), IV, 631–638. In 1609 the local definition of tributary was established as all men between the ages of 12 and 60 and all women between the ages of 10 and 60 inclusive; "Ordenanza de encomiendas de Sancho de Alquiza y de fray Antonio de Alcega de 30 de noviembre de 1609," published in Arcila Farías, *El régimen,* pp. 342–351.

12. AGN, Real Hacienda, leg. 12, fols. 166–167.

ENCOMIENDA, AFRICAN SLAVERY, AND AGRICULTURE

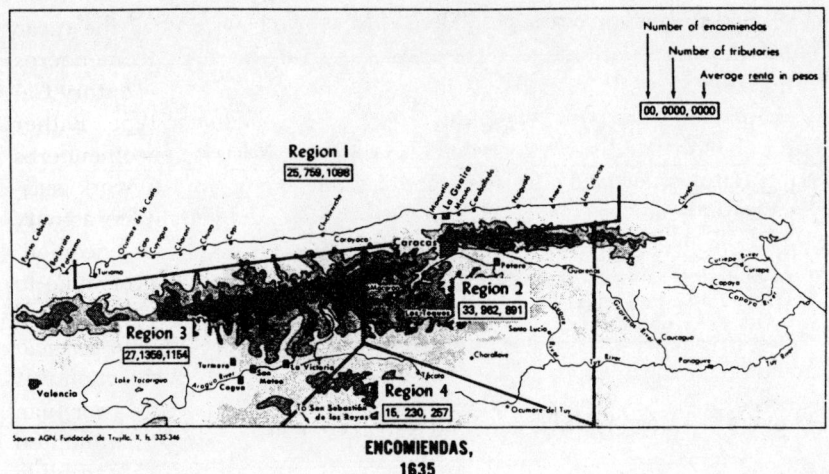

ENCOMIENDAS, 1635

27,450 pesos of total *renta* for the twenty-five encomiendas located along the *"costa del mar de la jurisdicción de Caracas,"* according to the 1635 list (Region 1 in Table IIa and Map).

At that time, 3,328 Indian tributaries were to be found divided unevenly among 100 encomiendas held by vecinos of three Spanish towns: Santiago de León de Caracas, Nueva Valencia del Rey, and San Sebastián de los Reyes. The surveyed area, comprising the legal jurisdictions of this triangle of towns, can be divided into four zones of agricultural production (Table II and Map). Region 1, designated Coastal Caracas, includes all the narrow Venezuelan coastline from present-day Chuspa to Puerto Cabello. Here the conditions are ideal for cacao cultivation. Prevailing winds release moisture accumulated during the Atlantic crossing as they collide with coastal mountains, providing a moderate, but constant, supply of water to the hot, cloud-covered valleys that rise steeply from the sea's edge. One-fourth of the encomiendas counted in 1635 were located in Region 1. In Region 2, Interior Caracas, a few Indians worked in isolated sugar trapiches on the banks of the Guarenas and Guaire rivers, but the majority of native laborers were second- and third-generation wheat farm and flour mill workers. Since the 1580s their toil had served a steady Spanish demand for bread, and in 1635 there were more encomiendas (33) in this temperate climate zone than in any other. Region 3, Interior Valencia, is the long, broad valley west of Caracas that forms at the inland base of the coastal mountains. In this area, described by archaeologist Cornelius Osgood as "probably the most hospitable site

TABLE IIA: Encomiendas ca. 1635 (*renta* is given in pesos).

Region:	(1) Coastal Caracas	(2) Interior Caracas	(3) Interior Valencia	(4) Interior San Sebastián	Province overall
No. encomiendas	25	33	27-	15	100
No. tributaries	759	962	1,359	230	3,310
Average no. of tributaries/encomienda	30	29	50	15	33
Total *renta*	27,450	29,403	31,158	3,855	91,866
Average *renta*/encomienda	1,098	891	1,154	257	908
Average *renta*/tributary	36.6	30.7	23.1	16.7	27.6

TABLE IIB: Encomiendas ca. 1635 with *renta* of at least 1,500 pesos.

Region:	(1) Coastal Caracas	(2) Interior Caracas	(3) Interior Valencia	(4) Interior San Sebastián	Province overall
No. encomiendas with *renta* of at least 1,500 pesos	7	8	8	0	23
No. of these encomiendas held by dons and doñas	1	5	5	0	11

Note: The seven coastal encomiendas (Region 1) may well have belonged to the eight unidentified vecinos described in 1628 as having "considerable wealth in haciendas de cacao." *Actas del Cabildo de Caracas*, 12 vols. (Caracas, 1943–75), VI, 266.

SOURCE: AGN, Fundación de Trujillo, leg. 10, fols. 335–346.

for a hungry population,"[13] the densest preconquest Indian population in central Venezuela had been divided into the province's largest encomiendas a decade before the foundation of Caracas. Thus concentrated, Valencia Indians survived as an ethnic group to become sharecroppers in that area in the nineteenth century.

On the other hand, many of those of the poor and distant San Sebastián district, Region 4, would be removed from their homeland during the early 1600s. Too few and too thinly scattered to be organized into

---

13. Cornelius Osgood, *Excavations at Tocorón, Venezuela* (New Haven, 1943), p. 49. For the literature on pre-Hispanic Venezuelan culture and civilization, see Mario Sanoja and Iraida Vargas, *Antiguas formaciones y modos de producción venezolanos* (Caracas, 1974). Climatic information is from J. Sánchez C. and J. García B., "Regiones meso-climáticas en el centro y oriente de Venezuela," *Agronomía Tropical* (Caracas), 18 (Oct. 1968), 429–439. A basic source for identifying Venezuelan place names is the *Gacetilla de nombres geográficos* (Caracas, 1974); Marco Aurelio Vila, *Antecedentes coloniales de centros poblados de Venezuela* (Caracas, 1978), is also helpful.

ENCOMIENDA, AFRICAN SLAVERY, AND AGRICULTURE

encomiendas, in many cases the most profitable use of the San Sebastián natives was their capture and sale to Spaniards whose agricultural enterprises were more directly integrated with the marketplace. Slave raids were still formally organized and carried out in this frontier zone in the 1640s.[14] The few small encomiendas in San Sebastián in 1635 were employed exclusively in cattle ranching.

The most profitable encomiendas in 1635 were the large grants located in the Valencia-Aragua Valley, Region 3. Three hundred and fifty tributaries, the largest native labor force in the province, worked a sugar trapiche at Turmero and produced an annual *renta* of 4,000 pesos for their encomendero, don Juan Martínez de Villegas. Only because their grants held more tributaries did the Valencia encomenderos receive an annual income greater than that of their counterparts whose Indians were laboring along the coast and in the immediate hinterland of Caracas. In terms of the value of each tributary's labor, average income from Region 3 encomiendas (23.1 pesos per tributary per annum) amounted to only two-thirds that of the coast (36.6 pesos) and three-fourths that of the Caracas region (30.7 pesos) (Table IIa). One of the several possible reasons for the low labor profitability of these larger encomiendas was the high cost of transport from the fertile, but landlocked, Valencia-Aragua zone. Regular commerce across the coastal mountains was impossible except at Caracas, where a break in the range provided relatively easy access to the sea. To reach Caracas from the west, agricultural produce was carried along the banks of first the Aragua and then the Guaire rivers. Except during the dry summer months from March to May, travel was frequently interrupted by floods and was therefore difficult and expensive. Even as late as 1841, carriage costs from the Aragua River to port at La Guaira were equal to shipping charges paid to transport the same commodities from La Guaira to Liverpool.[15]

---

14. The trade between Caracas and San Sebastián was too slight in 1609 to interest Hispanic muleskinners, and an exception to the general prohibition of Indian teamsters was made by Alquiza and Alcega in their ordenanza. So that San Sebastián encomenderos would not keep these drivers at work nearer the coast, the ordenanza required that Indian muleteers return from Caracas within fifteen days; Arcila Farías, *El régimen*, p. 347. At least until mid-century there was no permanent road between Caracas and San Sebastián; the cattle ranchers of the district offered encomendero Luis de Castro 1,000 pesos to open a trail from Paracotos to the Tuy River, midway between the towns; ARPC, Escribanías, Apr. 22, 1649. Evidence of Indian slavery is in AAC, Episcopales, Obispo Mauro de Tovar, 1641–51.

15. Travel in the Aragua Valley is described in Mariano Martí, *Documentos relativos a su visita pastoral de la diócesis de Caracas, 1771–1784*, 7 vols. (Caracas, 1969), II, 286–429. Shipping costs in 1841 are compared in John V. Lombardi, *The Decline and Abolition of Negro Slavery in Venezuela, 1820–1854* (Westport, Conn., 1971), p. 116 n.27.

By 1635 encomienda labor was clearly most profitable in the coastal cacao zone. Coastal encomiendas produced a *renta* of about six pesos more per tributary than those in the traditionally profitable wheat farming areas north and east of Caracas town (36.6 pesos per annum in Region 1 compared to 30.7 pesos in Region 2). With a similar population norm of 29 or 30 tributaries, the annual income from a coastal grant was therefore about 200 pesos, or 20 percent, more than a year's earnings from a typical grant near Caracas (Table IIa).

In itself this difference, something less than the value of one African slave, does not adequately reflect the impact that the cacao bean bonanza had already had on Caracas society by the 1630s. More revealing is the cabildo's lament in 1626 that the lure of greater profits from coastal cacao *arboledas* had left the wheat farm encomenderos without Spanish *labradores*, overseers who were willing to administer Indian labor in return for a share of the harvest.[16]

Cacao created fluidity at more than one level of the colonial society. Evidence from the 1635 document indicates that most of the richest coastal Indian grants were in the hands of recent immigrants to Caracas. Of twenty-three encomiendas with an annual *renta* of 1,500 pesos or more, incomes comparable to those of the most profitable cacao encomiendas in Central America at the middle of the sixteenth century,[17] eight were located in the Valencia-Aragua region, eight near the town of Caracas, and seven on the coast. A measure of the social standing of the holders of these high-income grants can be derived from the document compiler's recognition of some of them as "don" or "doña."[18] In the long-established regions of Valencia and Caracas, nearly two-thirds (10 of 16) of the encomiendas valued at 1,500 pesos or more *renta* belonged to a "don" or "doña" in 1635, but only one of the seven holders of high *renta* coastal encomiendas was distinguished by this suggestion of *hidalguía* (Table IIb). Who were these seventeenth-century parvenus with profitable encomiendas on the inhospitable coastal fringe of Caracas society?

Enough biographical data can be gathered on four of the seven to allow for a composite rendering of these first important cacao producers, the very first of the *grandes cacaos*, as they would be called by subsequent generations. One was Portuguese. There were several Portuguese among the poorer encomenderos, but Diego de Ovalle was the only individual of this nationality whose grant was valued at 1,500 pesos or

---

16. ACC, VI, 147 (Oct. 9, 1626).
17. MacLeod, *Spanish Central America*, p. 117.
18. The actual meaning in the Caracas context of this designation of gentility is not considered here. That the title held significance, although usage varied with time and place, is argued by James Lockhart, *The Men of Cajamarca* (Austin, 1972), pp. 31–33, 111, 208.

ENCOMIENDA, AFRICAN SLAVERY, AND AGRICULTURE

more annual income. Alejandro Blanco Ponte was a Canary Islander: his father was from Tenerife, his mother from Flanders. His brother Pedro Blanco Ponte had been in Caracas since at least 1619 when he declared himself a vecino "without estancia or land" and asked the cabildo for a plot near the La Guaira road where he could pasture his pack mules, but Alejandro did not leave Tenerife until after 1627, the year when the Casa de la Contratación formally prohibited direct trade between the Canaries and the Indies.[19] A third encomendero was Pedro de Liendo, a Vizcayan and the nephew of a retired admiral of the galleon fleet. Pedro and his brothers Santiago and Domingo de Liendo, who were not encomenderos, were perhaps Caracas's most ambitious cacao planters during the first half of the seventeenth century. The fourth grantholder for whom biographical information has been found, don José Rengifo Pimentel, was born in Santo Domingo, where his father was captain of the port and his grandfather had been a royal treasury accountant. The only encomendero on the coast whose rights to Indian labor earned him 1,500 pesos or more annually and who was also classified as a "don" on the 1635 list, Rengifo had in fact four encomiendas, three on the coast and one on the Guarenas River, east of Caracas. With a total of two hundred tributaries and an annual income from them of 6,250 pesos, don José was Caracas's foremost encomendero.[20]

The first of these men to settle in Caracas was Diego de Ovalle. In 1602 he married the youngest daughter of don Juan Vásquez de Rojas, a descendant of Caracas conquistadors in both maternal and paternal lines, and received an encomienda at Choroní as part of his wife's dowry. Ovalle was born in the Vila do Mojadoiro on the upper Douro River in the Portuguese province of Tras-os-Montes. There his family had vineyards and fruit orchards, the produce of which had been sent traditionally to Medina del Campo and Valladolid. With central Spain in depression at the end of the sixteenth century, Tras-os-Montes became part of the economic system of Seville, and there Ovalle established commercial relations that he maintained throughout his life. These connections, and

---

19. Ovalle's will is in ARPC, Testamentarías, 1650–53 *sin letra*. The Blanco Ponte family is traced in Carlos Iturriza Guillén, *Algunas familias caraqueñas*, 2 vols. (Caracas, 1967), I, 161–180. The first reference to the Blanco Ponte family in the Caracas documentation is ACC, leg. 4, fol. 310 (Sept. 9, 1619). Trade between the Canaries and the Indies is explored in detail in Huguette and Pierre Chaunu, *Séville et l'Atlantique (1504–1650)*, 8 vols. (Paris, 1955–58), VIII, pt. 1, 424–430.

20. Pedro de Liendo's will is in Universidad Central de Venezuela, *La obra pía de Chuao, 1568–1825* (Caracas, 1968), pp. 190–194. The Liendos and don José Rengifo Pimentel are also included in Iturriza's genealogical study, *Familias caraqueñas*, II, 451–452, 726. Multiple holdings of encomiendas were not unusual in seventeenth-century Caracas; Arcila Farías, *El régimen*, pp. 170–172.

those he evidently had with the Portuguese slavers who monopolized the supply of African labor to the Caribbean, were no doubt of some interest in Caracas. In 1607, the slave factor and six of the town's encomenderos, including Ovalle, were from Portugal.[21]

He was also a shrewd trader. In 1618 Ovalle sold one of the first lots of cacao ever to leave the Caracas coast to the Basque merchant Juan de Ibarra. Ibarra sent wheat flour and dry goods to Choroní and waited for his cacao to arrive at the La Guaira wharf. These were the infant years of cacao commerce, however, and this was Ibarra's first purchase. In the future, contracts would specify the location of delivery, either *"puesto puerto La Guayra"* or *"puesto costa del mar,"* but on this occasion Ovalle, who had not made arrangements to do so, refused to ship the beans at his cost to La Guaira. Ibarra, having already given merchandise for the much-wanted cacao, had no choice but to send for it and pay the transportation charges himself.[22]

Ovalle made a good deal of money selling cacao, and he used his wealth to protect himself from the problems that could arise on account of his Portuguese nationality. An appeal for funds by the crown or the cabildo would find him by far the most generous contributor.[23] Unlike others of his economic position, however, he did not acquire municipal office, and neither he nor his wife owned town property. They were permanent residents at Choroní, and it was there that they made their principal investment: African slaves. This was an investment against any misfortune and one that those who survived the prolonged depression of the 1650s and 1660s seemed to have made freely while cacao sales were brisk and profitable. To the 1602 encomienda work force of thirty-six Indian tributaries, Ovalle by 1626 had added fifty slaves. This number had risen to eighty-four in 1644, and by 1650, the year of his death, Ovalle was master of ninety-four slaves.[24]

The inventory of Ovalle's Choroní estate provides a glimpse of the material life constructed from the economic life of Indian and African

---

21. ARPC, Testamentarías, 1650–53 *sin letra*; Ovalle's wife's dowry, 12,565 pesos of 10 reales, is in ARPC, Escribanías, Feb. 21, 1602. His brother Antonio was corregidor in Santo Domingo; ARPC, Escribanías, Nov. 12, 1637. In the absence of other heirs, Diego's nephew Juan inherited the valuable Choroní estate, much to the disgust of Diego's wife's Caracas family. Ovalle heads the list of foreigners in "Relación de los estrangeros [Apr. 12, 1607]," Archivo General de las Indias, Seville (hereinafter AGI), Santo Domingo, leg. 193.
22. AAC, Episcopales, Obispo Gonzalo de Angulo.
23. His donation of 1,000 reales to aid the expedition against the Indian rebels in Nirgua, at a time when few benefits were to be had from such military activity, could not have been overlooked by his Spanish peers, whose contributions were 10, 25, or 40 reales; ACC, V, 389.
24. ARPC, Testamentarías, 1650–53 *sin letra*.

ENCOMIENDA, AFRICAN SLAVERY, AND AGRICULTURE

labor and fifty years of vigorous trading. No doubt primitive by Mexico City or Cartagena standards, the principal house at Choroní was comparable to any Caracas dwelling at that time. It was a simply furnished, two-story stone structure with a tile roof. Two gilded mirrors, a cedar table and chairs, and a silver service for six were the items of greatest value. A dozen books, including Antonio de Herrera's *Historia de las Indias*, occupied a shelf.[25] The upstairs bedroom contained just the touch of luxury in a style already common to Caracas cacao exporters: two carved cedar chests imported from New Spain and a four-poster bed covered with a Mexican quilt and pillows of Chinese silk reflected the importance of the market where Ovalle had made his modest fortune.[26]

Close by were the other buildings used for maintaining the estate and preparing cacao for shipment. A huge wooden storehouse doubled in the upper loft as a drying shed. Carpentry tools and a forge, both used by African artisans, were kept in a second structure. The slaves' quarters, because they represented no value to Ovalle, were not included in the inventory of his holdings, but the livestock and agricultural implements used to produce food to feed them were counted: seven milk cows, a hundred head of cattle, six teams of oxen with harnesses, several plows, hoes, and a variety of other tools filled out the list. In all, the material wealth of Choroní was predominantly utilitarian. The absence of expensive consumer goods may well have been the consequence of Ovalle's considerable investment in slaves. At 1650s prices, the Choroní Blacks were worth no less than 24,000 pesos.[27]

The last will and testament of a second cacao encomendero, Pedro de Liendo, has also survived. Like Ovalle, Liendo's principal expenditure was for African slaves. In 1635, his encomienda at Chuao was made up of thirty-five tributaries. In 1659, in addition to the estate house and tools, he owned a warehouse and two stores in La Guaira, a sloop to carry his cacao there, and 106 slaves.[28] For their part, the Blanco Ponte family participated directly in the slave trade; as buyers and brokers their credit was respected from Lisbon to Loanda to Cartagena. In May 1640, Pedro Blanco Ponte bought forty-eight slaves from Francisco de Aquilena, a vecino of Cartagena and agent to don Francisco de Vascon-

---

25. Antonio de Herrera's *Historia* was a common item in colonial Venezuelan libraries; see Ildefonso Leal, *Libros y bibliotecas en Venezuela colonial*, 2 vols. (Caracas, 1978), I, cix–cx. Ovalle's small collection does not appear in Leal's extensive inventory.

26. ARPC, Testamentarías, 1650–53 *sin letra*.

27. No peso value was assigned to the slaves on the Ovalle estate in the inventory of 1653, but the average value of slaves recorded in other inventories taken at about the same time was 250 pesos or more; ARPC, Testamentarías, 1648 RU; 1653–55 CL.

28. *La obra pía de Chuao*, pp. 191–194.

celo, governor of Angola, on credit at the bargain price of 160 pesos each (the current rate was about 250 pesos).[29] There is no record of slaves held or sold by don Rengifo Pimentel.

A Portuguese, a Basque, a Canary Islander, and a creole born in Santo Domingo—these were all men of seaborne and commercial experience. At the same time, they were not dependent on Seville and the traditional flota system, then in decline, which had provided a demand for Caracas wheat since the 1580s. Their entrepreneurial talents and trading contacts made them first welcome in Caracas and then part of its propertied elite. They all married well in that soon after their marriages they obtained encomiendas on the basis of the station and merits of their Caracas-born wives' families. Don Rengifo Pimentel married the Caracas cattle heiress doña Francisca Gámez and in her name held the four encomiendas already mentioned, including two on the coast worth 1,000 and 3,000 pesos per annum. Pedro de Liendo's wife was the wealthy doña Catalina Mexía. The granddaughter of Mérida (Venezuela) governor Suárez del Castillo, at her death—having outlived Liendo and two subsequent husbands—doña Catalina founded an *obra pía* on the encomienda and cacao groves at Chuao. While the encomienda was worth 1,600 pesos annually during the period of Liendo's possession, in time this estate became the most valuable ecclesiastical property in colonial Caracas. Alejandro Blanco Ponte's marriage to doña Francisca Infante, the granddaughter of town founders, brought him an encomienda at Caraballeda worth 1,500 pesos per annum. In 1632, the year of his sister's wedding, don Francisco Infante transferred a debt to the royal treasury of 5,403 pesos to his new brother-in-law, Alejandro Blanco Ponte, who immediately paid one-third of the outstanding amount in cash. In 1634, having already replaced Infante as encomendero at Caraballeda, Blanco Ponte received a 1,000-peso loan in Caracas guaranteed by 5,000 cacao trees in the coastal valley.[30] Perhaps entry into the Caracas elite in the 1630s required nothing more than one or two thousand pesos in gold coin. For the established families that provided these ambitious immigrants with wives, however, the acquisition of large numbers of African slaves assured them of a continuing return on their domestic investment. This is exactly what Ovalle, Liendo, and Blanco Ponte were able to provide.

They were not the first to bring Africans to Caracas. From early in

---

29. The total sale was for 7,693 pesos; ARPC, Escribanías, Sept. 21, 24, 1640.
30. The genealogies of these women's families are traced in Iturriza, *Familias caraqueñas*. The 1,000 pesos were loaned by the Rodríguez Santos family, wheat farmers and merchants; ARPC, Testamentarías, R, fols. 457–459.

ENCOMIENDA, AFRICAN SLAVERY, AND AGRICULTURE

the century, encomenderos had invested in slaves. Before 1628 the encomendero-merchant Alonso Rodríguez Santos had forty-seven slaves who farmed his wheat and herded his cattle. Francisco Castillo, encomendero and local tax collector for the Cartagena-based Inquisition court, directed a mixed-labor enterprise that was a variation on the common theme; the wheat grown by fifty African slaves in the Caracas valley was shipped to Cartagena in cotton bags woven by the Indians of his Valencia encomienda.[31] In the 1630s and 1640s, cacao sales gave a significant boost to the quantity of slaves that Caraqueños could purchase, and although the cacao encomenderos' Indians afforded them the means to obtain the largest slave holdings, other ways were found to form a slave gang. A considerable variety of these cases could be cited, but two examples will suffice. Baltasar de Escovedo came from Granada to Caracas expecting to inherit his uncle's estate, but when the elder Escovedo decided that there was much to be gained by the marriage of his daughter to the powerful Juan Rodríguez Santos, what Baltasar had hoped would be his became his cousin's dowry instead. He turned to making bricks and tiles for the houses of those more directly favored by cacao, and in time he was able to accumulate, in addition to the brick works, a respectable town house, a dozen slaves, and a small fortune in silver and gold coin. The Portuguese Agustín Pereira came to Caracas with a few slaves that he sold to buy two houses in the center of town. With this property as collateral, Pereira worked as a middleman, buying cacao with borrowed money guaranteed in Caracas and selling it for a profit at La Guaira. In 1684 his heirs owned a small cacao grove in the coastal valley of Osma.[32]

The slave trade and these secondary economic activities depended on cacao sales that most often took place at the gate or the wharf of the estate. Until about 1650 profits from these sales came particularly easily, for the greater part of the beans was collected from the virgin trees known as *árboles de la tierra*. The difficult labor of clearing the land and providing shade and irrigation for the young plants was not yet necessary, and there was no need to wait the five years or more after planting that it took for cacao trees to reach fruit-bearing maturity. A precise idea of the profitability of these groves is provided by accounts kept for the *arboledas* of Domingo de Liendo, brother of encomendero Pedro de Liendo.

---

31. Alonso Rodríguez Santos's will is in ARPC, Testamentarías, 1648 RU. Francisco Castillo de Consuegra's will is in ARPC, Testamentarías, 1614–34 CEFMSU.

32. Baltasar de Escovedo, ARPC, Testamentarías, 1634–37 MDPACVG; Agustín Pereira, ARPC, Testamentarías, 1656–57 *sin letra*; ARPC, Escribanías, Jan. 21 and Dec. 23, 1630, May 3 and Oct. 3, 1634. The property owned by Pereira's heirs in 1684 is listed in the document identified in n.48 *infra*.

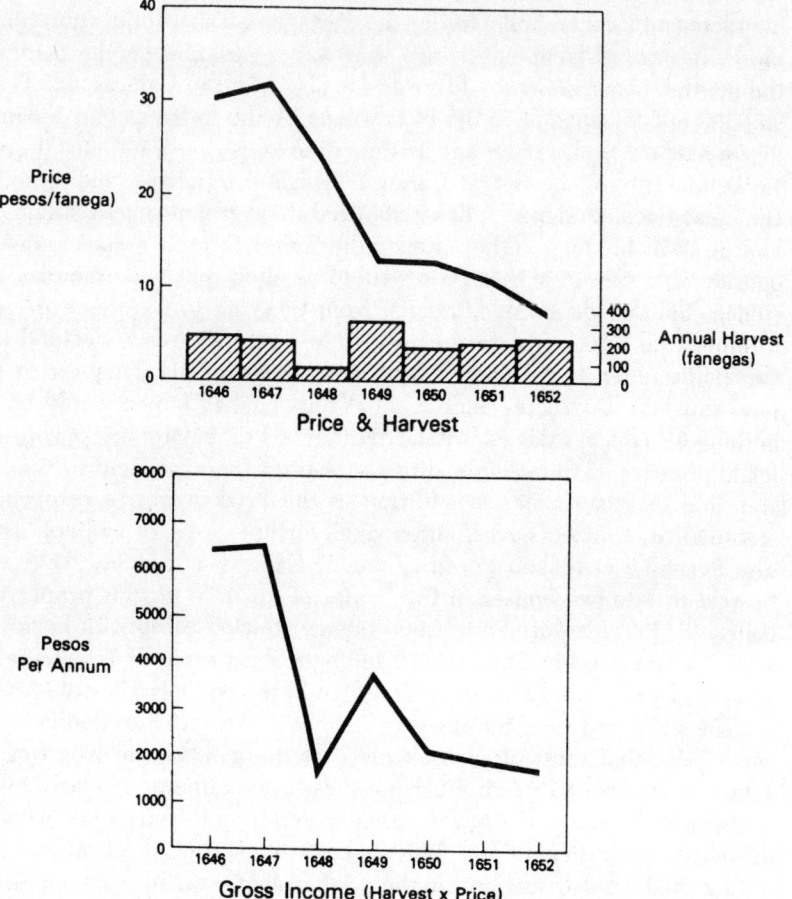

FIGURE 1

SOURCE: ARPC, Testamentarías, 1653–1655 CL.

More than 90 percent of Liendo's cacao in the coastal valley of Cepi was *de la tierra*. Since about 1640, his African slaves had planted some cacao trees in the valley, but in 1653 only 1,165 trees of a total of more than 12,000 were not indigenous to the area. More than half of all the trees were then stricken with a blight known as *alhorra*, but the Cepi groves had yielded more than 200 fanegas (a fanega of cacao weighed about 110 pounds) annually for five of the seven years before 1653. Although harvests varied somewhat from one year to the next, and prices were falling steadily, Liendo's gross income averaged nearly 3,500 pesos per year during the period 1645–52 (Figure 1).

That fishhooks were imported to Cepi and small amounts of maize were exported from the valley suggests that food costs were low or nil and that the plantation was largely self-sufficient. Since cacao beans were sold directly from the groves, there were no freight or transportation charges. The value of the Cepi land in 1650 is unknown, but the half of the valley not owned by Liendo was sold for 1,000 pesos in 1630 by someone who had no slaves to someone who did. The thirty-two slaves who formed the Cepi work force were worth a total of 8,550 pesos in 1658.[33]

Thus the cacao groves and slaves at Cepi represented an investment of probably no more than 10,000 pesos, and, therefore, it seems reasonable that in an average year Liendo's gross earnings equaled about 35 percent of the value of his capital, with each adult slave annually earning, on the average, about 40 percent of his or her market value. Considering in addition that many groves were first harvested by "cheap" encomienda labor and not slaves, and that sales were made at high prices from the late 1620s until about 1650, the dynamic profitability of Caracas's first cacao bonanza becomes clear.[34]

At mid-century there coincided at Caracas several events that brought this period of remarkable prosperity to an abrupt end. The problems of supply at this critical juncture are much more evident than those that caused the price paid for cacao in Caracas to drop to one-fifth of its customary level (Figure 2). A principal factor in the crisis was the blight known as *alhorra*. Beginning in the late 1630s on the plantations upwind of La Guaira, within a decade the *alhorra* had destroyed more than half of all the cacao trees on the coast, leaving many groves without a single fruit-bearing plant. Newly planted trees would begin to reach maturity by 1660, but the coastal groves would not recover their original productivity until the eighteenth century.[35] A second disaster took the form of a surprisingly destructive earthquake, which reduced modest Caracas to rubble in 1641. At one blow much of the urban property that had been used to finance the purchase of slaves and cacao was eliminated. The ecclesiastical establishment, its temples and convents in ruins and the

33. ARPC, Escribanías, Jan. 15, 1630; ARPC, Testamentarías, 1653–55 CL.
34. ARPC, Testamentarías, 1656–57 *sin letra*, will of Elvira de Campos, states that 1,800 fanegas of cacao worth an estimated 50,000 pesos were harvested from the 22,000-tree coastal estate of Juan Navarro during an unspecified number of years before 1637.
35. Ibid. The Navarro groves were completely destroyed by the *alhorra* blight. See n.39 *infra*.

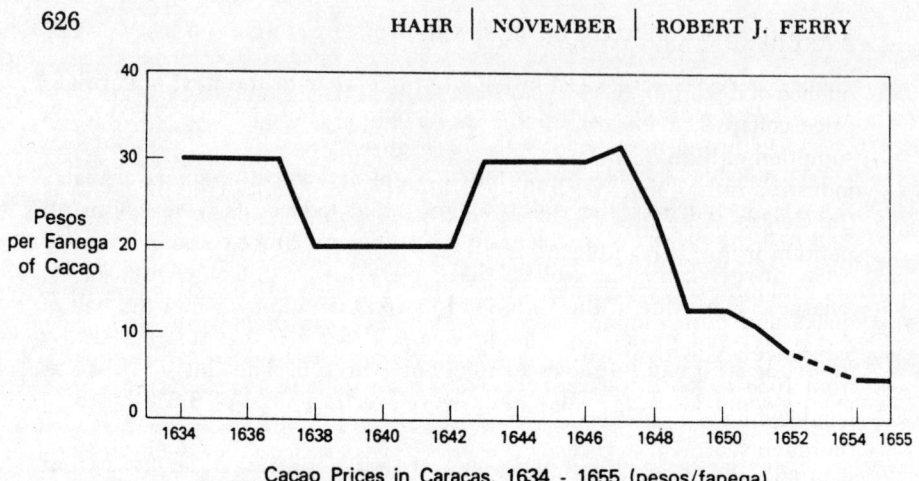

Cacao Prices in Caracas, 1634 - 1655 (pesos/fanega).

FIGURE 2

SOURCE: ARPC, Testamentarías, 1646–1648 MLJERC, will of Rodrigo Gallegos (years 1634–1642); Testamentarías, 1653–1655 CL, will of Domingo de Liendo (years 1643–1652); Testamentarías, 1656 CLR, will of Blas Correa (years 1654–1655).

mortgages it held rendered worthless by the destruction, suffered the most. Exacerbated by the truculence of Bishop Mauro de Tovar, who refused the vecinos' request to move the town (his approval would have nullified all claims held by the Church against the owners of the destroyed property), these reverses initiated a rancorous epoch marked by spectacular quarrels between the bishop and many of Caracas's first citizens.[36]

As yet there is no clear explanation why cacao prices, at twenty and thirty pesos the fanega until 1647, collapsed to five pesos by 1654. The answer will come from a fuller understanding of the Mexican market than is possible from the vantage point of Caracas. MacLeod documents falling prices for cacao in Santiago de Guatemala after 1651 and attributes the decline to successful competition from Guayaquil and Venezuela: "In all the price history of cacao in Middle America, the main factor seems to have been the question of supply."[37] If this was indeed the case, the high prices paid for cacao in the Mexican highlands and elsewhere (including Caracas) until the late 1640s may have been in part the conse-

---

36. The best description of the earthquake is Bishop Mauro de Tovar to the King, Aug. 14, 1641, AGI, Santo Domingo, leg. 218. The bishop's colorful career is described in Andrés F. Ponte, *Fray Mauro de Tovar* (Caracas, 1945), and in Manuel Guillermo Díaz, *El agresivo obispado caraqueño de don Fray Mauro de Tovar* (Caracas, 1956).

37. MacLeod, *Spanish Central America*, p. 251.

ENCOMIENDA, AFRICAN SLAVERY, AND AGRICULTURE 627

quence of declining production from Caracas because of the *alhorra*. The price collapse in Caracas at that time could then be linked to the resumption of high output in Guayaquil after 1640.[38] The diseased Caracas industry, rather than contributing to the seventeenth-century crisis of the Central American cacao economy, was like it overwhelmed by competition from Guayaquil.

Undoubtedly much impressed by the combined disasters of earthquake and crop blight, Caraqueños appealed to divine providence for assistance. A festival with the Virgin Mary as patron was held annually from 1638 to 1670 to plead for relief from the *alhorra*.[39] On the other hand, while the documents abound with references to these events, a thorough search uncovered no comment whatsoever about the low prices paid for cacao. What little evidence pertaining to Caracas's commercial woes exists suggests that cacao growers did not understand their economic problems in terms of competition from other suppliers. With silver production in decline, and forced to depend on the Mexican peso by the crown's decision to devalue the debased Peruvian currency, would-be buyers in New Spain preferred to hoard their sound coins rather than let them go out to distant markets such as Venezuela from whence they might not circulate back. From 1650 to the 1670s a general crisis in the supply of currency choked trade in the Spanish Caribbean.[40] To aggravate matters further, from 1620 to 1650 the Mexican Inquisition tried more than 200 Portuguese for practicing Judaism, many of them cacao traders who had been the primary buyers of Caracas cacao.[41] By eliminating these middlemen and reducing the opportunity to barter cacao for Africans, Caracas growers were forced to depend on cash sales to Mexican merchants who were generally unwilling to make such purchases. The growers were aware of the currency shortage and, even as the price dropped to five pesos the fanega, of the lack of buyers for their cacao beans. In 1649, Santiago de Liendo paid the priest at Cepi in devalued copper currency, the unwanted *vellón*, because "there was nothing else." In 1654, for the first time cacao had to be brought at Liendo's cost from Cepi to La Guaira, "to sell it and give it a market because there was none in Cepi and the cacao was about to be lost, rotting in the storehouse."[42]

38. Ibid., pp. 242–244.
39. Archivo del Consejo Municipal del Distrito Federal, Caracas (hereinafter cited as ACM), Actas del cabildo, Originales, 1669–72, Aug. 6, 1670.
40. MacLeod, *Spanish Central America*, pp. 280–287.
41. Millions of pesos were confiscated by the inquisitors; Hordes, "The Crypto-Jewish Community," p. 153.
42. ARPC, Testamentarías, 1653–55 CL.

The better to weather this combination of events, Caracas residents withdrew to the countryside. Urban functions and services came to a standstill. For years the cabildo was without its full complement of regidores and the governors' secretaries had to double as *escribanos públicos*.[43] By the 1670s the preoccupations of property and status had returned to give renewed importance to urban life. Plans to construct a new seminary were presented and discussed in the cabildo. Of more dubious benefit, but a clear indication of the reemerging social order, the jail, destroyed in the 1641 earthquake, was rebuilt in 1674. Frightened by marauding French corsairs and by raids on Maracaibo by the Englishman William Henry Morgan, townsmen expressed faith that Caracas was once again worth protecting and weighed the possibilities of persuading the crown to finance a new fort to be built on the road to La Guaira.[44]

It is reasonable to suppose that the opportunistic Dutch, then the Atlantic's leading merchants, provided planters with an alternative outlet for their cacao during the years of sluggish legal commerce. Indeed, fewer than half of all the men designated for the Caracas militia, encomenderos and planters among them, responded in 1639 when an expedition was formed to dislodge the Hollanders from the nearby island of Curaçao.[45] What took place along the miles of open coast is as difficult to account for now as it was to control then, but the Dutch were no doubt quite willing to replace the Portuguese in the lucrative slaves-for-cacao exchange. By 1671, the role of Curaçao in the supply of slaves was so commonplace that the Caracas cabildo issued a matter-of-fact warning that slaves afflicted with the "Dutch evil" (*mal 'olanda*) were entering from the island.[46] Yet it may be that Curaçao did not immediately become "little more than a gigantic slave pen" after the Dutch occupation, as one historian recently suggested.[47] Between 1659 and 1671, the slave population at Chuao underwent changes in demographic composition that suggest that few, if any, African slaves had been introduced to that coastal

---

43. New appointments to the cabildo are in ACM, Actas del cabildo, Originales, 1673–76, Feb. 6, May 16, 1675.

44. The construction of the new seminary is in ACM, Actas del cabildo, Originales, 1673–76, Oct. 25, 1673; the jail is mentioned on Sept. 2, 1674; and the fort was discussed in the sessions of Nov. 16, 27, and Dec. 1, 1673.

45. Lucas Guillermo Castillo Lara, *Las acciones militares del gobernador Ruy Fernández de Fuenmayor (1637–1644)* (Caracas, 1978), pp. 35, 52–58.

46. ACM, Actas del cabildo, Originales, 1669–72, May 21, June 20, 1671. If the *Mal 'olanda* were the same disease later described as the *mal de Loanda*, then it probably meant scurvy, an infirmity associated with overloading and undersupplying the slave ships. See Joseph C. Miller, "Mortality in the Atlantic Slave Trade: Statistical Evidence on Causality," *The Journal of Interdisciplinary History*, 11 (Winter 1981), 412–413.

47. MacLeod, *Spanish Central America*, p. 363.

valley. The overall population declined from 110 to 100 individuals, and of those who can be identified by place of birth with certainty (76 percent in 1659; 79 percent in 1671), the number of African-born dropped from 36 to 26. A sharp decrease in the number of single men, from 30 to 11, and a significant rise in married or widowed men and women, from 20 to 27 men and from 26 to 34 women, are further indicators of adjustment to a reduced slave supply. More conjugal pairs and a greater sexual balance in 1671 occurred as single men found creole wives. The majority (14 of 27) of married and widowed men in 1671 were African-born, but the majority of women of the same status (17 of 34) were creoles. At least in the Chuao case, the immediate presence of Dutch traders was unimportant and, from the point of view of the slave community, the stagnant African trade had beneficial demographic effects (Table III).

By the 1680s the Caracas economy was vigorous once again. In 1684, the crown ordered a survey of rural property to determine whether non-collection of the alcabala, which had been suspended by royal favor since 1631, was still justified. Acknowledged by contemporaries as the first such census of the town's rural domain, the 1684 *padrón* is a crucial document for the history of seventeenth-century Caracas.[48] Caraqueños owned 373,250 cacao trees distributed on 167 estates, 18 wheat farms, 26 sugar trapiches, and 28 cattle *hatos* in 1684. Wealth accumulated during the first years of booming commerce made it possible for the families of many of the first encomendero-slaveholders to survive the several decades of economic slump. A close study of the 146 owners of cacao groves listed in the 1684 *padrón* reveals that at least half were descendants of encomenderos named in 1635, and of the 38 encomenderos who had then been holders of grants of 1,000 pesos or more *renta*, the heirs of at least 28 were growing cacao in 1684.[49] The moderate size of many of the cacao groves belonging to children and grandchildren of the first planters, and the location of these *arboledas* in sites different from their ancestors' encomiendas, suggests that the majority of the 1684

---

48. This document is in AGI, Contaduría, leg. 1613; it was published in *Revista de Historia* (Caracas), 28 (Aug. 1970), 63–81. The auction of the alcabala was held on Sept. 25, 1673, and again on June 5, 1675, but no one was willing to bid. ACM, Actas del cabildo, Originales, 1673–76.

49. Only four of the thirty-eight encomenderos who held encomiendas with a *renta* of 1,000 pesos or more have not been identified by place of birth and with a descending kindred network. The genealogical studies used are those by Carlos Iturriza Guillén, *Algunas familias caraqueñas*, already cited, and *Algunas familias valencianas* (Caracas, 1955). José Antonio de Sangroniz y Castro, *Familias coloniales de Venezuela* (Caracas, 1943) is serviceable, but lacks the detail and completeness of Iturriza's work. The marriage registry for the Cathedral parish has been published by the Instituto Venezolano de Genealogía, *Matrimonios y velaciones de españoles y criollos blancos celebrados en la catedral de Caracas desde 1615 hasta 1831* (Caracas, 1974).

TABLE III: Demography and Regional Origins of Slaves at Chuao, 1659 and 1671.

	1659		1671	
	No.	% Total	No.	% Total
Married and widowed men				
West Africa	4		4	
Central Africa	9		10	
Creoles	3		9	
Unknown	4		4	
Total	20	18	27	27
Married and widowed women				
West Africa	2		0	
Central Africa	9		10	
Creoles	2		17	
Unknown	13		7	
Total	26	24	34	34
Single men				
West Africa	6		1	
Central Africa	5		0	
Creoles	12		7	
Unknown	7		3	
Total	30	27	11	11
Single women				
West Africa	0		1	
Central Africa	1		0	
Creoles	0		0	
Unknown	1		1	
Total	2	2	2	2
Children under 12	32	29	26	26
Total slaves	110	100	100	100

SOURCE: *La obra pía de Chuao*, pp. 191–193, 227–229. My thanks to Professors George Brooks and Phyllis Martin of Indiana University for their assistance in solving problems of nomenclature and establishing the ethnicity of these slaves.

groves had been planted since the 1640s. Started during the period of *alhorra* blight and low prices, this was expansion that already purchased slaves could execute. Unfortunately, the only estimate of the slave labor force is a figure by Bishop González de Acuña, who claimed that there were 16,000 slaves in the Caracas region in 1674.[50] That more than 10 percent of the cacao trees counted in 1684 were new plants located along the Tuy River south and east of Caracas, however, is proof that there had

---

50. Bishop Antonio González de Acuña to the King, June 15, 1675, in Guillermo Figuera, ed., *Documentos para la historia de la iglesia colonial en Venezuela*, 2 vols. (Caracas, 1965, 1967), II, 101–104.

ENCOMIENDA, AFRICAN SLAVERY, AND AGRICULTURE

been labor enough to increase production. Once planted, an *arboleda* required only a maintenance crew of waterers and weeders during most of the year, and this meant that a gang of the younger, sturdier slaves could be kept constantly at work bringing new land under cultivation. Such were the profits from cacao that favored planters came to have three sets of slaves: in addition to their domestic servants, they had a caretaker crew and an expansion crew. In 1658, the heirs of Domingo de Liendo had 24 men and boys and 8 women and girls on the original coastal estate at Cepi, 7 men and boys and 7 women and girls in domestic service in Caracas, and 8 men and 3 women at work planting cacao trees at Santa Lucía in the Tuy Valley.[51]

Here we have perhaps the most salient characteristics of colonial Caracas economy and society: the ready availability of well-watered land and the mobility of the slave gang encouraged constant expansion. In 1720, before the establishment of the Guipuzcoana Company in Venezuela, there were more than two million cacao trees in the Caracas province, and by 1744 the number of trees had risen to over five million; in each case more than half of the trees were located in the Tuy River region. The multitude of small holdings that resulted, brought and held together by a complex system of carefully arranged marriages, kept elite status and profits in the hands of generations of select Caracas families.[52] Even the humbler growers, indigent Canary Islanders for the most part, who came to Caracas in a steady stream after 1680,[53] could make a bid

---

51. ARPC, Testamentarías, 1653–55 CL. The average age and cash value of the Liendo slaves sixteen years old and older are as follows:

	Cepi		Caracas (domestic)		Sta. Lucía	
	Males	Females	Males	Females	Males	Females
Avg. Age	39	33	27	29	35	30
Avg. Value (pesos)	258	294	343	293	369	350

52. The strategies of kinship and their relationship to elite status and property are studied in Robert James Ferry, "Cacao and Kindred: Transformations of Economy and Society in Colonial Caracas" (Ph.D. Diss., University of Minnesota, 1980), chap. 4. *Capellanías* were not used as capital depositories for leading families in Caracas as they were in other regions of Spanish America. See Marta Espejo-Ponce Hunt, "The Process of the Development of Yucatán, 1600–1700" in Ida Altman and James Lockhart, eds., *The Provinces of Early Mexico* (Los Angeles, 1976), p. 38. For Caraqueños, African slaves were much preferred as a form of investment and accumulation of capital, a fact that helps explain why there was little planter indebtedness to the Church in colonial Venezuela; see Ermila Troconis de Veracochea, *Las obras pías en la iglesia colonial venezolana* (Caracas, 1971).

53. Leopoldo de la Rosa, "La emigración canaria a Venezuela en los siglos xvii y xviii," *Anuario de Estudios Atlánticos* (Tenerife), 20 (1976), 617–631.

for gentry status with cacao and the few slaves they were able to buy. Before 1700 these hopeful immigrants had begun to harvest quantities of cacao in the valleys at the remote edge of the province. They also settled in Caracas town, in the Candelaria parish on the settlement's eastern fringe, which in time became a Canario barrio. This was a labor force of a somewhat different order, although as artisans, grocers, vegetable farmers, overseers, and, perhaps more than any other occupation, muleteers who served the increasingly distant *arboledas*, they were as closely dependent on the production and sale of cacao as the estate owners who bought their wares and hired their services.[54]

Before 1700 the population structure of the Caracas region had become a racial-class continuum, with white masters and Black slaves at opposite poles and Canarios, other poor Spaniards, and the mixed-blood pardos and mulattoes scattered in between. In this world that his forced labor had helped create, the Indian seems to have had no clearly defined place. At best, it was an uncertain and unstable order. With a growing slave population and at a time when several decades of economic hardship had ended and a new, poor white class had begun to fill the town and countryside, Caracas seems to have become quite race conscious. The marriage of a Valencia favorite son to the daughter of an Aragua Valley cacique provoked near hysteria in the Hispanic community.[55] In an effort to freeze melting social-racial distinctions, Caracas Bishop González de Acuña declared that he would no longer ordain anyone who had as much as one-fourth Indian or African ancestry—a challenge to the Church's evangelical mission that was quickly disallowed by Pope Innocent XI.[56]

In this tense atmosphere, the crown's decision in 1691 to insist on the termination of the *servicio personal* encomienda and thereby put an end to more than 150 years of de facto defiance on the part of the Caracas encomenderos met surprisingly little resistance. Definitely not the outcome of easy mestizaje and racial toleration, the encomienda's demise was finally allowed because the immobile, legally circumscribed institution had little place in the dynamic cacao economy now based on African slavery. Once freed from encomienda service, it was decided that Indians would be required to pay a tax for the religious instruction that they were to continue to receive. Bishop Baños y Sotomayor's *Ynforme* of 1690, on which the tax was to be based, allows a glimpse at the remnant of Indian

---

54. The concentration of Canarios in the Candelaria parish is reflected in the "Matrículas de las parroquias de Caracas y demás pueblos de su diócesis, 1759," a bound manuscript located in the Biblioteca Nacional, Caracas.
55. Castillo Lara, *Materiales para la historia de Aragua*, pp. 240–244.
56. Guillermo Figuera, ed., *Documentos para la historia de la iglesia*, II, 119–120.

ENCOMIENDA, AFRICAN SLAVERY, AND AGRICULTURE 633

society that was then very much at the margin of Caracas's Hispano-African community.[57]

The bishop's agents found 7,464 Indians on 64 encomiendas in the jurisdictions of Caracas, Valencia, and San Sebastián.[58] Of these the majority, 5,278 Indians and 37 encomiendas, were in the possession of Caracas vecinos. Comparison with the agricultural property owned by the encomenderos who held these 37 grants—as the 1684 *padrón* permits—makes it possible to account for the specific activities that employed Indian labor at the end of the century. More than three-fifths, 3,262, were in 9 encomiendas located near sugar mills owned by their encomenderos; 1,085 Indians in 11 encomiendas were at work farming wheat; and 931 were spread thinly among 17 encomenderos whose grants were located near their coastal cacao groves. More revealing is the portion of all agricultural estates that benefited from encomienda labor; while a third of the sugar trapiches and fully two-thirds of the wheat farms were owned by encomenderos, only about one cacao planter in ten was also the holder of an Indian grant in 1690.

The bishop's *Ynforme* also indicates that for some time many encomenderos had allowed their grants to revert to the crown rather than pay fees for the confirmation of their titles. Twenty-four of the province's sixty-four encomiendas were already royal possessions in 1690. Legal rights had been allowed to lapse in both the case of the small coastal encomiendas, now of only minimal value, and in the case of the large, still important grants in the Valencia-Aragua sugar region. Only encomenderos with Indians in the wheat fields near Caracas, in Petare, Baruta, and La Vega, had uniformly paid their taxes and kept their titles in order. Why this was so can be inferred from the demographic composition of the encomiendas recorded in the *Ynforme*.

By 1690, cacao cultivation and the forced immigration of Black slaves had reduced the coastal Indian communities to fragments of their original form. African men without African women were fortunate if, as occurred in Chuao, they were provided with creole women; otherwise they found substitutes among the women from the coastal encomiendas. Many of the husbands of the women in the coastal encomiendas at Maiquetía, Caraballeda, and Naiguatá were listed as anonymous *negros* in the *Ynforme*, and most of the encomienda-born children were recognized as

---

57. The 1690 encomienda census is in AGI, Santo Domingo, leg. 197-B. A copy is in the Archivo de la Academia Nacional de la Historia, Caracas, Traslados, Sección Caracas, vol. 138.

58. All encomienda Indians, whether tributaries or not, were counted in 1690.

Afro-Indian *zambos*.[59] For their part, the Indian men from these encomiendas were frequently married to women who belonged to populous inland encomiendas near the sugar trapiches at Turmero and Guarenas. These groups were still stable in 1690, sexually balanced and headed by an aged patriarch, quite unlike the truncated, often caciqueless encomiendas on the coast.[60]

In the Aragua Valley, where the original population had been dense and indigenous culture was able to survive the seventeenth century more nearly intact than it had on the coast or in Caracas, it may have simply made good tactical sense to allow the encomiendas to return to the crown. In this area, eliminating the encomienda altogether assured access to Indian labor to those whose encomienda privileges had expired; in this way, there would be no chance of one's competitors gaining exclusive control of a needed work force. With two of the four trapiches located in the Aragua pueblo of Turmero, the powerful and wealthy Tovar family had no desire to part with the skilled Indian workers who had made sugar for them for decades. The end of the encomienda also put an end to this possibility, and the Tovar dominion in Aragua continued. In 1710, Turmero Indians complained that Tovar canefields were encroaching on their village, and in 1730 these same Indians were still identified as "the people of Don Antonio Tovar," their last encomendero—who had been dead for more than forty years.[61]

Finally, although now nearly insignificant as a source of income alongside slavery and cacao, the prestige value of the wheat farm encomienda had not diminished during the course of the seventeenth century. Mario Góngora has shown how the encomienda in agrarian Chile was absorbed by a "class of owners" who derived their power from various sources and who were therefore not dependent on Indian tribute for wealth and prestige.[62] This was all the more true in Caracas, where the monetary value of an encomienda frequently depended on the market skills of the encomendero, and where, especially after 1650, ownership of large, movable gangs of slaves was the decisive factor in the maintenance of aristocratic status. In 1684, all eleven of the encomenderos whose Indians grew wheat near Caracas were also owners of either sugar or cacao prop-

---

59. Classified "zambos" by the *Ynforme* compilers, these children of Indian and African parentage were considered Indians for tax purposes.
60. There were virtually no Indians remaining on the coast by 1719. Archivo de la Academia Nacional de la Historia, Caracas, Misiones de Capuchinos, Trinidad, Guayana y los Llanos de Venezuela, leg. 2, no. 36, fol. 81.
61. ARPC, Civiles, 1730.
62. Mario Góngora, "Urban Social Stratification in Colonial Chile," *HAHR*, 55 (Aug. 1975), 430–431.

erties. Born several generations after the time when wheat exports and Indian labor had determined membership in the local elite, these grantholders could liken themselves to those earlier encomenderos and thereby call attention to the tenure of their families' importance in Caracas. For a boom-and-bust society of slavers and smugglers, it was a useful symbol of stability and the nobility of their origins, and for this reason the titles of these encomiendas were nearly all in order in 1690.

During the half century after 1630, Caracas completed the transformation from a minor supplier of a European food crop to Spanish sailors to a major supplier of a tropical quasi-drug to Indian and European consumers alike. In the process, African slavery rendered the encomienda a nonessential anachronism. The Mexican demand for cacao initiated this transformation, and it would appear that, once committed to cacao and slavery, Caracas continued to be very responsive to shifts in the economy of New Spain. The Mexican market for cacao surged in the late 1620s and lasted for about two decades thereafter. New Spain merchants offered considerably less for Caracas cacao from the late 1640s until the late 1660s, a period that corresponds closely to the general depression of Zacatecas silver mining. Silver production reached new highs from 1675 to 1690 as the supply of mercury was made certain once again,[63] and it was at this time that cacao exports resumed and Caracas's prosperity returned. To what degree these trends are coincidental must await a study of the New Spain cacao market and its merchants, but such a study might well reveal that the rise of seventeenth-century Caracas was closely tied to the simultaneous development of a diversified capitalist economy in Mexico. Increasingly self-sufficient generally, New Spain continued to depend on external suppliers for cacao, and when commercial capital was unavailable, as was the case from about 1650 until about 1670, Mexicans invested their money in local enterprises and probably did without cacao beans. For their part, growers might have taken Dutch slaves and merchandise in exchange for their harvests during these long years, but only after 1670, when "the discriminating customer"[64] from across the Caribbean was again willing to buy, did Caracas, permanently established as New Spain's cacao colony, enjoy a resurgence of wealth and well-being.

63. Peter J. Bakewell, *Silver Mining and Society in Colonial Mexico: Zacatecas, 1546–1700* (Cambridge, 1971), pp. 208–220.
64. Ibid., p. 229.

6
Slaves in Piedmont Virginia, 1720–1790
Philip D. Morgan & Michael L. Nicholls

ON Monday, May 25, 1772, a young African woman, one of many slaves aboard the *Polly*, stepped onto Virginia soil at Bermuda Hundred, a Chesterfield County village close to the confluence of the James and Appomattox rivers. Perhaps she caught a glimpse of Shirley, the imposing brick manor house majestically sitting beyond the James. Would her fate fall there? she may have wondered. Had she arrived in Virginia a half century earlier, she would probably have landed at a wharf along the York River or the Rappahannock, perhaps destined to live the rest of her life in the tidewater region. But in the years after 1750 most Africans brought to Virginia were taken up the James to be sold at ports like Bermuda Hundred. Most were then marched into the interior, where planters eagerly sought their labor on newly settled piedmont plantations and quarters. In this regard, the fate of this African woman was typical.

No doubt, many planters and merchants were drawn to Bermuda Hundred on May 25 by newspaper notices advertising the *Polly*'s 450 "fine healthy SLAVES." Among them was Paul Carrington, holder of several local offices and a member of the House of Burgesses for Charlotte County. He bought 50 slaves with intent to resell them in the Southside. As the king's attorney in several counties and a professional lawyer, Carrington traveled regularly in this rapidly expanding subregion. He must have been aware of the Southside's insatiable demand for labor. Perhaps he also found encouragement in the slave prices. Richard Hanson, a Petersburg area merchant, expressed surprise at the owners of this consignment, Burnley and Braikenridge, "breaking the price so low £60 and £65 privilege . . . as the People expected to give £65 to £67.10 privilege. . . . They likewise abated £3 pr inch in the small slaves." Nevertheless, "considering the sum large and a considerable risque in the health & life of the Slaves," Carrington took on three silent partners. With their financial support he could proceed more securely in the resale of his purchases. Carrington led the Africans to his plantation near the junction of the Roanoke and Little Roanoke rivers in southern Charlotte County. He ultimately kept only

---

Mr. Morgan is a member of the Department of History at Florida State University. Mr. Nicholls is a member of the Department of History at Utah State University. The authors wish to thank Stanley Engerman, Brady A. Hughes, Sarah S. Hughes, and Lorena Walsh for their helpful criticisms, and the Department of History, Utah State University, for the financial support for the map and figures.

one of them for his own use; he estimated her age as eighteen years; he named her Kate.[1]

Initially, Kate lived at Carrington's home plantation. She was accompanied for at least some months by other Africans from the *Polly,* since the credit crisis of 1772, the effects of which were felt in Virginia that summer, undoubtedly hampered sales. By the fall, however, Carrington had apparently sold about thirty-five of his Africans, for he then bought winter clothing for only fourteen. What ties Kate developed to her shipmates cannot be determined, nor can the number of men, women, or children in this dwindling group. More critical to Kate's future were the five adult slaves who already could call this plantation home. Two were men, both named Will. One, aged about forty-two, had been purchased about a decade earlier; the other, at least half his age, only two years before. The three adult women included Barbara, acquired five years earlier and, at age twenty-four, the mother of a two-year-old daughter named Sarah; Amey, given as a young girl to Carrington by his father-in-law Clement Read, now twenty-five years of age and the mother of three boys, Lewis, Hampton and Amos; and Nell, born in July 1754, now almost eighteen, about the same age as the African newcomer. Kate lived among these slaves—some Africans, some Virginians—for about a year and a half.[2]

By residing on the manor plantation Kate came into close contact with whites. When she arrived, Carrington, a widower, was raising five children—two girls and three boys—who ranged in age from eight to sixteen years. It is impossible to gauge the impact on Kate of these early encounters with her owner, his children, and white visitors. What is clear is that this close contact with her master and his family was short-lived.

By November 1773 Carrington had transferred Kate to his Twitty's Creek plantation a mile or so up the Little Roanoke from his home. There she joined eight other adults: five men and three women. One of the

[1] *Virginia Gazette* (Purdie and Dixon), May 14, 21, 1772. For a succinct account of Carrington's life see Richard L. Morton's sketch in *Dictionary of American Biography,* II (New York, 1958), 522; "Deposition given between Gordon v. Lockhead, Mar. 5, 1801," Paul Carrington Papers, MSS1 C2358 c5-7, Virginia Historical Society, Richmond; and Hanson to Thomas Deane, July 27, 1772, Roger Atkinson Letterbook, 1769-1776, Alderman Library, University of Virginia, Charlottesville. The term "privilege" slave referred to those few Africans shipped free of freight charges by the captain. We assume, therefore, that they would have been the cheapest slaves on offer and would have set the standard for subsequent sales.

[2] The details of Kate's life presented in this and the following paragraphs have been constructed from Paul Carrington's Account Books 1755-1779, MSS1 C2358 ff. 3, 13, 18, 22, Va. Hist. Soc., and Charlotte County Deed Book 5, 1782-1787, f. 172, microfilm, Virginia State Library, Richmond. Unfortunately, the Carrington accounts provide little information on the social, economic, and material dimensions of slave life. Unless otherwise noted, county records cited here are microfilm copies on deposit at the Va. State Lib.

women was an African named Dicey, who, in the spring of 1763, had been judged by the Lunenburg court to be twelve years old. Since that time, she had given birth to three children, two of whom had died. Perhaps she helped Kate bring her first child into the world, a son named Byas, born in February 1774, twenty-one months after his mother's arrival in Virginia, the father unknown. Another woman, Belinda, had turned seventeen in the fall of 1773. She, too, was pregnant and would give birth to her first child, Ryly, one month before Kate. The age of Tabb, the remaining woman, was never recorded, but she was probably middle-aged, since she was exempt from field labor (being put instead to spinning) and never bore a child while a Carrington slave. She had been purchased, along with two other Twittys Creek slaves—Robin, the thirty-three-year-old head of the quarter, and Toby, aged twenty-five—from Benjamin Harrison in 1770. The other men at this quarter included Hampton, about twenty-eight years old, purchased from Edward Branch in late 1770; George, nearly thirty-four, who came into Carrington's possession in the fall of 1767; and Jack, in his early thirties, possibly an African, bought from the merchant Thomas Tabb in September 1762. All nine adults therefore had been in Carrington's service no more than a few years, a decade or so at most.

Kate's transfer to Twittys Creek apparently removed her from continual contact with whites. With the possible exception of one year, Carrington did not employ a white overseer at this quarter. In fact, he soon referred to it as "Robins." When he gave one thousand acres to his youngest son, Paul Carrington, Jr., in July 1786, the land included a parcel on the east side of the Little Roanoke above Horsepen Creek, including the mouth of Twittys Creek "and the plantation called Robins." Robin himself became the possession, but not yet the property, of the younger Carrington. By January 1782 Kate, who seems to have remained at Robin's, was the mother of three more children: two girls, Lucy and Anitta, and a boy, Abram. Together with Byas, Kate and her three younger children most likely passed into the ownership of Paul Carrington, Jr., at his marriage in August 1785. At that point she disappeared from the senior Carrington's records.

The story of Kate's first decade in piedmont Virginia is tantalizingly inconclusive at its most critical junctures. How had this young African woman withstood the horrors of the Middle Passage? How did she view her new surroundings and master? How did she react to the sale of her shipmates? What friendships did she strike up with the slaves on Carrington's home plantation and Twittys Creek quarter? Did she form a special relationship with Nell at the former residence and Dicey at the latter? Was it an African from the *Polly*, one of the two men named Will at the home quarter, or some other person who fathered her first child? Who fathered her next three? There are no answers to these questions. It is impossible to penetrate the veil of silence.

We can, however, place the known facts of Kate's life in broader context. For one thing, Kate was part of a large stream of people flowing

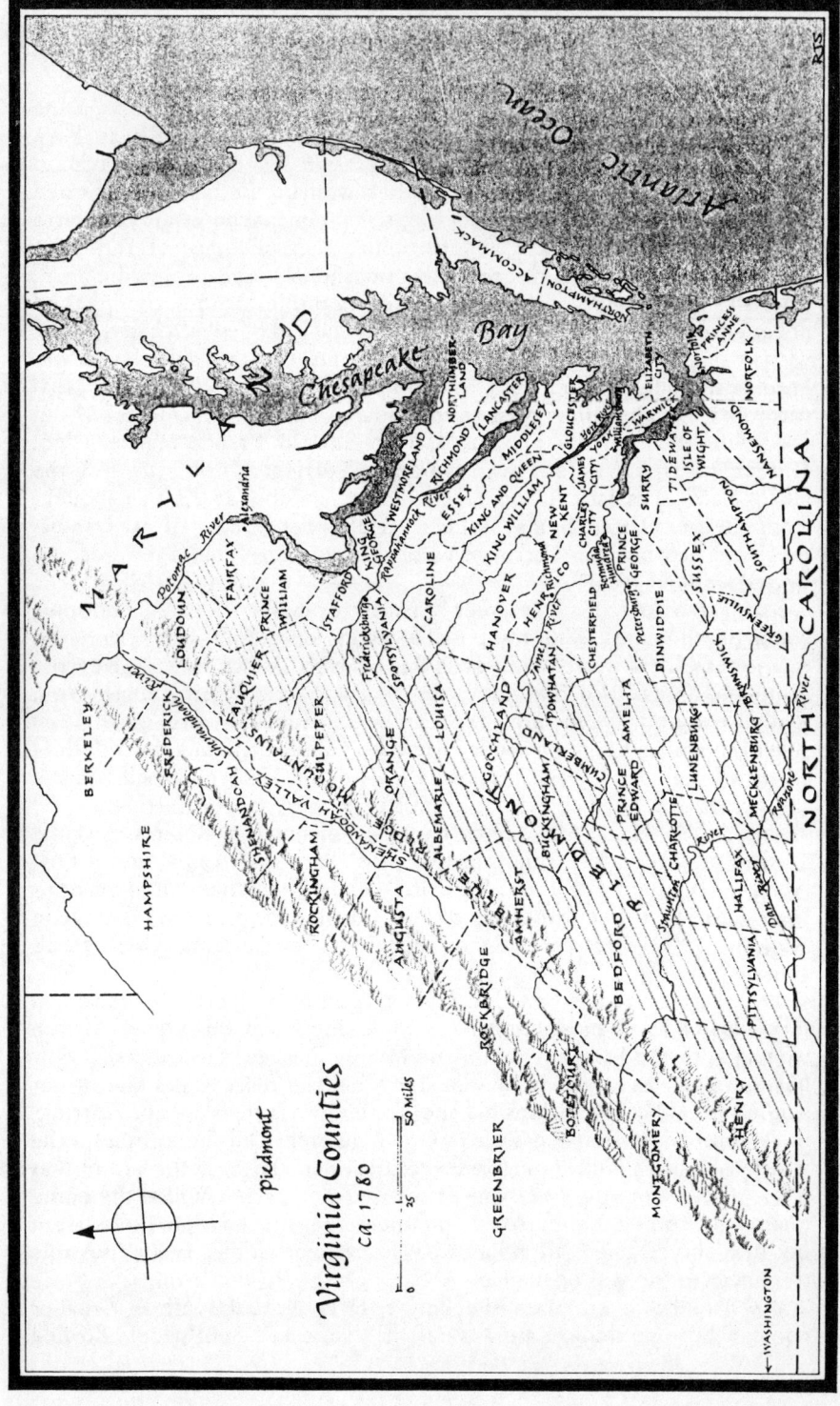

Virginia Counties ca. 1780

SLAVES IN PIEDMONT VIRGINIA 215

TABLE I
DISTRIBUTION OF TITHABLES IN VIRGINIA, 1729-1773

	1729 %	1749 %	1755 %	1773 %
Tidewater	92	68	60	50
Piedmont	8	28	35	44
Shenandoah Valley		4	5	6
Number of Tithables	51,195	91,864	103,318	157,325

Note: In Virginia in this period a tithable was any freeman and all slaves over the age of 16. The 1729, 1749, and 1755 figures come from Evarts B. Greene and Virginia D. Harrington, *American Population before the Federal Census of 1790* (New York, 1932). Those figures have been modified, and gaps filled in, by searching all extant county order books for reported tithables. Fortunately, Peter Bergstrom, formerly of the Colonial Williamsburg Foundation, had already assembled many of these data, including those for 1773. We are grateful to him for sharing this information.

into the piedmont region. An understanding of this remarkably rapid settlement, which shifted the center of black life from tidewater to piedmont about the time Kate arrived in Virginia, will enlarge her tale. Second, it is not too farfetched to speak of the Africanization of the piedmont during the third quarter of the eighteenth century. Kate played a small role in that broader story. Third, Kate's life encourages us to explore the demographic profile of the piedmont slave population. While large numbers of women and children appeared late among tidewater slaves, they were important to piedmont settlements from the first. Finally, the setting that awaited most of the region's black settlers will be investigated. It will be argued that remarkably early in piedmont settlement most slaves found themselves on middle-sized estates, with a fair degree of autonomy from whites. Kate's situation was not unusual. Thus we move beyond Kate's story to explore the peopling of a region and the emergence of black community life.

Settlers had been trickling into the piedmont since the turn of the eighteenth century, but it was not until 1721 that the first three predominantly piedmont counties—Spotsylvania, Brunswick, and Hanover—were authorized by the assembly. By the end of the decade about 8 percent of Virginia's population resided beyond tidewater, mostly in the northern and central piedmont. In another twenty years the proportion exceeded one quarter and settlement had begun to spread farther south. On the eve of the Revolution the piedmont could claim almost 45 percent of Virginia's population, and settlement stretched diagonally from the Potomac River, down the Blue Ridge, to the North Carolina line (see Table I). Within a half century, then, around two hundred thousand settlers swarmed beyond the fall line to take up residence in the Old

Dominion's rolling hills.[3] When war broke out, more of Virginia's population lived in the piedmont and the Shenandoah Valley than in the tidewater.

A number of developments contributed to the rapid expansion of the piedmont. The white population of the tidewater, which had been growing from natural increase since the turn of the century, began to put pressure on local resources. Younger sons, in particular, were forced to seek new opportunities; many moved to the piedmont frontier, where land was readily available. In contrast to the seventeenth century, when settlers secured headrights through purchase, fraud, or the importation of servants, prospective piedmont residents could more easily patent crown lands by purchasing treasury rights at a fee of five shillings sterling for each fifty acres. This ready sale of crown lands encouraged land speculation, which in time attracted more settlers to the region. Most people came from the well-settled peninsulas of the tidewater, but, by the 1740s, their ranks were swelled by immigrants sweeping south from Pennsylvania.[4]

In addition to these demographic forces, economic changes accelerated piedmont expansion. European demand for tobacco, which had been stagnant around the turn of the century, expanded from about 1710 onward. This demand pushed up tobacco prices. Moreover, as British merchants penetrated new European markets, the price differential narrowed between the two varieties of tobacco, the more expensive sweet-scented and the more widely available oronoco. The piedmont's soils were well suited to oronoco cultivation; as the region's planters developed better grades of oronoco and as the French market, which favored oronoco, grew ever more important, the piedmont's farmers benefited from rising prices. In addition, a new system of stores, run by Scottish merchants, spread throughout the region, providing not only a ready market for piedmont produce but a source of supplies and credit. Pioneer planters eagerly took advantage of these new opportunities—so much so that by the 1770s the Upper James Naval District, which encompassed the James River Basin above Williamsburg, accounted for about half the tobacco exported from the colony.[5]

As tobacco cultivation expanded, so did slavery. By the early eighteenth

---

[3] In 1773 the piedmont returned about 70,000 tithables, which translates to a population of almost 200,000.

[4] For a general description of piedmont expansion see Richard L. Morton, *Colonial Virginia. Vol. II: Westward Expansion and Prelude to Revolution, 1710-1763* (Chapel Hill, N.C., 1960), 539, and *passim*.

[5] Jacob M. Price, *France and the Chesapeake: A History of the French Tobacco Monopoly, 1674-1791, and of Its Relationship to the British and American Tobacco Trades*. 2 vols. (Ann Arbor, Mich., 1973), I, 658-677; Russell R. Menard, "The Tobacco Industry in the Chesapeake Colonies, 1617-1730: An Interpretation," *Research in Economic History*. V (1980), 109-177; Darrett B. Rutman and Anita H. Rutman, *A Place in Time: Explicatus* (New York, 1984), 3-7; Allan Kulikoff, *Tobacco and Slaves: The Development of Southern Cultures in the Chesapeake, 1680-1800* (Chapel Hill, N.C., 1986), 45-161.

century the association between this staple crop and slave labor had become firmly established. When piedmont planters sought slaves to meet their laboring needs, they were following well-worn footsteps. Thus the piedmont's extraordinary growth was fueled by a spectacular rise in black numbers. White settlers seem to have carried slaves into the region from the first, but, before long, the addition of blacks outpaced that of whites. Amelia, for example, was the seventh county to be formed in the piedmont. It was created in 1735, largely from the western portion of Prince George County. One year after its establishment, white men outnumbered black men by three to two; thirty years later, the proportions were almost reversed. Similarly, in mid-century Lunenburg County (four years after the county's creation), slaves formed just one-third of tithables, but by 1769 they constituted 60 percent.[6]

In the space of two generations the center of gravity of Virginia's slave system shifted from tidewater to piedmont. Slavery's first significant expansion into the piedmont occurred in the second decade of the eighteenth century. Within one generation, about 40,000 slaves, one-third of the colony's total, resided in the region. Within a second generation, this number almost trebled. Virginia emerged from the Revolutionary War with more slaves living in the piedmont than in the tidewater (see Table II). A remarkable transformation had occurred in just fifty years: the center of black life now lay beyond the fall line.[7]

Part of this striking rise in black numbers can be attributed to movement of both tidewater slaves and African arrivals. Because the latter can be measured fairly directly, it makes sense to begin with them. Historians have noted that a marked shift occurred in the marketing of Africans in Virginia over the course of the eighteenth century. By mid-century, most Africans were sold on the upper James rather than in the York and Rappahannock river basins as in the past. The spread of tobacco into the Southside and central piedmont, and the consequent rising demand for labor, accounts for this dramatic relocation of African sales.[8]

Another piece of evidence—the numbers of African children brought into piedmont county courts to have their ages adjudged—can help determine the size of the influx. If children made up roughly a sixth of all slave cargoes—a standard proportion in the trade as a whole, as well as in

[6] Amelia County Tithable Lists, 1736 and 1767, MS, Va. State Lib. (all subsequent tithable lists cited are also manuscripts on deposit at the Va. State Lib.); Richard R. Beeman, "Social Change and Cultural Conflict in Virginia: Lunenburg County, 1746 to 1774," *William and Mary Quarterly*, 3d Ser., XXXV (1978), 464. It should be noted that the 1769 listing for Lunenburg is incomplete.

[7] Richard S. Dunn, "Black Society in the Chesapeake, 1776-1810," in Ira Berlin and Ronald Hoffman, eds., *Slavery and Freedom in the Age of the American Revolution* (Charlottesville, Va., 1983), 58.

[8] Kulikoff, *Tobacco and Slaves*, 336. See locations of sales in Walter Minchinton, Celia King, and Peter Waite, eds., *Virginia Slave-Trade Statistics, 1698-1775* (Richmond, Va., 1984).

TABLE II
DISTRIBUTION OF VIRGINIA'S SLAVE POPULATION, 1755-1790

	1755 %	1782 %	1790 %
Tidewater	66	46	44
Piedmont	33	51	51
Shenandoah Valley	1	3	5
Number of Slaves	119,996	229,088	292,717

Note: For published sources see Greene and Harrington, *American Population*, 150-155; *Heads of Families at the First Census of the United States Taken in the Year 1790: Reviews of State Enumerations: 1782 to 1785. Virginia* (Baltimore, 1966), 9-10; and Robert E. Brown and B. Katherine Brown, *Virginia 1705-1786: Democracy or Aristocracy?* (East Lansing, Mich., 1964), 73 (Table 1). We have estimated Loudoun's population in 1755 and subtracted it from that of Fairfax. The 1782 figures were assembled from the individual county personal property assessments in the Virginia State Library, Richmond. We also compiled all figures for 1783-1787 to check on the reliability of those for 1782 and to supply an occasional missing county return. We are therefore convinced that the 1782 figure is fairly trustworthy.

those few Virginia cargoes for which ages can be determined—then piedmont planters seem to have purchased just under half of the 59,000 newcomers imported into the colony between 1725 and 1775. Moreover, in the last quarter century of the trade, this region's planters garnered almost all the incoming Africans (see Table III). Compared to the tidewater at this time, the piedmont experienced a continous, if uneven, infusion of Africans into its black population.[9]

In some piedmont counties, at particular times, the adult slave population was heavily African. In Amelia County, which probably received more Africans than any Chesapeake county in the forty years before Independence, about 60 percent of the adult slaves in 1755 were Africans. In 1782, although the proportion had dropped, it still stood at around one-fifth. In Chesterfield County, established fourteen years after Amelia, the proportion of adults who were Africans reached almost 40 percent in 1755 and remained at about 27 percent in 1782 (see Table IV).

What sort of Africans were piedmont planters buying? Direct evidence exists only for children. Still, the annual totals of African children registered by the region's planters correspond closely with known yearly imports. Indeed, the synchronization seems surprisingly exact, suggesting that slave children were a significant component of slave cargoes or of

[9] We do not mean to suggest that Africans were in a majority in piedmont slave populations or that African, rather than creole, immigration was critical to the region's slave population growth. Neither is true. Rather, we mean only that the slave population of the piedmont comprised a surprisingly large number of Africans in the late colonial era—certainly more than the tidewater region. For children in slave cargoes see next note.

SLAVES IN PIEDMONT VIRGINIA

TABLE III
AFRICAN IMMIGRANTS TO THE
VIRGINIA PIEDMONT AND TO VIRGINIA AS A WHOLE, 1725-1775

Years	Piedmont[a]	Virginia[b]
1725-1729	390	7,741
1730-1739	2,868	16,226
1740-1749	3,852	12,113
1750-1759	7,302	9,197
1760-1769	8,766	9,709
1770-1775	4,020	3,932
Totals	27,198	58,918

[a]For the calculations on which these numbers are based see Appendix 1.
[b]These figures are derived from Walter Minchinton, Celia King, and Peter Waite, eds., *Virginia Slave-Trade Statistics. 1698-1775* (Richmond, Va., 1984). Some adjustments have been made for missing data. See Appendix 2.

piedmont planter purchases. Furthermore, when one applies the standard multiplier of six to the slave child registrants so as to arrive at the total number of African immigrants to the piedmont, the resulting estimates actually exceed all known imports into Virginia from 1750 to 1774; indeed, in the 1770s they exceed even the revised import figures. It therefore seems likely either that piedmont planters were buying a disproportionately large share of children or that slave cargoes in late colonial Virginia were composed of slightly more than one-sixth children.[10]

Both explanations are plausible. The costs of setting up plantations on the Virginia frontier, along with higher than average expenditures for transporting farm products great distances over land, would certainly have reduced funds available for purchasing adult slaves. Buying children may well have been the piedmont planter's most practical option. On the other hand, African imports into Virginia declined from a peak of over 16,000 in the 1730s to about 9,000 in both the 1750s and 1760s. The increasing marginality of Virginia to the burgeoning Atlantic slave trade may well have resulted in a larger proportion of women and children in the region's slave cargoes. We know, for instance, that Chesapeake-bound slavers generally arrived late in the trading season on the African coast. Indeed, a comparison of voyage times between British ports and Chesapeake

[10] Allan Kulikoff, "A 'Prolifick' People: Black Population Growth in the Chesapeake Colonies, 1700-1790," *Southern Studies.* XVI (1977), 398-399. It should be noted that the statistical basis for the multiplier of six is exceedingly slender—just five ships that entered the Chesapeake in the early 18th century. But the same proportion has been found among 40,000 Africans imported into colonial South Carolina (Philip D. Morgan, *Slave Counterpoint: Black Culture in the Eighteenth-Century Chesapeake and Lowcountry* [Chapel Hill, N.C., forthcoming]).

TABLE IV
AFRICANS AMONG ADULT SLAVES IN TWO PIEDMONT COUNTIES, 1735-1782

Years	African Immigrants	Surviving African Immigrants	Adult Slaves	% of Adults Who Were African
Amelia				
1734-1754	1,860			
1755		983	1,652	59
1755-1774	1,820			
1782		880	4,059	22
Chesterfield				
1749-1754	790			
1755		439	1,198	37
1755-1774	1,575			
1782		733	2,681	27

Notes: The adult populations of the two counties are derived from the 1755 tithable lists and from the 1782 personal property assessments. African immigrants were calculated from age-registration data using the standard multiplier of 6. The survivor schedule is adapted from Allan Kulikoff, "A 'Prolifick' People: Black Population Growth in the Chesapeake Colonies, 1700-1790," *Southern Studies*. XVI (1977), 421. In one year or less after importation we expect .755 of Africans to have survived; after 2 years, .649; after 3-5 years, .625; from .620 to .517 for the 6th to the 10th year; then .453 for 11-20 years; .200 for 21-30 years; and .050 for 31-40 years. It should be emphasized that these calculations are little more than educated guesses.

wharves from the 1720s to the 1760s indicates that, on average, slavers took an extra month to deliver their cargoes by the latter decade. Arriving late on the coast, and spending more months to fill their manifests, late colonial Chesapeake slavers could hardly have found it easy to obtain choice adult males.[11] We also know that those involved in the Virginia slave trade recognized the value of adolescent Africans, "men-boys" and "women-girls" in their parlance. Thus in 1751 Charles Steuart thought Africans "from 14 to 18 years . . . most saleable," while twelve years later

[11] Herbert S. Klein, *The Middle Passage: Comparative Studies in the Atlantic Slave Trade* (Princeton, N.J., 1978), 121-140; James A. Rawley, *The Transatlantic Slave Trade: A History* (New York, 1981), 181-182; W. E. Minchinton, "The Slave Trade of Bristol with the British Mainland Colonies in North America, 1699-1770," in Roger Anstey and P.E.H. Hair, eds., *Liverpool. the African Slave Trade. and Abolition* ([Liverpool], 1976), 39-59. Analysis of the patterns of Chesapeake voyages (based on 34 in the 1720s and 28 in the 1760s, when average sailing times from Britain to Virginia, via Africa, were 10 and 11 months respectively) was made possible by information reported in Minchinton *et al.*. eds., *Virginia Slave-Trade Statistics*.

William Taylor declared that "planters in general Chuse men Boys and women Girls before grown slaves."[12]

The sex ratios of African children registered in piedmont county courts support the contention that the composition of slave cargoes changed over time. In the late 1720s and 1730s boys outnumbered girls by 145 to 100 in piedmont registrations, a more balanced sex ratio than was common in the slave trade perhaps, but certainly not markedly out of line.[13] However, for much of the remaining time in which slaves were imported into colonial Virginia, the ratio was much lower. And we need only recall that, during this period (1750-1775), most of the Africans were apparently being purchased in the piedmont. It is difficult to see how the piedmont planters' purchases could have been atypical, since these planters were the primary buyers. Rather, the shift was almost certainly in the composition of the slave cargoes themselves. Boys and girls were fairly equally matched from the 1740s on, most particularly in the 1750s and 1760s. Indeed, in the Southside and northern piedmont in the 1750s, and in the central piedmont in the 1760s, girls actually outnumbered boys among registrants (see Table V). Apparently, slave ships were carrying more girls than ever before.[14]

Most African children brought into the piedmont county courts were judged eight to twelve years of age. There seems to be no discernible trend over time. On average, both boys and girls were ten or eleven years when taken into court. Girls tended to be a little older than boys, generally by about six months (see Table VI). Since most African girls were about eleven years of age when they arrived in the piedmont, this region's planters could expect them, with any luck, to be contributing to the natural increase of the black population within six to seven years.

Natural increase would also have been aided, of course, by an influx of creole slave women and girls. We now know that the tidewater's black population began to increase by natural means in the early eighteenth

[12] Charles Steuart to Messrs. Blackman and Adams, July 5, 1751, Steuart Letterbook, microfilm, Colonial Williamsburg Foundation Library, Williamsburg, Va.; William Taylor to Samuel and William Vernon, July 5, 1763, Slavery Collection, New-York Historical Society, New York City. Taylor was the captain of the *Little Sally* that consigned 60 Africans to Richard Adams in late June. Taylor wrote from the town of Richmond on the James River.

[13] Kulikoff reports a sex ratio of 200 for children aged 10-14 years imported into the Chesapeake ("A 'Prolifick' People," *So. Studies.* XVI [1977], 399). Maryland inventories in the first decade of the 18th century reveal a sex ratio of 142 among children (many of whom must have been Africans) (Russell R. Menard, "The Maryland Slave Population, 1658 to 1730: A Demographic Profile of Blacks in Four Counties," *WMQ.* 3d Ser., XXXII [1975], 32). See also David W. Galenson, "The Atlantic Slave Trade and the Barbados Market, 1673-1723," *Journal of Economic History.* XLII (1982), 505.

[14] The reversal of the trend in the early 1770s is somewhat puzzling, but it should be remembered that this sex ratio is based on a smaller number of observations: just over 500 in the 1770s, as contrasted with over 2,000 for the 1750s and 1760s.

TABLE V
SEX RATIOS OF AFRICAN CHILDREN
REGISTERED IN PIEDMONT COUNTY COURTS, 1725-1774

	Central Piedmont	Southside	Northern Piedmont	Total
1725-1729	142			142
1730-1739	132	165		145
1740-1749	119	125		121
1750-1759	108	97	95	101
1760-1769	92	116	131	106
1770-1774	142	126	122	131
Numbers of Children	1,586	1,613	163	3,362

Note: Ratios are expressed as boys per 100 girls. For the sources of these figures see Appendix 1. Totals of children reported in this table vary somewhat from those reported in the appendix because not all children could be differentiated by sex.

century. By mid-century some large planters in the older areas had little need to buy more slaves because the laborers they inherited were now reproducing at a rate sufficient to supply—and sometimes oversupply— their labor needs. They therefore began to transfer or sell their surplus slaves westward. It is not possible to measure the size and composition of this creole diffusion in the same way as for African importations. However, by a set of complex calculations based on admittedly flimsy data, it has been estimated that about 17,000 tidewater slaves were moved to the piedmont between 1755 and 1782. This is slightly more than the number of Africans who arrived in the piedmont over the same period.[15] Data for the composition of the transferred slaves are fragmentary at best. Planters who held slaves in both tidewater and piedmont provide one set of clues. Thus a listing of William Byrd III's slaves in 1757 reveals that, at Westover in the tidewater, men outnumbered women heavily, while children outnumbered women by fewer than 2 to 1. Further west, on the other hand, Byrd's slaves exhibited relatively equal adult sex ratios, while children outnumbered women by 2.3 to 1 on his Roanoke River quarters and 2.6 to 1 at his quarters near the falls of the James River. Similarly, on Landon Carter's home quarters in the tidewater in 1779 males outnumbered females by 113 to 100 while on his piedmont quarters the ratio was almost reversed. Large planters apparently kept more male slaves on their tidewater or home quarters, in part to fulfill their artisanal needs, and

[15] Philip D. Morgan, "Slave Life in Piedmont Virginia, 1720-1800," in Lois Green Carr, Philip D. Morgan, and Jean B. Russo, eds., *Colonial Chesapeake Society* (Chapel Hill, N.C., 1988), 436-437.

SLAVES IN PIEDMONT VIRGINIA 223

TABLE VI
AGES OF AFRICAN CHILDREN
REGISTERED IN PIEDMONT COUNTY COURTS, 1725-1774

Years	Central Piedmont		Southside		Northern Piedmont		Numbers	
	B	G	B	G	B	G	B	G
1725-1729	10.3	10.2					17	12
1730-1739	11.0	11.4	10.4	10.9			196	135
1740-1749	10.8	11.2	9.9	10.1			265	218
1750-1759	10.6	11.1	10.7	10.9	11.6	11.2	473	467
1760-1769	10.3	10.7	10.2	10.7	9.7	11.5	561	520
1770-1774	10.3	10.0	10.4	10.8	10.7	10.9	304	225
Totals	10.5	10.9	10.4	10.7	10.6	11.3	1,816	1,577

Note: B = boys, G = girls. For the sources of these figures see Appendix 1. Totals of children reported in this table vary slightly from those reported in that appendix because not all children's ages were given or are decipherable.

transferred female and more youthful slaves to labor in the newly opened tobacco fields.[16]

In only a couple of instances can we catch clearer glimpses of tidewater planters moving slaves westward. In the late 1750s or early 1760s Peyton Randolph set up a plantation in Lunenburg County; in 1764 he reported sixteen black tithables on this quarter. Nine of these sixteen can be identified from a listing by Peyton's son, taken twenty years later. Almost all of the nine (five men, four women) were under the age of twenty when they were moved west, and the women generally had small children in tow. In 1787 Robert Carter of Nomini Hall shunted twenty-eight slaves to a new quarter in the Shenandoah Valley. Although this transfer was not to the piedmont, the type of slaves chosen is instructive. Twenty-four were under the age of twenty, while fully half were youths aged between nine and seventeen, often separated from their immediate families. Females outnumbered males.[17]

Our final measure is somewhat more indirect but based on many more cases. Profiles of slaves remaining in the tidewater from 1720 to 1780 may suggest the types of slaves being transferred out of the region. For

[16] William Byrd III Memorandum Book, Library of Congress; inventory of Landon Carter, 1779, Carter Family Papers, College of William and Mary, Williamsburg, Va.

[17] Accounts of the estate of Peyton Randolph (listings of 1775/6 and 1784), microfilm, Colonial Williamsburg Foundation Research Department, Williamsburg, Va.; Landon C. Bell, ed., *Sunlight on the Southside: Lists of Tithes, Lunenburg County. Virginia. 1748-1783* (Philadelphia, 1931), 223 (listing of 1764); Robert Carter to George Newman, Oct. 23, 1787, Robert Carter Letterbook, VIII, 18, Duke University, Durham, N.C.

comparative purposes, two counties were chosen where out-migration was thought to be minimal. As expected, male-female sex ratios were quite equal in these two counties—in Northampton County, this was true throughout the period; in Isle of Wight, certainly from around mid-century—while children outnumbered women by more than 2 to 1 for most of the period (see Figure 1).[18]

For tidewater counties from which transfers of slaves westward seem to have been commonplace, two trends can be discerned. First, adult sex ratios were especially unbalanced in the first half of the century. Undoubtedly, this extreme male majority reflected the impact of African importation, but perhaps it also owed something to transfers of women into the piedmont. However, from 1750 onward there is little evidence that seaboard planters moved disproportionate numbers of slave women west, since there was no significant male surplus in the tidewater. Indeed, adult sex ratios were little different between exporting and non-exporting counties. The second trend does differentiate markedly between exporting and non-exporting counties, and became more noticeable over time. In exporting counties the ratio of children to women never reached 2 to 1, something that could be expected in a naturally growing population and certainly characterized predominantly non-exporting counties, as we have seen. Morever, a significant decline in that same ratio occurred over the course of the century. It thus seems that tidewater planters singled out young slaves when they opened up quarters further west. Moreover, this practice seems to have become more widespread over time (see Figure 2).[19]

[18] Figure 1 is based on Isle of Wight County Will and Inventory Books 2-9 (1715-1785), wherein 2,324 slaves are listed, and Northampton County Will and Inventory Books 15-26 (1717-1783), wherein 2,631 slaves are listed. Tithable lists surviving for 18th-century Northampton Co. reveal remarkably similar adult sex ratios to those found in inventories. Thus, in 1720, a tithable list reveals an adult sex ratio of 115 men per 100 women (249 slaves); in 1727 it was 111 (352 slaves); in 1744 it was 113 (567 slaves); and in 1769 it was 110 (971 slaves). These and other tithable lists are to be found in loose court papers at the county courthouse, and findings are reported in Joseph Douglas Deal III, "Race and Class in Colonial Virginia: Indians, Englishmen, and Africans on the Eastern Shore during the Seventeenth Century" (Ph.D. diss., University of Rochester, 1981), 206.

[19] Figure 2 is based on York County Wills and Inventories, 15-22 (1716-1783), wherein 3,819 slaves are listed, and Essex County Wills, Inventories, Settlements of Estates, 3-13 (1717-1785), wherein 5,062 slaves are listed. Apparent corroboration of the drop in woman-child ratios around mid-century comes from Middlesex Co. data kindly supplied by Darrett and Anita Rutman. From the early 1720s to the early 1740s, children outnumbered women by about 1 1/2 or 2 to 1. But in the mid-1740s, where the Rutmans' analysis ends, a significant shift occurred: women actually outnumbered children. The Rutmans wonder whether this reflects the continuing effects of a children's epidemic but also suggest that it may reflect the opening up of new plantations in the piedmont by medium and large planters who presumably transferred disproportionate numbers of young couples.

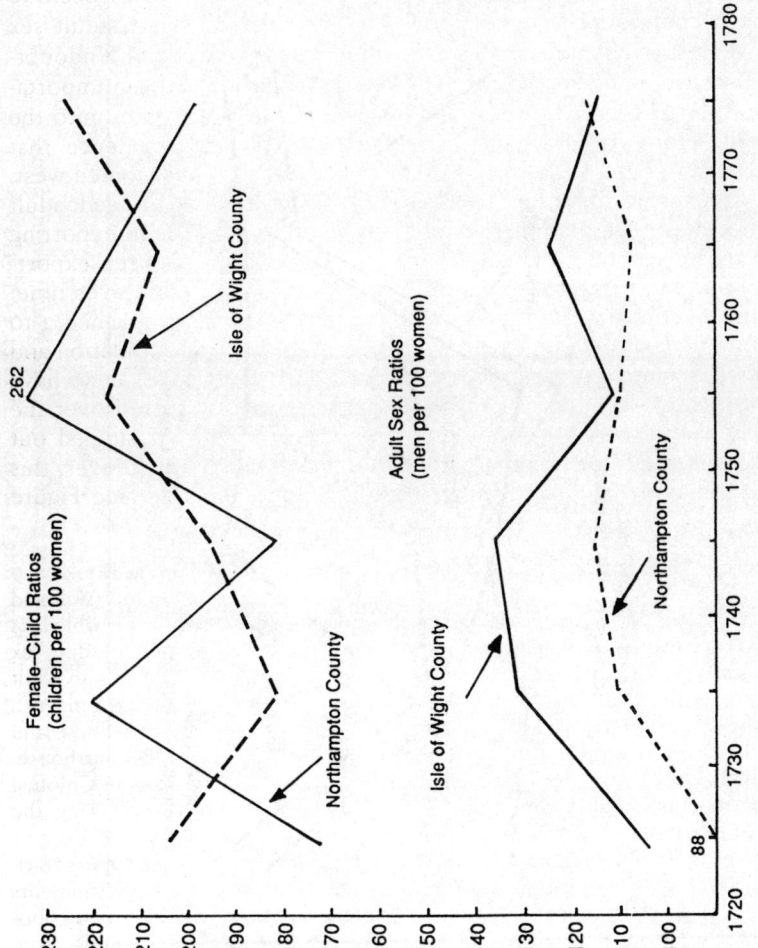

Figure 1. *Adult Sex Ratios and Female–Child Ratios among Slaves in Two Predominantly Non-Exporting Tidewater Counties, 1720–1779*

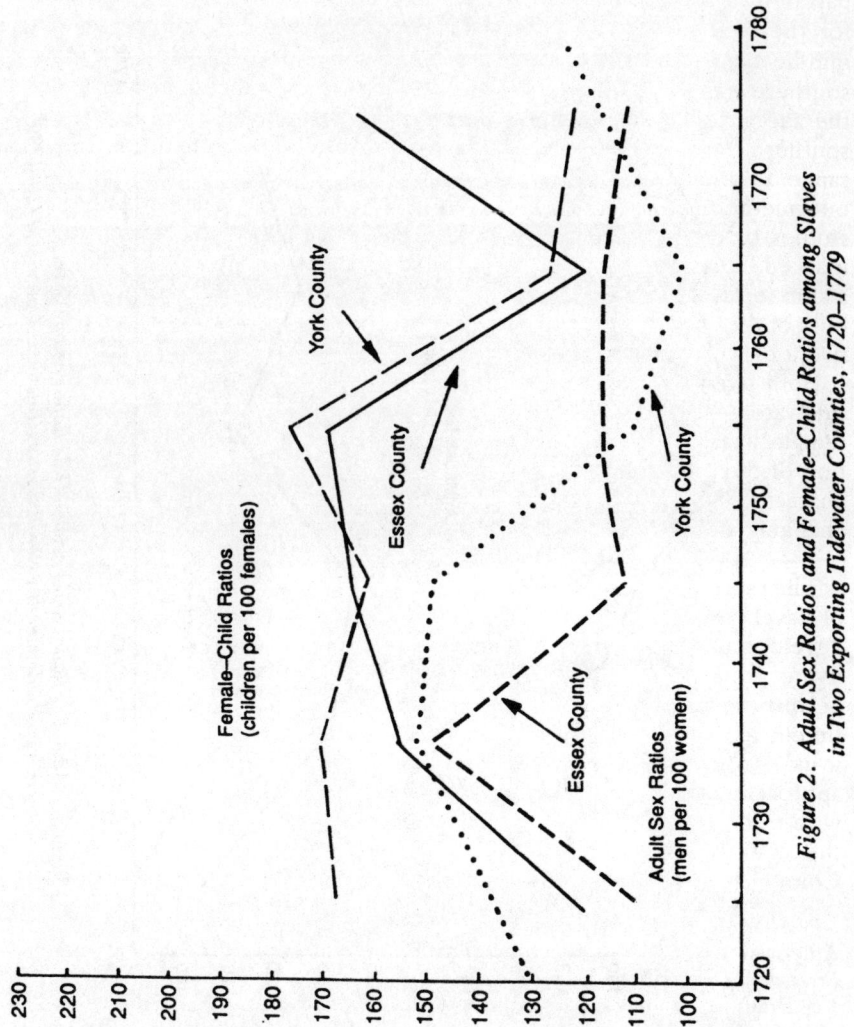

Figure 2. *Adult Sex Ratios and Female–Child Ratios among Slaves in Two Exporting Tidewater Counties, 1720–1779*

These trends are quite consistent with those found among the out-migrations from states of the upper South in the late eighteenth century and the nineteenth. In the late eighteenth century, planters moving to the Southwest generally took as many female as male slaves and twice as many children as women. When Leonard Covington of Maryland transferred part of his slave complement to Mississippi in 1809, he singled out youths for the new venture and left a disproportionately large number of his middle-aged slaves behind.[20] In the decade 1810-1820 planters in the six southern states that imported slaves possessed more females than males in the age group 14-25, whereas their counterparts in the three exporting southern states (including Virginia) had a surplus of males in the same age range. In other words, among the young slaves who were prime targets for out-migration, planters removed more females than males.[21] Although studies of the domestic slave trade in the antebellum era reveal a less marked female predominance, one analysis of over 3,000 slaves has found a slim female majority. Young unmarried slaves certainly constituted the bulk of the antebellum domestic slave trade. Slaves in their teens and early twenties were most at risk.[22]

Both streams of immigrants, then—those from Africa and those from tidewater—seem to have been disproportionately composed of young and female slaves. Is it therefore possible to speak not just of the Africanization of the piedmont slave population but also of its female and youthful character? To explore this subject further, we will need to investigate the changing structure of the resident slave population.

The rapidity with which the sexes approached parity is a striking feature of slavery in the piedmont. Tithable lists are our most reliable guide to this development, since they canvass large numbers of adult slaves among a living population. However, few such lists have survived; even fewer permit the reconstruction of sex ratios; and, consequently, change over time is not easily pinpointed. Yet the surviving evidence, meager though it is, strongly suggests that the numbers of slave men and women approached equality quickly in the piedmont, particularly when compared with trends in the tidewater. Furthermore, it would seem that the newer the county, the more quickly this equality was achieved. In Amelia County, at its establishment in 1735, slave men were twice as numerous as

[20] Allan Kulikoff, "Uprooted Peoples: Black Migrants in the Age of the American Revolution, 1790-1820," in Berlin and Hoffman, eds., *Slavery and Freedom*. 143-171, esp. 156-161.

[21] Brady A. Hughes, "Migration of Virginia Slaves Based on the Census of 1820," *Virginia Social Science Journal*. XX (1985), 86-92. This essay reveals a more complicated picture in terms of male and female surpluses at the county level.

[22] Michael Tadman, "Slave Trading in the Ante-Bellum South: An Estimate of the Extent of the Inter-Regional Slave Trade," *Journal of American Studies*. XIII (1979), 195-220, esp. 198-200; Robert William Fogel and Stanley L. Engerman, *Time on the Cross: The Economics of American Negro Slavery*. 2 vols. (New York, 1974), I, 49, 79.

women; this ratio dropped to 147 in 1749, 122 in 1767, and 115 in 1778-1779. At mid-century, Lunenburg County's slaves had a lower sex ratio than Amelia's, even though their county was then only four years old. The slaves of Prince Edward County, which was formed from Amelia County, achieved a sex ratio similar to that of the parent county as early as 1767, only thirteen years after the county had come into existence. In the same year, at its establishment, Pittsylvania County had a more balanced adult sex ratio than either Amelia or Prince Edward. Indeed, in 1777, female slaves outnumbered their male counterparts in Pittsylvania. Finally, in Amherst, another frontier county formed in the 1760s, the numbers of slave men and women were in rough balance by 1775, more equal, in fact, than the county's white population.[23]

Inventories are another source for measuring sex ratios. They are less reliable than tithable lists since they provide information only on slaves belonging to decedents, who were generally older and wealthier than the average householder. Yet they have one advantage over listings: they capture demographic changes in the slave population. They show, for the piedmont region as a whole, the same rapid downward trend in adult sex ratios, though punctuated by periodic upward shifts. Adult males heavily outnumbered females during the first decade or so of piedmont settlement. But this imbalance was short-lived. In the late 1730s and early 1740s there were fewer than 120 men for every 100 slave women. More skewed imbalances periodically recurred, presumably as waves of Africans and creoles moved into the region, most notably in the late 1740s, when adult sex ratios climbed back to 141, and the late 1750s and early 1760s, when they reached 130. However, the overall trend accelerated downward. By the late 1760s and 1770s, the piedmont had almost as many slave women as men, a result of the cumulative effect of transferring and importing significant numbers of women and children along with natural population growth within the region (see Figure 3).[24]

A county-by-county analysis reveals two distinct patterns within the broader regional trend, centered not only on when, but also on where, a county was formed. Thus those piedmont counties where immigrants from both tidewater and Africa were more readily available generally reveal a greater preponderance of men. In the early years of Spotsylvania,

[23] Amelia County Tithable Lists, 1736, 1749, 1767 and 1778-1779; Lunenburg County Tithable Lists, 1750; Prince Edward County Tithable Lists, 1767 (incomplete listing, but containing slightly over half the tithables taxed that year); Pittsylvania County Tithable Lists, 1767 and 1777; William Cabell's Diary, Vol. 5, Mar. 24, 1776, Va. Hist. Soc. (in Amherst as reported in this diary, the adult sex ratio for blacks was 104 and for whites 111). See also Joan Rezner Gundersen, "The Double Bonds of Race and Sex: Black and White Women in a Colonial Virginia Parish," *Journal of Southern History*. LII (1986), 355-356.

[24] Figures 3-5 are based on information from all extant inventories (up to the year 1775) of Amelia, Chesterfield, Cumberland, Goochland, Halifax, Loudoun, Lunenburg, Orange, Pittsylvania, Prince Edward, and Spotsylvania counties (4,115 adults, 3,527 children). All inventory books are on microfilm, at the Va. State Lib.

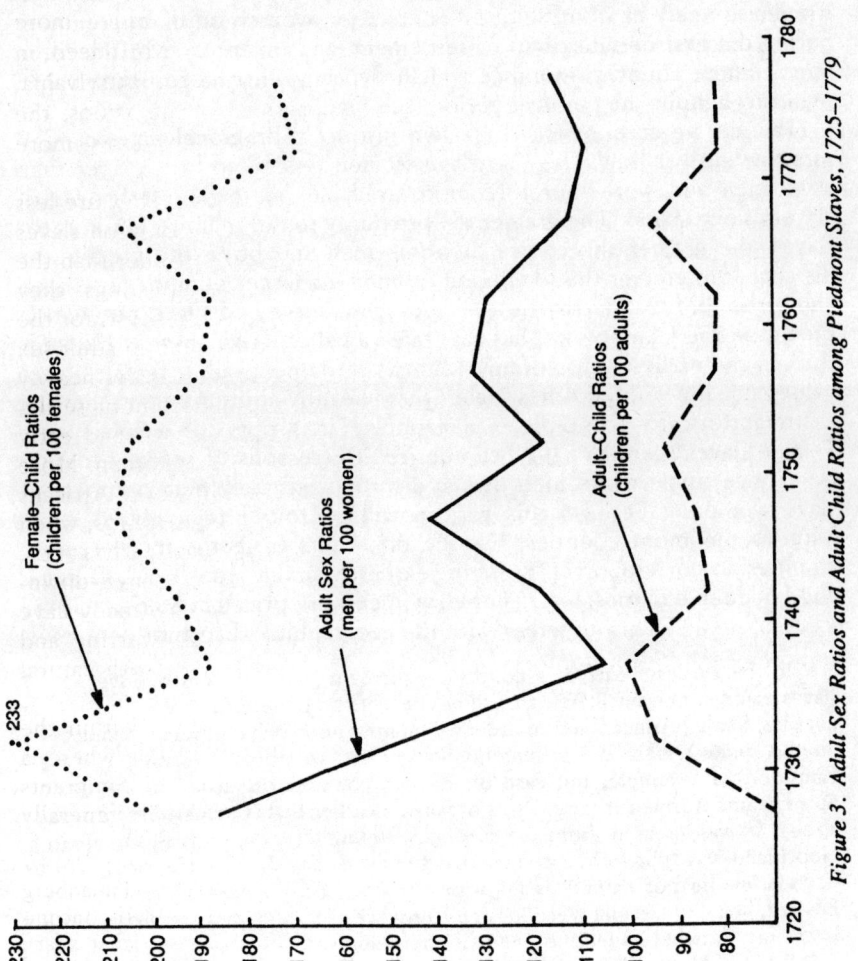

Figure 3. Adult Sex Ratios and Adult Child Ratios among Piedmont Slaves, 1725–1779

created in the 1720s, Amelia in the 1730s, Lunenburg and Chesterfield in the 1740s, men outnumbered women extremely heavily, though the sexes soon arrived at a rough parity in each county (see Figure 4). Those piedmont counties on the outer fringes of settlement—Goochland in the 1720s, Orange in the 1730s, Prince Edward in the 1750s, and Pittsylvania in the 1760s—generally exhibit a more equal balance between men and women. Indeed, in all these outlying counties, women outnumbered men during the first decade or so of settlement, and inventories of the three most remote counties—Orange and Pittsylvania—list more women than men throughout the colonial period (see Figure 5).[25]

How do we account for these two distinct subregional patterns? An intuitive answer might be that slave women tended to be cheaper than slave men and were therefore more available on the outer fringes of slavery's expansion. The greater the proximity to the regional markets for slaves, the greater the choice of prime men and boys; the greater the distance, the greater the likelihood of finding a larger number of women and girls. P.M.G. Harris, in a wide-ranging survey of black population growth in the Chesapeake, has suggested an alternative answer. He finds that, in the earliest years of any county's settlement, adult females were valued equally with (at times higher than) healthy adult males. Eschewing a straightforward economic explanation, Harris opts for a social one. Female slaves seem to have been desired for reasons of sexual exploitation, since single white men owned disproportionately high numbers of slave women.[26] Perhaps this pattern will be found to apply to some Virginia piedmont counties, but the price data from the more remote counties do not support it. From the earliest years in Goochland, Orange, and Loudoun counties, for example, women were priced at 80 percent the value of men.[27] This, together with the geographical distribution of adult

---

[25] In the earliest years of a county's settlement, the numbers of inventoried slaves tended to be small. The trend lines become more determinate in succeeding decades. Some justification is in order for identifying some counties as having been on the outer fringes of settlements and others as more accessible to slaves. Consider, for example, the case of the two counties organized in the 1720s. Spotsylvania (formed in 1721) was organized earlier than Goochland (formed in 1728); it was created from a thoroughly tidewater county (Essex), whereas Goochland was split from a county that straddled the fall line (Henrico); and its boundaries did not extend as far west as those of Goochland. Orange, Prince Edward, and Pittsylvania were located to the west of, and were generally settled later than, Amelia, Lunenburg, and Chesterfield.

[26] P.M.G. Harris, "The Spread of Slavery in the Chesapeake, 1630-1775" (paper presented at the "Maryland, A Product of Two Worlds" conference, St. Mary's City, May 1984).

[27] The earliest inventories of six piedmont counties reveal prices of slave women within a 74%-88% range of those for men: in Goochland Co. (1730-1739), men averaged £26.10 while women averaged £20.10 (18 observations); in Orange Co. (1738-1744), the differential was £27.10 for men and £23.10 for women (49 observations); in Prince Edward Co. (1754-1763), it was £59.15 for men and £44 for women (28 observations); in Halifax Co. (1756-1765), £53 for men and £47

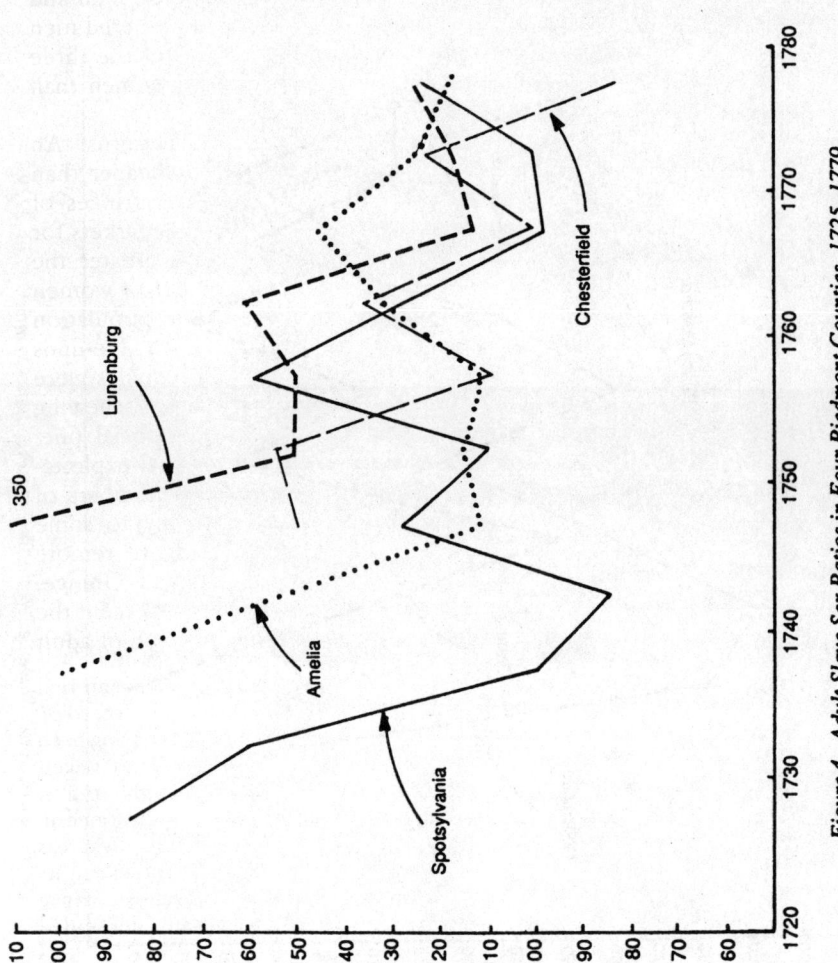

Figure 4. Adult Slave Sex Ratios in Four Piedmont Counties, 1725–1779

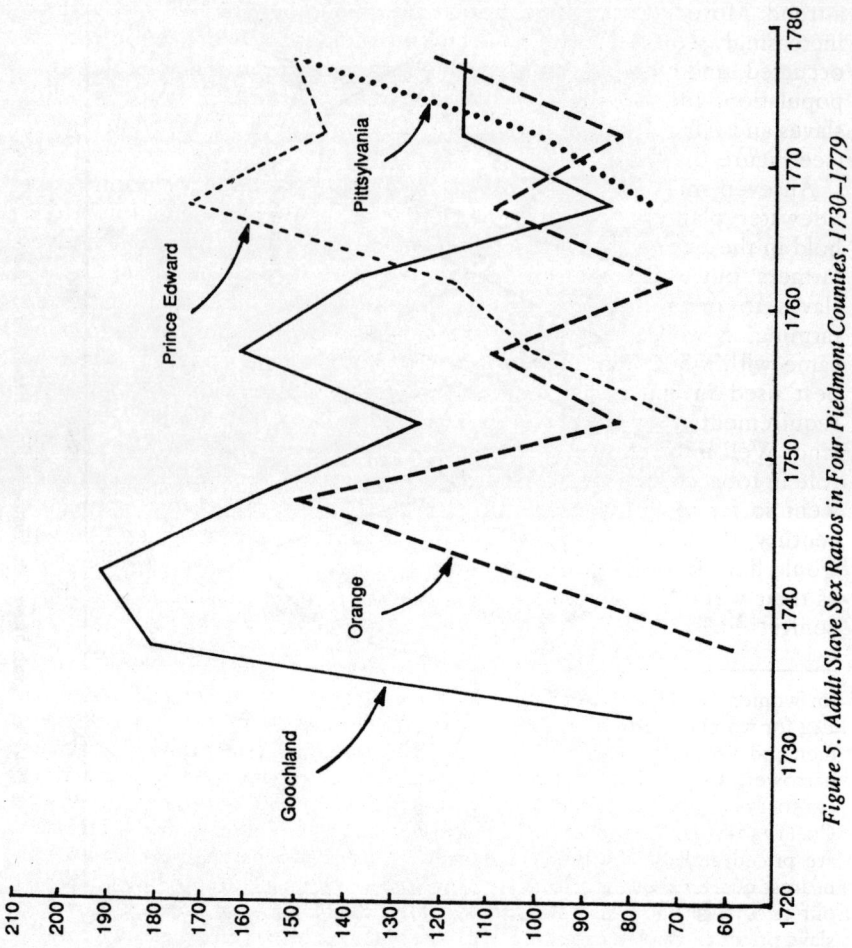

Figure 5. Adult Slave Sex Ratios in Four Piedmont Counties, 1730–1779

slave sex ratios, emboldens us for the moment to hold to our first, admittedly rather obvious, answer.

One reason why females were more available on the margins of slavery's expansion may arise from their increasing proportion in the African slave cargoes arriving in late colonial Virginia. As we have seen, there is no direct evidence to demonstrate that this in fact happened, but the extreme rapidity with which the adult sex ratios approached equality suggests such a trend. Moreover, the importation of large numbers of children, with an increasingly equal distribution between the sexes, which we do know occurred, undoubtedly contributed to lowered sex ratios among the adult population. Indeed, the close parallel trends in the sex ratios of adult slaves and African children underline the importance of this contribution (see Figure 6).[28]

An even more important reason may reside in the labor priorities of tidewater planters. As diversification into small-grain agriculture took hold in the tidewater, the region's planters, now often styling themselves farmers, put a premium on experienced male labor. They wanted male slaves to meet the new artisanal responsibilities attendant on mixed farming, as well as to perform the plowing, carting, and harvesting that came with wheat farming. Conversely, women (and children) could be best used in traditional tobacco and corn culture with its high labor requirements, low degree of mechanization, and high ratio of labor to land. Well into the twentieth century, women and children played a full role in tobacco cultivation. One twentieth-century female informant even went so far as to say, "Men ain't no good in 'bacca." In the eighteenth century, then, it seems quite logical that slave women and their children would have followed the course of expanding tobacco culture, particularly as their work roles diminished in areas where traditional tobacco culture contracted.[29]

---

for women (41 observations); in Loudoun Co. (1758-1764), £55.10 for men and £44 for women (37 observations); and in Pittsylvania Co. (1768-1775), £68.15 for men and £57.10 for women (26 observations). In one county the gap was much narrower, and in another women averaged more than men: in Lunenburg Co. (1750-1759), men averaged £37 and women £35.10 (48 observations); in Amelia Co. (1737-1746), men averaged £23 and women £25 (18 observations—three men are priced at £15 or less in this sample). All prices have been rounded to the nearest quarter of a pound and are expressed in Virginia currency. We omit from our calculations any adult slave valued at less than £9. For an excellent survey of slave prices throughout the New World see Manuel Moreno Fraginals, Herbert S. Klein, and Stanley L. Engerman, "The Level and Structure of Slave Prices on Cuban Plantations in the Mid-Nineteenth Century: Some Comparative Perspectives," *American Historical Review*, LXXXVIII (1983), 1201-1218.

[28] Figure 6 draws on two sources: adult sex ratios are based on inventory data, while African child sex ratios rely on age registration material.

[29] Pete Daniel, *Breaking the Land: The Transformation of Cotton, Tobacco, and Rice Cultures since 1880* (Urbana, Ill., 1985), 28-29, 198-200, 263-267 (quotation on p. 198). For the impact of diversification on slave work patterns see Morgan, *Slave*

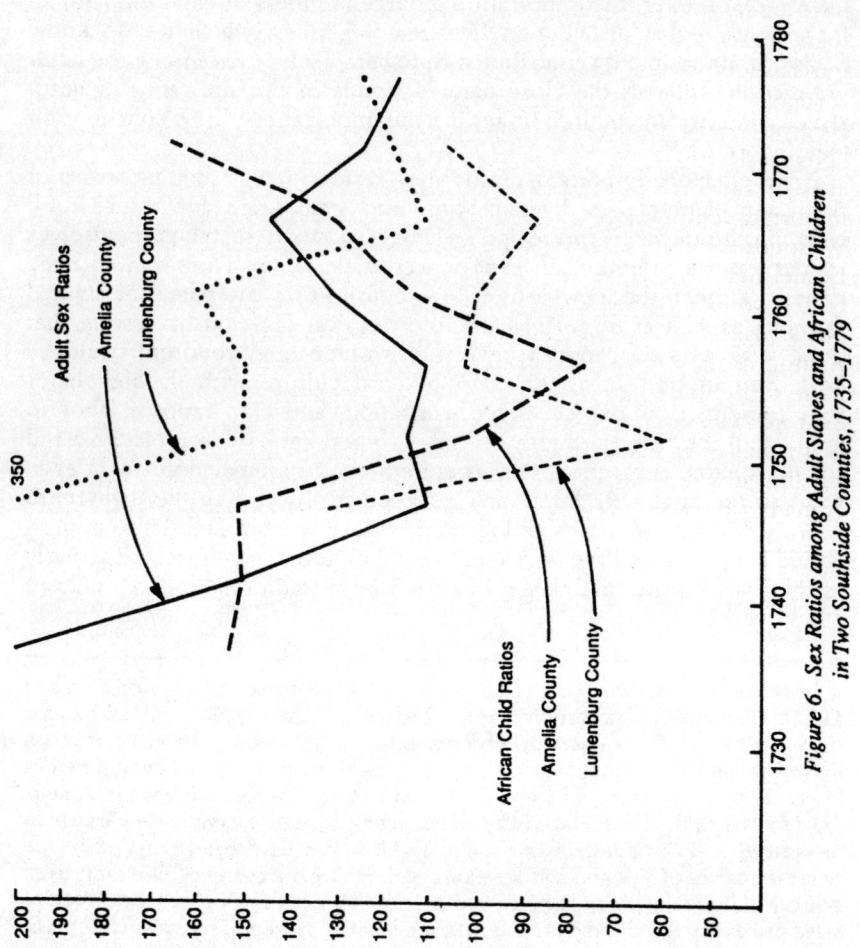

Figure 6. *Sex Ratios among Adult Slaves and African Children in Two Southside Counties, 1735–1779*

Moreover, some piedmont planters, unlike their tidewater counterparts of the seventeenth and early eighteenth centuries, recognized the reproductive value of slave women. For the seventeenth century, Russell R. Menard has shown that wealthier planters in tidewater Maryland—the very people with the means to equalize sex ratios among their slaves—actually created the most unbalanced slave complements. In tidewater Virginia, well into the eighteenth century, planter perceptions seem to have undergone little change. Most testators, when they specified how their legacies were to be expended on human property, ordered the purchase of slaves irrespective of gender. In York County this was certainly the case. Alexander Anderson hoped his son might "purchase a Negroe"; John Smyth ordered his land in England to be sold to buy "Negro slaves"; Joseph White gave his two daughters a slave between the ages of sixteen and twenty-six, suggesting that he thought age a more important consideration than gender.[30] Much the same was true in Westmoreland County. Buy ten slaves from "any Negro ship," ordered Daniel McCarty; purchase "a negro boy or girl," stipulated John Baker; buy "two young negroes," commanded Thomas Arrington. Northern Neck planters, too, placed a greater priority on age than gender. Henry Ashton desired that seven slaves, "none to be over forty years old nor none to be under ten years old," be purchased for his heirs, and Joseph Sanford willed that "two young Negro slaves" be bought for his son. Nathaniel Pope offered a rare contrary view. "Nothing is more to the advantage of my son," he wrote in 1719, "th[a]n young breeding negroes." A man of his word, he purchased two fifteen-year-old girls for this heir.[31]

Counterpoint, and Lois Green Carr and Lorena S. Walsh, "Economic Diversification and Labor Organization in the Chesapeake, 1650-1820," in Stephen Innes, ed., *Work and Labor in Early America* (Chapel Hill, N.C., 1988), 144-188. For general accounts of women's work under slavery see Jacqueline Jones, *Labor of Love, Labor of Sorrow: Black Women, Work, and the Family from Slavery to the Present* (New York, 1985), 11-43, and Deborah Gray White, *Ar'n't I a Woman? Female Slaves in the Plantation South* (New York, 1985), 66-67, 114, 120-121, 128-129.

[30] Menard, "Maryland Slave Population," *WMQ*, 3d Ser., XXXII (1975), 44; wills of Alexander Anderson, Jan. 9, 1687, John Smyth, Dec. 22, 1687, and Joseph White, Feb. 27, 1706, York County Deed, Order, and Will Books, 7, 324; 8, 92-94; 14, 76-77. All York Co. wills from 1675-1770 were searched; no testator expressed a preference for female slaves. Essex Co. wills were checked selectively. Only one testator, late in the colonial period, expressed a preference: he willed that "as young negroes of the female kind are much wanting on my estate I desire that my negro fellow Cupid be sold by my executors for the most that can be got as soon as conveniently may be and the money owing thereby laid out in purchasing a couple of Negro girls" (Will of John Richards, Feb. 12, 1773, Essex County Wills 12, 529-530).

[31] Wills of Daniel McCarty, Mar. 29, 1724, John Baker, Nov. 19, 1727, Thomas Arrington, Oct. 16, 1715, Henry Ashton, Feb. 26, 1731, Joseph Sanford, Apr. 3, 1741, and deed of gift of Nathaniel Pope, May 26, 1719, Westmoreland County Deeds and Will Books, 8, Part 1, 18, 91; 5, 498; 8, Pt. 1, 160-161; 9, 155-156; 6,

Piedmont planters seem to have agreed with Pope. When providing for slave purchases in their wills, they almost always stated a preference for women over men. In 1751 William Watson commanded his executors to spend all his available cash on "Neger Wenches or Girls." Eleven years later, John Sandford ordered his land to be sold so that slave women could be bought. In 1773 John Speed, Jr., directed his executors to hire out his slaves annually, to sell his land at three years' credit, and to invest the proceeds "in young Negroes chiefly wenches," who were to be hired out. In the same year Benjamin Wimbish of Prince Edward County requested that his executors buy "a young wench" if his runaway mulatto man, Toba, could be caught and sold. On the eve of the Revolution, Sion Spencer, former sheriff of Charlotte County, desired his executors to purchase "young Negro wenches for the benefit of the increase of my Estate and Children."[32] Piedmont planters seem to have been well aware that young slave women could "increase" an estate.

If the proportion of adult women in the black population of the piedmont is immediately striking, even more so is the proportion of children. From the first years of piedmont settlement children outnumbered women by at least 2 to 1. This ratio dropped somewhat in the late 1760s and 1770s, but was still around 1.75 to 1 (see Figure 3). By the early 1780s, when it is possible to compare adult-child ratios systematically across the state of Virginia, there was a clear difference. In over four-fifths of piedmont counties the ratio was over 110 in 1784 but below that figure in almost half of tidewater counties and two-thirds of Shenandoah Valley counties.[33] Slavery in the piedmont exhibited a more youthful profile than that in either of the other two Virginia regions.

480-481. All Westmoreland Co. wills from 1712-1773 were searched; most expressed no preference for female slaves; in just two cases, females were ordered to be bought for a daughter or granddaughters—wills of Joseph Sanford, Apr. 3, 1741 and Jemima Spence, Westmoreland County Deeds and Wills, 9, 155-156; 12, 284-285. Sanford ordered that four slave girls or women be purchased "out of a Negro Ship as soon as Conveniently they can after my said Daughter shall marry, the said four negroes to be as near as can be adjudged of her own age."

[32] Will of William Watson, Dec. 3, 1751, Amelia County Will Book 1, 78-79; will of John Sandford, Mar. 1762, Lunenburg County Will Book 1, 340-341; will of John Speed, Jr., Oct. 1773, Mecklenburg County Will Book 1, 186-187; will of Benjamin Wimbish, July 1773, Prince Edward County Will Book 1, 168; will of Sion Spencer, July 1775, Charlotte County Will Book 1, 129-130. Piedmont planter awareness of the reproductive capacities of slave women did not translate into higher prices for females than males.

[33] This state-wide survey is based on the Property Tax Recapitulations, 1784, Miscellaneous Auditor's item no. 296, Va. State Lib. together with certain county personal property assessments to fill in missing data. In 1786, another state-wide survey shows that adult-child ratios had declined throughout Virginia, but most notably in the tidewater. In the piedmont, 70% of counties registered adult-child ratios over 110, whereas this was true of fewer than half of tidewater counties. See

A number of reasons can be adduced to explain this difference. First, there is the character of the piedmont's creole immigrants. Tidewater planters seem to have favored young slaves for their piedmont quarters, in part because they could be shifted with least disruption to families, in part because they might be thought good pioneers, and in part because they, like women, could provide useful service in traditional tobacco culture. Second, there is the character of the region's African immigrants. African cargoes to the late colonial Chesapeake seem to have comprised more children (and females) than was typical in the trade generally. Since most of these Africans ended up in the piedmont, this influx also enhanced the youthfulness of the piedmont slave population. Finally, the more benign environment of the piedmont as compared to the tidewater encouraged the reproductiveness of both creoles and Africans, while the relocation of many native-born slave women from tidewater to piedmont meant that they could contribute to the natural growth of the region's black population almost from the first. The childbearing pattern of the slave women with whom Kate came into contact lends support to these explanations. Both Belinda and Barbara had a first child at least by age eighteen; Dicey, an African, had given birth to three children by her early twenties; Amey was a mother of three children by her twenty-fifth birthday. Kate herself became a mother at age twenty and could boast four children before her thirtieth year.[34]

Although the ratio of children to women among piedmont slaves was surprisingly high for most of the colonial period, a notable decline occurred in the late 1760s and 1770s. This downward trend can also be explained in many ways. First, it took place just when the piedmont became the receiver of virtually all Africans imported into the colony. And, while these African cargoes may have comprised more children than was typical, adults still predominated. As with most immigrant populations, the upheaval of moving disrupted family formation. Some women had to wait a number of years—valuable childbearing years—before finding a mate. Moreover, it seems probable that African women nursed their children for longer periods than did their creole counterparts, a practice that further depressed overall fertility.[35] Second, even though young slaves undoubtedly composed a significant proportion of creole

---

Property Tax Recapitulations, 1786, Miscellaneous Auditor's item no. 49, Va. State Lib. These lists report adults and children; it is no longer possible to isolate child-female ratios.

[34] More age-at-conception data, which support these observations, are available in Morgan, "Slave Life in the Virginia Piedmont," in Carr et al., eds., *Chesapeake Society*. 445. See also Gundersen, "Double Bonds of Race and Sex," *Jour. So. Hist.*, LII (1986), 360-361.

[35] Herbert S. Klein and Stanley L. Engerman, "Fertility Differentials between Slaves in the United States and the British West Indies: A Note on Lactation Practices and Their Possible Implications," *WMQ*. 3d Ser., XXXV (1978), 357-374.

immigrants, adults predominated among their numbers, too. This influx of creole adults depressed female-child ratios in the short run. Finally, the fertility of slave women may well have declined during the late colonial era owing to new work requirements. The shift toward a more diversified economy, which occurred even in the predominantly tobacco-growing areas of the piedmont, may well have had adverse effects on pregnant women. As slave men assumed new skilled positions as carters, plowmen, and tradesmen—activities directly related to the increased cultivation of grains and raising of livestock—the burden of additional field work fell disproportionately on the shoulders of women. Declining child-female ratios in the third quarter of the eighteenth century may therefore reflect a harsher work regimen for slave women.[36]

Although large numbers of children and a rough balance of men and women occurred remarkably early in the history of the piedmont's slave population, these characteristics are generalizations drawn from the region as a whole. Aggregate statistics can often mask immense variations. Numbers of slave men and women may approach equality in regional terms but prove remarkably unequal on individual plantations. An investigation of the plantation setting, particularly of variations in size of estate and in the presence or absence of whites, can tell us much about the quality of life among piedmont slaves.

Most piedmont slaves, like Kate, resided on small to middling plantations or quarters. In the early years of a piedmont county, a clear majority of slaves lived on quarters comprising fewer than ten slaves. Equally clearly, the size of these units increased remarkably quickly. Generally speaking, it took no more than one or two decades after a piedmont county was formed before a majority of its slaves resided on plantations with more than ten slaves (see Table VII). Twitty's Creek quarter, it will be recalled, housed well over ten slaves by the mid-1770s. In the following decade, about 40 percent of slaves in most piedmont counties lived on plantations of more than twenty slaves. By the Revolutionary era, the concentration of slaves on medium-sized estates with eleven to twenty-nine slaves was little different between the piedmont and tidewater regions of Virginia.[37] Although large estates (thirty or more slaves) were rare in the late eighteenth-century piedmont, most of the region's slaves lived on plantations of sufficient size to support a measure of community life (see Table VIII).

But what about family life? In a newly formed piedmont county, where plantations were small, one might expect that adult slaves would find

---

[36] That work levels did have an impact on slave fertility see John Campbell, "Work, Pregnancy, and Infant Mortality among Southern Slaves, *Journal of Interdisciplinary History*. XIV (1984), 793-812. See also Gundersen, "Double Bonds of Race and Sex," *Jour. So. Hist.*. LII (1986), 366-367.

[37] Allan Kulikoff, "The Origins of Afro-American Society in Tidewater Maryland and Virginia, 1700 to 1790," *WMQ*. 3d Ser., XXXV (1978), 246-248.

SLAVES IN PIEDMONT VIRGINIA

TABLE VII
PLANTATION SIZES IN THE PIEDMONT, 1720-1779

County/Year	Proportion of Slaves Living on Units of				Number of Slaves
	1-5	6-10	11-20	21+	
Spotsylvania					
1720s	30	35	35		37
1730s	6	42	20	32	209
1740s	18	26	29	27	292
1750s	11	21	25	43	618
1760s	12	22	35	31	592
1770s	19	26	23	32	352
Goochland					
1730s	49	30	21		71
1740s	38	23	39		128
1746*	21	34	22	23	2,246
1750s	32	38	13	17	180
1760s	32	45	10	13	160
1770s	13	25	38	25	464
Orange					
1730s	50	50			28
1740s	18	17	41	24	226
1750s	8	3	39	50	238
1760s	15	33	40	12	211
1770s	12	40	40	8	255
Amelia and Prince Edward					
1730s	100				8
1736	30	35	28	7	698
1740s	31	16	53		107
1749	23	35	25	17	2,842
1750s	19	11	13	57	311
1755 (P.E.)*	22	45	21	11	482
1760s	12	20	31	37	699
1767*	13	28	30	29	7,098
1770s	13	18	44	25	511
Lunenburg, Charlotte, Mecklenburg					
1740s	100				4
1750	30	34	13	23	1,540
1750s	39	36		25	114
1760s	20	19	13	48	359
1764*	20	32	27	22	4,546
1770s	10	19	55	16	482

TABLE VII (CONTINUED)
PLANTATION SIZES IN THE PIEDMONT, 1720-1779

County/Year	Proportion of Slaves Living on Units of				Number of Slaves
	1-5	6-10	11-20	21+	
Chesterfield					
1750s	14	41	45		345
1756*	17	29	27	27	2,420
1760s	8	13	44	35	518
1770s	12	25	37	26	878
Halifax and Pittsylvania					
1750s	51	49			37
1755	15	44	14	27	254
1760s	42	32	26		95
1767 (Pitt.)	30	34	23	13	646
1770s	32	29	24	15	144
Loudoun					
1760	23	29	25	28	992
1760s	42	8	50		102
1770s	16	34	30	19	274

*Incomplete listing

Notes: Information for a whole decade is derived from inventories. Inventory data are particularly useful when determining plantation size, because the slaves are occasionally listed by quarter. We have measured unit, rather than estate, size wherever possible. Sources are the Will Books of Amelia County 1-2 (1734-1780), Charlotte County 1 (1765-1791), Chesterfield County 1-3 (1750-1785), Goochland County 1-12 (1728-1779), Halifax County 0-1 (1753-1782), Loudoun County A-B (1757-1782), Lunenburg County 1-2 (1746-1778), Mecklenburg County 1 (1765-1782), Orange County 1-3 (1735-1801), Pittsylvania County 5 (1767-1780), Prince Edward County 1 (1754-1785), and Spotsylvania County A-E (1727-1798), Va. State Lib.

Information for a single year is derived from tithable lists. All are in the Va. State Lib. except for the Loudoun listing, which is analyzed in Donald M. Sweig, "Northern Virginia Slavery: A Statistical and Demographic Investigation" (Ph.D. diss., College of William and Mary, 1982), 42-43; the Lunenburg listing of 1764, which is found in Landon C. Bell, *Sunlight on the Southside: Lists of Tithes, Lunenburg County Virginia, 1748-1783* (Philadelphia, 1931), 212-268; and the Prince Edward listing of 1755, which is found in *Tyler's Quarterly Historical and Genealogical Magazine*, XVIII (1937), 50-54. We have doubled tithable numbers to arrive at approximate plantation sizes.

The Southside counties have been arranged in three contiguous groups in order to increase the number of slaves analyzed in any one decade. By combining an older county with its offshoots, the concentration of slaves in the more established county is not fully captured, but the misrepresentation is not marked.

We are impressed by the general congruence between tithable and inventory data in Virginia. Cf. Jean Butenhoff Lee, "The Problem of Slave Community in the Eighteenth-Century Chesapeake," *WMQ*, 3d Ser., XLIII (1986), 346n.

SLAVES IN PIEDMONT VIRGINIA

TABLE VIII
PLANTATION SIZES IN THE PIEDMONT, 1782-1785

County	Proportion of Slaves Living on Units of				Number of Slaves
	1-5	6-10	11-20	21+	
Amelia	10	16	28	46	8,749
Cumberland	12	21	24	43	3,934
Chesterfield	11	19	30	40	5,961
Goochland	15	20	24	41	3,840
Albemarle	15	19	26	40	4,409
Louisa	15	19	29	37	4,485
Orange	16	15	30	38	2,848
Amherst	18	18	28	36	3,852
Fluvanna	22	21	19	37	1,330
Halifax	22	24	32	22	3,290
Pittsylvania	23	32	25	20	1,835
Totals	15	19	27	39	44,533

Note: Individual county listings are from personal property assessments, Va. State Lib., or *Heads of Families—Virginia*.

partners only with difficulty. This was undoubtedly the case. In fact, in a number of counties, about three-quarters of the adult slaves lived on plantations where members of only one sex resided or where men heavily outnumbered women. But more propitious conditions developed quite quickly. In most piedmont counties, within a couple of decades of their formation, at least a half and, in some cases, almost two-thirds of adult slaves resided on units with quite evenly balanced numbers of men and women. Indeed, in piedmont counties formed late in the century, the initial period of marked sexual imbalances hardly occurred at all (see Table IX).

Opportunities for finding a spouse or mate varied with plantation size, but not markedly. In Amelia County in 1736 men predominated on plantations with fewer than three adult slaves; much more balanced ratios characterized the larger plantations. However, at mid-century, this differential between large and small plantations was much smaller in Lunenburg and Chesterfield counties, and in Goochland County it was virtually nonexistent. Indeed, in counties formed later in the century, such as Prince Edward and Pittsylvania, and in long-settled counties like Louisa, there was even something of a reverse phenomenon. There, the sex ratio was actually lower on the smaller plantations (see Table X).

Opportunities for a more stable family and community life were enhanced by the growing weight of black numbers in the piedmont. As Richard S. Dunn has recently demonstrated, of the ten Virginia counties that had black majorities in the middle of the eighteenth century only three were located in the piedmont. A generation later, the piedmont

TABLE IX
PROPORTION OF SLAVES ON PLANTATIONS WITH
VARYING ADULT SEX RATIOS

County/Years	Single-Sex Estates	Near-Balanced Estates	Unbalanced Estates	Number of Adults
Spotsylvania				
1720s	15	10	75	20
1730s	3	63	34	107
1740s	10	37	53	119
1750s	7	32	61	292
1760s	9	50	41	256
1770s	8	59	33	213
Goochland				
1730s	27	18	57	23
1740s	22	27	51	68
1746	13	44	43	1,123
1750s	21	32	47	92
1760s	41	43	16	73
1770s	9	53	38	233
Orange				
1730s	45	55		11
1740s	14	51	35	114
1750s	5	69	26	110
1760s	15	55	30	115
1770s	13	51	36	128
Amelia and Prince Edward				
1730s		40	60	5
1736	28	37	35	348
1740s	13	62	25	39
1749	19	46	35	1,413
1750s	14	67	19	165
1760s	9	56	35	396
1767	10	54	35	3,523
1770s	7	47	46	272
Lunenburg, Charlotte, Mecklenburg				
1750	19	45	36	752
1750s	19	35	46	63
1760s	16	59	25	136
1770s	7	50	43	195

SLAVES IN PIEDMONT VIRGINIA 243

TABLE IX (CONTINUED)
PROPORTION OF SLAVES ON PLANTATIONS WITH
VARYING ADULT SEX RATIOS

County/Years	Single-Sex Estates	Near-Balanced Estates	Unbalanced Estates	Number of Adults
Chesterfield				
1756	14	43	43	1,210
1750s	8	38	54	164
1760s	7	58	35	238
1770s	11	51	38	308
Halifax and Pittsylvania				
1750s	14	57	29	21
1760s	15	50	35	72
1767	21	58	21	320
1770s	16	47	37	87
Cumberland				
1783	10	57	33	1,670
Louisa				
1783	9	61	30	1,423

Notes: Near-balanced estates are ones where the adult sex ratios range from <67 to >150. Unbalanced estates have adult sex ratios above 151 and below 66. It should be noted that the proportion of adult slaves on estates with women only or with a marked female majority (that is, less than 66) rose substantially, so that by the late colonial period it stood at 15%-20%. Since men outnumbered women in the general population, women on such estates presumably could find mates in the neighborhood.

Sources are the same as for Table VII except for the use of two personal property assessments of 1783 that list names of all slaves, though the Louisa listing is incomplete in this respect. These are on film at the Va. State Lib.

region had increased its share to ten counties in the state total of twenty-four.[38] In 1782 one-third of the counties between the fall line and Blue Ridge, as compared to 47 percent in the tidewater, contained more blacks than whites.

But did the large size of piedmont counties not dilute the impact of rapidly growing slave numbers and black majorities? To some degree, this was so. Thus in 1755 Orange, Louisa, Chesterfield, Dinwiddie, Amelia, and Southampton counties contained roughly 220 acres of taxed land for each tithable slave. By comparison, Gloucester, a tidewater county with the colony's largest number of tithable slaves, taxed approximately fifty-three acres for every black tithable, a density four times that of these piedmont counties. On the other hand, on average, a tidewater county had

[38] Dunn, "Black Society in the Chesapeake," in Berlin and Hoffman, eds., Slavery and Freedom. 56-58.

TABLE X
ADULT SLAVE SEX RATIOS BY PLANTATION SIZE

County/Year	Number of Adults						
	1	2-3	4-5	6-7	8-9	10+	Total
Amelia, 1736							
R	430	296	156	150	117	171	203
% [N=349]	15	26	24	7	15	13	100
Goochland, 1746							
R	159	140	145	148	168	155	150
% [N=1,123]	7	27	21	6	13	26	100
Amelia, 1749							
R	172	158	166	131	105	136	147
% [N=1,409]	9	26	22	15	9	19	100
Lunenburg, 1750							
R	191	139	155	117	200	121	143
% [N=752]	14	28	22	9	3	25	100
Chesterfield, 1756							
R	119	141	125	148	145	133	135
% [N=1,210]	7	22	19	13	11	28	100
Amelia, 1767							
R	122	135	116	115	131	124	122
% [N=3,080]	4	18	17	16	11	34	100
Pr. Edward, 1767							
R	100	125	109	77	150	173	127
% [N=443]	5	22	30	7	8	28	100
Pittsylvania, 1767							
R	71	122	98	117	118	148	106
% [N=320]	18	19	27	12	8	16	100
Cumberland, 1783							
R	76	107	99	103	89	100	99
% [N=1,668]	5	18	14	19	10	33	1,668
Louisa, 1783							
R	88	88	110	89	124	107	102
% [N=2,041]	6	15	18	14	12	35	100

Note: R = adult sex ratio; % = percentage of adult slaves in groups of that size; N = number of adult slaves. Sources are the same as for Table IX.

a black population density 2.5 times that of a piedmont county in 1755. This difference narrowed over the course of the next quarter century, so that in 1787 it had been reduced to 1.7 times that of the average piedmont county.[39]

[39] Taxable acres are a rough proxy for settled land. They are reported for colonial Virginia counties in C.O. 5/1327-1330, Public Record Office, and have been compiled by George H. Reese, MS in authors' possession. The county-wide lists of black tithables in 1755 are reported in R. A. Brock, ed., *The Official Records*

SLAVES IN PIEDMONT VIRGINIA

Moreover, even within a large piedmont county, the density of the slave population varied considerably. In Amelia County in 1767, the ratio of taxable acres per tithable slave ranged from 92 in the area between Deep and Flatt creeks in the north central part of the county to 146 in part of Nottoway Parish in the southern half. Nearly 70 percent of this county's adult slaves lived in neighborhoods where the ratio of taxable acres to black tithables was 115 or less. This concentration of blacks approached average tidewater levels of just a decade earlier. By 1787 the northern half of Amelia County averaged 75 acres for each tithable slave, while the southern half, soon to become Nottoway County, averaged only 85 acres for each of its tithable slaves.[40]

This growing concentration of blacks raises the question of how closely piedmont slaves rubbed shoulders with masters and other whites. In the very earliest years of piedmont counties, a large proportion of slaves lived on quarters, usually with a white overseer, but sometimes with no white present. These quarters contained from one-fourth to just over a half of adult slaves in newly formed piedmont counties. When Amelia County was in its infancy, at least a third of its tithable slaves lived on units whose owner lived in another county. In Loudoun County, six years after its formation, over a half of all tithable slaves lived on such quarters (see Table XI).[41]

Over time, however, units without masters became more infrequent. In Chesterfield County in 1756, only one-fifth of adult slaves lived on estates where the master was absent; in longer-settled counties like Amelia, Cumberland, and Louisa later in the century the proportion was closer to one-tenth. As more piedmont whites became slaveholders late in the century, more slaves came under the direct influence of resident masters. Indeed, this soon became the overwhelming pattern. At the same time, the proportion of slaves living on quarters close to a home plantation began to rise. A quarter of Amelia County slaves in 1767 and a sixth of Cumberland County slaves in 1783 lived on a complex of units comprising home plantation and adjacent quarters (see Table XI). Kate's eventual

---

of Robert Dinwiddie, Lieutenant-Governor of the Colony of Virginia, 1751-1758..., Virginia Historical Society, *Collections*, IV (Richmond, Va., 1884), II, 352-353. A comparable set of figures is available for 1787 (when the tithable age was the same as in 1755). Taxable acres are reported in county land books in the Va. State Lib. (for 9 tidewater and 15 piedmont counties this figure was totaled). The corresponding county figures on adult blacks (aged 16 and above) can be found in a 1787 listing, Auditor's Item no. 49, Va. State Lib. In 1755 the average taxable acre/black tithable was 108 in tidewater counties and 275 in piedmont counties (median 96 and 220 respectively), based on 37 county returns. In 1787 it was 85 in tidewater counties and 145 in piedmont counties (median was 85 and 130 respectively), based on 24 county returns.

[40] Amelia County Tithable list, 1767, and personal property tax list, 1787.

[41] See also Kulikoff, *Tobacco and Slaves*, 97, 153. In early Orange Co. (1739), 24% of tithables lived on quarters: "Orange County Tithe Lists," *WMQ*, 2d Ser., XXVII (1918), 22-27.

TABLE XI
PROPORTION OF ADULT SLAVES ON PLANTATIONS AND QUARTERS

County/Year	Resident Masters	Resident Masters and Outlying Quarters[a]	Quarters with White Overseers[b]	Quarters with Blacks Only[b]
Amelia				
1736 [N=349]	61	3	23	13
1740 [N=578]	42	16	29	13
1749 [N=1,420]	57	12	25	5
1767 [N=3,081]	60	26	10	2
Goochland				
1746 [N=1,123]	48	14	31	7
Lunenburg				
1750 [N=774]	60	4	30	6
Chesterfield				
1756 [N=1,210]	66	15	15	5
Loudoun[c]				
1760 [N=496]		c48	46	6
Prince Edward				
1767 [N=443]	57	17	24	2
Pittsylvania				
1767 [N=323]	76	5	18	1
Cumberland				
1783 [N=1,677]	70	16	10	4
Louisa				
1783 [N=4,485]	78	11	5	6

[a] If a slaveowner is credited with 10 or more tithables, of whom at least one seems to be a white overseer, it can be inferred that outlying quarters existed. For example, in 1756 Archibald Cary of Chesterfield Co. returned 38 tithables, 4 of whom were whites with surnames other than Cary. It seems highly likely that many of Cary's 33 adult slaves wre placed under the supervision of these whites on outlying quarters within the county. Occasionally, a listing refers to an "overseer" in a household with a resident master.

[b] Tithable lists refer to quarters directly, as in "William Bolling's Quarter," or indirectly, as in "William Bolling's list." In such a case, William Bolling is not counted as a tithable by the enumerator; rather, the unit is either headed by a white overseer, sometimes stated explicitly, or is composed only of slaves.

[c] Sweig, "Northern Virginia Slavery," 39-40. Sweig did not distinguish between those estates with and without outlying quarters.

Sources are the same as for Table IX.

residence on a quarter beyond the close control of her master was therefore not unusual, though living with a resident master became the typical slave experience in the late eighteenth-century piedmont.

What, then, is the import of Kate's story when placed in this larger context? In some respects, Africans like Kate who reached the piedmont in the third quarter of the eighteenth century replicated many of the experiences of their predecessors in the tidewater during the previous half-century. As before, these newcomers contributed to extremely rapid population growth; in certain places at certain times, the adult slave population was heavily African; as waves of immigrants moved into the region, imbalances between men and women occurred, making stable family life difficult to achieve; and, in the early years of settlement, most new arrivals resided on small plantations and quarters.

Thus there were similarities between this later piedmont and earlier tidewater slave experiences, yet the differences are more striking. Even the composition of African immigrants seems to have changed. African girls, and perhaps women, formed a larger proportion of newcomers than ever before. Because of this, and because of a similar trend among creoles, the numbers of slave men and women moved rapidly toward equality. In addition, children formed a significant proportion of the piedmont's slave population from the beginning. The opportunities for family life were therefore not negligible. Moreover, women and children were important to production in the piedmont at an early stage in the region's development. Whether they gained any material advantages from the early assumption of these field work responsibilities remains an open question. Finally, before long, most Africans and creoles found themselves on middle-sized, not small, estates, often in neigborhoods where many other slaves resided. At the same time, the chances to live beyond the control of a white master were greatest in the earliest years of piedmont settlement, in marked contrast to the tidewater pattern.

Kate's situation can hardly be described as fortunate. Nevertheless, she and many other immigrants like her at least enjoyed more favorable demographic and social conditions than had their predecessors in the tidewater. Newcomers like Kate discovered it much easier to find a mate; they had more children; they were less isolated from fellow slaves; they played a central role in the primary productive activity of the region, which may have afforded them some advantages; and some even lived beyond the purview of whites. The demographic constraints that existed for generations in the oldest tidewater counties disappeared in a matter of decades on the frontier.

APPENDIX I

AFRICAN IMMIGRANTS TO THE PIEDMONT

A Virginia statute of 1680 stipulated that all recently imported African children had to be brought into county court to have their ages adjudged. At that time, children were judged to be twelve years or under for the purposes of this act; in 1705 this age limit was raised to sixteen.[42] A search of all extant piedmont county court order books, with adjustments for

[42] William Waller Hening, *The Statutes at Large. Being a Collection of All the Laws of Virginia* ... (Richmond, Va., and Philadelphia, 1809-1823), II, 479-480, III, 258-259, VI, 40-41.

AFRICAN CHILDREN BROUGHT INTO PIEDMONT COUNTY COURTS,
1725-1774

Years	Central Piedmont	Southside	Northern Piedmont	Total
1725-1729	65 [29]			65 [29]
1730-1739	306 [193]	138 [138]	34 [0]	478 [331]
1740-1749	383 [282]	228 [206]	31 [0]	642 [488]
1750-1759	561 [426]	506 [439]	150 [76]	1,217 [941]
1760-1769	692 [479]	662 [530]	107 [73]	1,461 [1,082]
1770-1774	259 [185]	367 [329]	44 [20]	670 [534]
Totals	2,266 [1,594]	1,901 [1,642]	366 [169]	4,533 [3,405]

Notes: Bracketed figures are the totals based on extant returns. To calculate missing returns, we plotted projections based on the data of adjoining counties.

The central piedmont comprises the counties of Spotsylvania, Hanover, Goochland, Orange, Louisa, Albemarle, Cumberland, Chesterfield, Amherst, and Buckingham. The records of two of these counties (Hanover and Buckingham) are missing entirely; three are incomplete (Louisa for 1748-1765, Albemarle for 1749-1774, and Amherst for 1761-1765 and 1770-1773); the other five are complete.

The Southside comprises the counties of Brunswick, Amelia, Lunenburg, Halifax, Dinwiddie, Prince Edward, Bedford, Charlotte, Mecklenburg, and Pittsylvania. The records of Dinwiddie Co. are missing entirely; those of Brunswick are incomplete for 1743-1745 and 1761-1763; the other eight are complete.

The northern piedmont comprises the counties of Prince William, Culpeper, Loudoun and Fauquier. The records of Prince William and Culpeper are incomplete (1731-1751, 1758-1759, 1764-1765, 1770-1774, and 1749-1762, 1764-1774 missing respectively); the other two are complete.

missing data, produced the numbers displayed below. It is generally thought that children formed about one-sixth of slave cargoes (although, elsewhere in this article, we wonder whether this proportion is too low, particularly for the later slave trade to Virginia). However, for the purposes of calculating the total number of African immigrants to the piedmont, this multiplier will suffice, since we can probably assume that a good number of newly imported African children went unnoticed by the county courts. The numbers recorded in Table III are therefore derived by multiplying the totals in the table below by six.

Trends in African child registrations correspond closely to known import figures (see Figure 7).[43] Similarly, the months in which children were registered and slaves imported also move in tandem. A purchaser of a newly imported slave child was required to bring his or her charge into court within about three months of arrival in Virginia.[44] Allowing for an inevitable lag

[43] African child registrations are the adjusted yearly totals reported in this appendix; slave imports are the adjusted yearly totals reported in Appendix 2, based on Minchinton et al., eds., *Virginia Slave-Trade Statistics*.

[44] See laws cited in n. 42 above.

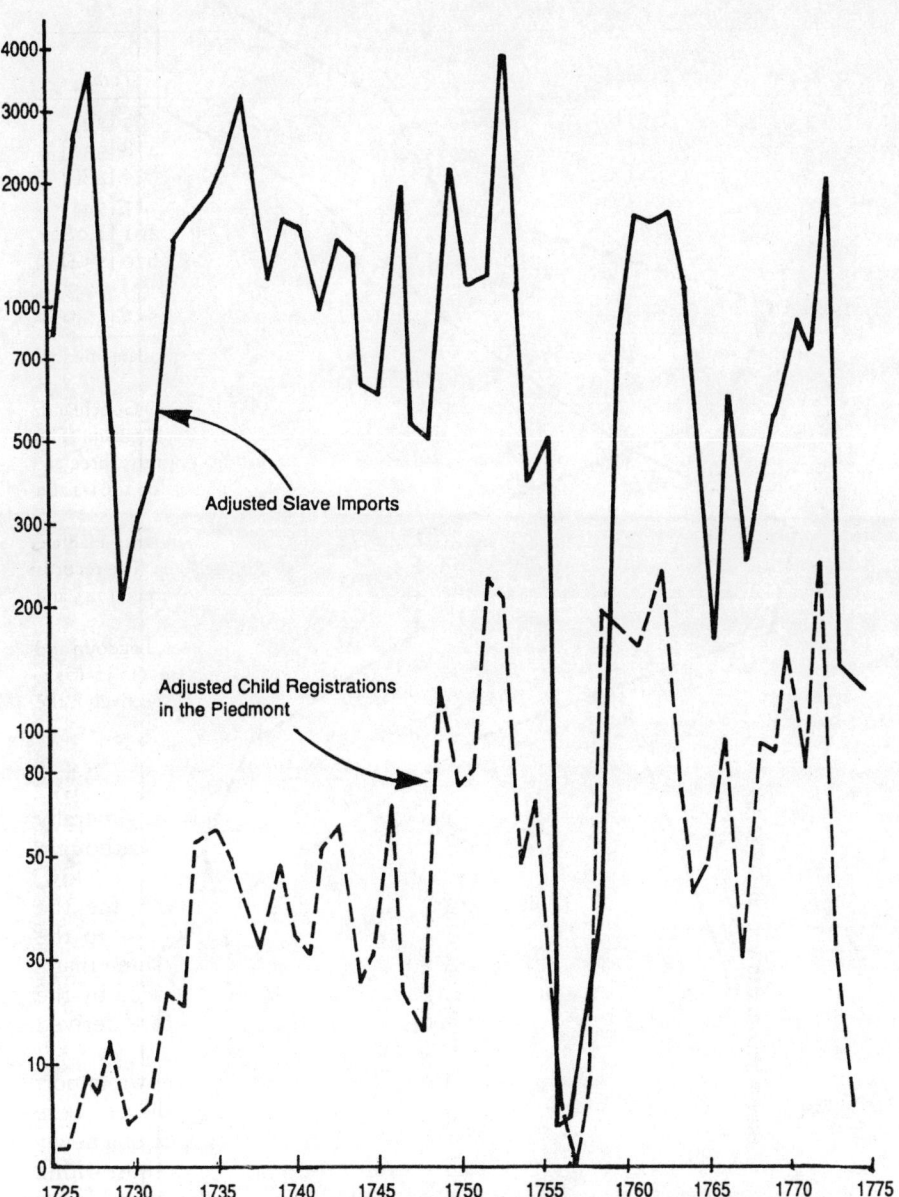

Figure 7. Slave Imports into Virginia and African Child Registrations in Piedmont Counties, 1725–1775

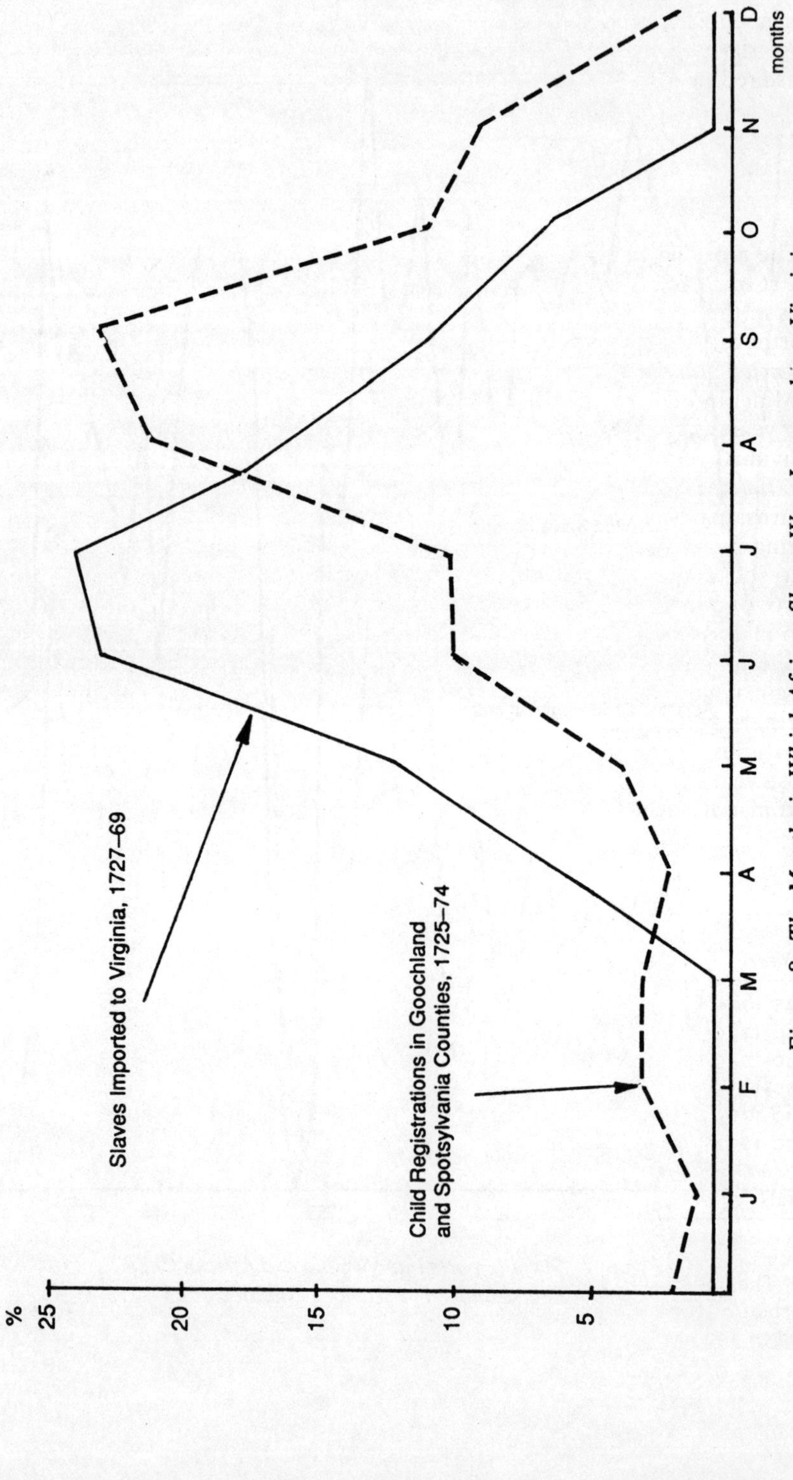

Figure 8. The Months in Which African Slaves Were Imported into Virginia and Slave Children Registered in Two Piedmont Counties, 1725–1774

SLAVES IN PIEDMONT VIRGINIA

between the time an African was imported and the appearance of the newly purchased African child in county court, the synchronization is close to exact. Most slaves arrived in Virginia in June and July. Most children were registered in August, September, and October (see Figure 8).[45]

APPENDIX 2

SLAVES IMPORTED INTO VIRGINIA, 1725-1774

The most reliable listing of slave cargoes to Virginia is that of Minchinton et al., eds., *Virginia Slave-Trade Statistics*. Unfortunately, there are gaps in the records and perhaps a measure of smuggling, particularly into Northern Virginia. The known figures and our revisions are set out below.

Known Numbers: these figures concur with those presented in Tables 3-6 of Minchinton et al., eds., *Virginia Slave-Trade Statistics*, xiv-xv, except for two years, 1749 and 1768, when there was a slight undercounting of the individual returns.

Estimated Numbers: making allowances for missing Naval Office shipping returns, particularly in customs districts normally receiving slaves and during quarters of the year when slaves could be expected, we estimate that 500 additional slaves were imported in 1728-1729, 690 in 1730-1736, 950 in 1746-1748, 200 in 1750-1751, 1,000 in 1753, 100 in 1755, 500 in 1759, 550 in 1766-1769, 150 in 1770, and 450 in 1772-1774. In addition, on the basis of some evidence of smuggling into Northern Virginia, we estimate that an additional 1,000 slaves were brought into the northern piedmont in the late 1750s and early 1760s from Maryland, since the duties on imported Africans were much lower in that colony during those years. See Donald Sweig, "The Importation of African Slaves to the Potomac River, 1732-1772," *WMQ*, 3d Ser., XLII (1985), 507-524.

SLAVES IMPORTED TO VIRGINIA

Years	Known Numbers	Estimated Numbers
1725-1729	7,241	7,741
1730-1739	15,346	16,226
1740-1749	11,163	12,113
1750-1759	6,997	9,197
1760-1769	8,559	9,709
1770-1774	3,332	3,932
Totals	52,638	58,918

[45] The months in which African slavers arrived in Virginia are taken from Herbert S. Klein, "Slaves and Shipping in Eighteenth-Century Virginia," *JIH*, V (1975), 396.

7
Plantations, Paternalism, and Profitability: Factors Affecting African Demography in the Old British Empire

Daniel C. Littlefield

HISTORIANS HAVE OFTEN OBSERVED THAT THE BLACK POPULATIONS in the West Indies and Latin America, unlike those in North America, failed to reproduce themselves.[1] Indeed, the striking lack of population growth among blacks in the West Indies during the eighteenth century actually prompted the view that blacks were less fertile than whites.[2] The difference between North America and other plantation areas in this regard has caused scholars to reconsider previous theories about the nature of slavery in various New World societies, particularly those which posited the notion that slavery in North America was harsher than in Latin America. Clearly, if North American slavery were more stringent than any other, the evidence on slave reproduction was an inexplicable anomaly. Eugene D. Genovese has refined the argument on treatment, indicating that academic disputes often resulted because the word was used with divergent meanings,[3] and he and other scholars have discredited the hypothesis, advanced most cogently by Frank

---

[1] See for example C. Vann Woodward, "Southern Slaves in the World of Thomas Malthus," in Woodward, *American Counterpoint: Slavery and Racism in the North-South Dialogue* (Boston and Toronto, 1971), 78-106; Franklin W. Knight, *Slave Society in Cuba During the Nineteenth Century* (Madison, Milwaukee, and London, 1970), xvii; Philip D. Curtin in, among other places, "The Slave Trade and the Atlantic Basin: Intercontinental Perspectives," in Nathan I. Huggins, Martin Kilson, and Daniel M. Fox, eds., *Key Issues in the Afro-American Experience* (2 vols., New York and other cities, 1971), I, 92.

[2] Henry Ellis to Lord Hawkesbury, March 31, 1788, Liverpool Papers, CCXXVII, Additional Manuscripts 38416, ff. 69-70 (British Museum, London, England).

[3] Genovese, "The Treatment of Slaves in Different Countries: Problems in the Application of the Comparative Method," in Genovese, *In Red and Black: Marxian Explorations in Southern and Afro-American History* (New York, 1971), 158-72; also in Laura Foner and Genovese, eds., *Slavery in the New World: A Reader in Comparative History* (Englewood Cliffs, N. J., 1969), 202-10.

Tannenbaum, that varying types of slave institutions were preeminently a function of the culture of the master class.⁴ Yet the reasons for the phenomenal growth of North America's slave population, although receiving increasing attention, have not been adequately explained or explored. A host of factors have been considered, including the demographic configuration, family stability, nutrition, epidemiological conditions, the type and severity of work, and, most recently, lactation practices.⁵ Many of these considerations are interrelated or bear upon still other important factors, such as the age of menarche, but none alone is sufficient explanation of the natural population growth on North American plantations and its failure elsewhere. This phenomenon is probably, however, an aspect of the character of the society as a whole and not one particular facet of it, a notion this essay seeks to advance. In addition, it will adduce a consideration that has not been adequately explored, the influence of economics in the shaping of a social attitude.

Models already exist which would seem to encompass the essential differences between the variant plantation systems. Gerald W. Mullin has contrasted the paternal-autarkic nature of Virginia plantation society with what could be styled the modern-industrial nature of South Carolina slave society.⁶ This typology ought, perhaps,

⁴ See Genovese, "Materialism and Idealism in the History of Negro Slavery in the Americas," *Journal of Social History*, I (Summer 1968), 371–94; also in Genovese, *In Red and Black*, 23–52; and Foner and Genovese, eds., *Slavery in the New World*, 238–55; Marvin Harris, *Patterns of Race in the Americas* (New York, 1964), whose whole book practically is an attack on Tannenbaum; David B. Davis, *The Problem of Slavery in Western Culture* (Ithaca, 1960), *passim*; and Knight, *Slave Society in Cuba*, xvi, among others.

⁵ Recent works dealing with this issue include Michael Craton, "Hobbesian or Panglossian? The Two Extremes of Slave Conditions in the British Caribbean, 1783 to 1834," *William and Mary Quarterly*, 3d Ser., XXXV (April 1978), 324–56; Richard S. Dunn, "A Tale of Two Plantations: Slave Life at Mesopotamia in Jamaica and Mount Airy in Virginia, 1799 to 1828," *ibid.*, XXXIV (January 1977), 32–65; Herbert S. Klein and Stanley L. Engerman, "Fertility Differentials Between Slaves in the United States and the British West Indies: A Note on Lactation Practices and Their Possible Implications," *ibid.*, XXXV (April 1978) 357–74. Craton deplores the tendency to regard West Indian plantation societies as equally destructive of human life and has compared the Rolle plantation on Great Exuma Island in the Bahamas with the Worthy Park plantation in Jamaica. The Bahamas plantation was notable for the ability of its population to reproduce. In this instance, though, high fertility was correlated with a nuclear family structure and a pyramidal demographic pattern, configurations that did not exist on most West Indian plantations in either the eighteenth or the nineteenth century. Moreover, the slave force in the Bahamas had come from North America and the circumstances that made it exceptional might have been at least partially a result of that experience. Dunn has illustrated that many of the common assumptions made about contrasting conditions on plantations in the West Indies and North America, particularly in terms of sex ratios, while valid for the eighteenth, may not have been true for the nineteenth century but that they still did not alter the facts of reproduction.

⁶ Mullin, "Religion and Slave Resistance," paper prepared for the American Historical Association Meeting, New Orleans, 1972, pp. 2–7. Also see Mullin, *Flight and Rebellion: Slave Resistance in Eighteenth-Century Virginia* (New York, 1972), *passim*; and Michael

PATERNALISM AND SLAVERY

to be extended to include the West Indies (and other, similar societies) at one end of the scale and the Chesapeake at the other, with South Carolina somewhere in the middle and containing elements of both. This dichotomy is comparable to Eric R. Wolf's distinction between "old-style" plantations, which have a traditional, paternalistic ethos, and "new-style" plantations, which have a free-labor, capitalistic ethos.[7] These varying *ethea,* in association with material and environmental circumstances—climate, economy, epidemiology—might be crucial determinants of a situation in which a servile population could or could not reproduce itself. Abstractly, the *ethea* might be considered apart from the environment but in fact would be largely created by it. They could be used, therefore, as succinct expressions of the host of circumstances which brought them about.

The variance can be seen in concrete terms by considering two important aspects of slave reproduction, an adequate sex ratio and family stability. The paternalistic system would be characterized by a great interest—extending even to interference—in the personal life of the servant which tightly circumscribed his autonomy and hampered his initiative but compensated for this by a greater concern for his material well-being and respect for familial cohesiveness, which, among other things, dictated a more equal sex ratio. The industrial system would be typified by a comparative lack of concern about the private affairs of the bondsman, his household social structure, or his material welfare; its preeminent consideration would be business efficiency; and it would result in a relatively greater amount of individual freedom but also a less equal sex ratio. The more equal sexual distribution, in association with other aspects of the more paternal system, would encourage reproduction; the less equal sex ratio, in association with other aspects of the industrial system, would discourage reproduction.[8]

Mullin, ed., *American Negro Slavery: A Documentary History* (New York and other cities, 1976), *passim.* Mullin used the term "autarkic" but not the words "modern" or "industrial" in the paper and so should not be blamed for this typology, which may or may not express his meaning.

[7] See Eugene D. Genovese, *The World the Slaveholders Made: Two Essays in Interpretation* (New York, 1969), 11; and Wolf, "Specific Aspects of Plantation Systems in the New World: Community Sub-Cultures and Social Classes," in Pan American Union, *Plantation Systems of the New World (Social Science Monographs,* No. 7: Washington, D. C., 1959), 136-46.

[8] Peter H. Wood has outlined the retrogression in South Carolina from one type of system to the other, where largely West Indian settlers were encouraged by the Society for the Propagation of the Gospel to exercise more supervision over their slaves than had previously been their practice, eventuating in greater restrictions on the servitor's liberty and cultural vitality. See Wood, *Black Majority: Negroes in Colonial South Carolina from 1670 Through the Stono Rebellion* (New York, 1974), 131-42. The material benefits, therefore, might have been offset by greater psychological stress. A capsule series of comparisons can be found in Foner and Genovese, eds., *Slavery in the New World.*

The argument here is not that masters were more or less humane in one type of society as opposed to the other but that they perceived their interests differently and thereby acted in such a way that, in the one case, the slave received greater physical benefits. This conjunction can clearly be seen in East Florida, where the British attempted, between 1763 and 1783, to establish a plantation colony modeled somewhat after that of South Carolina. Governor James Grant requested permission from London to purchase state slaves to be used for provincial construction.[9] One of his correspondents, in supporting his plan, wrote in 1765 that "it must manifestly appear to every person to be a saving to the Publick, as the Slaves will rather increase than diminish in Value or number, and will always at least be worth the original purchase, in so healthy a Country as East Florida, the young ones will even do more than to keep up the number, a few likely young Wenches must be in the parcell, & [he added in a phrase which calls attention to the sexual predisposition of some white colonials] should their Husbands fail in duty, I dare say my friend Sweetinham & other publick Spirited Young Men, will be ready to render such an essential service to the Province, as to give them some help."[10] The use of the word husband in this case could have been rather casual, and the desire to have the slaves reproduce did not necessarily oblige the creation of an equal sex ratio, force a regard for monogamous family units, or mandate stability. Certainly, the attitude of the above writer indicates a concern for progeny without any sensitivity to human considerations requisite for procreation. (This might not, in fact, mean that he had no such awareness, as the offending phrase could simply be an example of the kind of gross jocularity with which men probably have contemplated sex since time immemorial.) But however else the response is interpreted, it explicitly recognizes a regard for natural increase and from that a need for women.

The men actually engaged in building plantations had myriad reasons for wishing a more equal balance between male and female. Richard Oswald, writing from London, notified the governor in 1767 that he was taking advantage of a ship going to St. Augustine via the African coast to send some females to the colony as "it might be of bad Consequence if the Men Slaves now on the Plantation, remained longer unprovided with Wives."[11] The Earl of Egmont wrote in 1769 that he hoped soon to reap a profit from the slaves who already occupied his property, out of which he proposed

---

[9] James Grant to the Board of Trade, July 16, 1765, Grant of Ballindalloch MSS, 0771/Letterbook (National Register of Archives, Edinburgh, Scotland). Grant addressed the letter to the "Lords of Trade."
[10] John Graham to Grant, July 19, 1765, *ibid.*, 0771/401.
[11] Oswald to Grant, May 20, 1767, *ibid.*, 0771/295.

PATERNALISM AND SLAVERY

more Negroes to be bought. And I Shd. be obliged to Yr. Excellency, for advising those who may be concerned for me, in such Encrease of Negroes, not so much to Consult my most Immediate Profit, as to render the Negroes I now have happy and contented, wch I know they cannot be without having each a Wife. This will greatly tend to keep them at home and to make them Regular and tho the Women will not work all together so well as ye Men, Yet Amends will be sufficiently made in a very few years by the Great Encrease of Children who may easily [be] trained up and become faithfully attached to the Glebe and to their Master.[12]

As if to confirm that the solicitude of these gentlemen was not idle worry, Grant's overseer sent the departed governor a report in 1781. He advised that £500, which was planned to be used in another way, be spent instead in buying young women for the plantation, which suffered by not having enough. Both the labor and increase of slaves was lost, since the men, rather than work, ran away to town in search of female companionship and caused trouble by enticing young women to run away. He had had, he said, "frequent complaints on this head, and their excuse is what must they do for Wife."[13]

The concern for permanence and stability as well as the foresight implicit in these statements, entirely aside from the benefits accruing to the slave, were uncharacteristic of the exploitation mentality associated with the industrial plantation model. Rather than trying to make as much as possible in the least imaginable time, regardless of consequences, as was typical of planters in the West Indies and in areas of Latin America, the posture expressed here is a willingness to defer immediate, temporary earnings for measured, long-term gains. This outlook, of course, was entirely pragmatic and reflected a realistic appraisal of continental possibilities. Richard Oswald, for instance, admitted to Governor Grant in 1768 that many of the worthies involved in the East Florida venture had become discouraged because they "lookt upon having a Tract of Land in Florida, as having a Plantation; & without giving themselves the trouble of further enquiry, were too easily amused with the idea of a Similarity to that respectable denomination when applied to a West India possession." But when they received a host of bills for implements, slaves, and supplies, he went on, with little prospect of a ready return on their outlay, their enthusiasm waned. He concluded by expressing the hope that a common acquaintance might "be cautious, & not overburden his scheme with too great preliminary expense, Since Circumstances in North America are so different from

[12] Earl of Egmont to Grant, May 14, 1769, *ibid.*, 0771/264.
[13] David Yeats to Grant, February 3, 1768, *ibid.*, 0771/250.

the West Indies, that it will in any event be a long time before they can be fetched up, by the fruits of the Settlement, & if half these late [discouraging] reports are true," he noted with pessimism, they "never will. . . ."[14] It is possible, therefore, that the more paternal attitude was determined by the economics of the situation, that the return on some crops was such that their production would not be profitable if the labor force had to be continually replaced.

A judgment of the extent to which this was true will ultimately depend upon a comparative analysis of the economics of production in the three major agricultural regions of the old empire, an endeavor beyond the scope of the current study. It is certain that rice, Carolina's primary crop, in association with its subsidiary, indigo, yielded much higher returns than did Chesapeake tobacco. It created, in fact, the greatest plantation wealth on the continent, though the riches of Carolina planters paled in comparison to that of their compatriots in the West Indies. But, as Edmund S. Morgan asserts, "To make a profit, [island] sugar planters worked their slaves to death . . . [while mainland] planters did not have to" do so.[15] Morgan argues, indeed, that the development of slavery in Virginia was inhibited initially in part because of the high death rates among whites and blacks alike.[16] Slaves were more expensive than servants (they cost about twice as much), and there was no reason to pay more for one than the other if a black man might live no longer than a white man could be indentured. When, among other things, survival rates improved, slavery took hold, marking a clear recognition of longevity as a consideration in economical plantation management.

Rice cultivation was more strenuous than raising tobacco, but it does not seem to have been so hazardous to life as sugar production. The life expectancy of a slave imported into the sugar islands was from five to seven years, it being considered more economical to use the slave in such a way that he died within that period than to improve or create conditions wherein his life could be longer preserved. Even if, as one scholar asserts, slaves managed to survive longer than that, the attitude is instructive.[17] In South Carolina, by contrast, planters estimated that a slave paid for himself within four

[14] Oswald to Grant, February 19, 1768, *ibid.*, 0771/259.

[15] Morgan, *American Slavery, American Freedom: The Ordeal of Colonial Virginia* (New York, 1975), 301.

[16] *Ibid.*, 297–98.

[17] Herbert S. Klein, *The Middle Passage: Comparative Studies in the Atlantic Slave Trade* (Princeton, 1978), 246, rejects this statement on the survival rate of slaves, but somewhat ambiguously, since his comment refers to all the Americas whereas the traditional statement refers specifically to the West Indies, and modern studies seem to support it. See note 21 below.

PATERNALISM AND SLAVERY

or five years so that the real profit from his labor came after that span. They looked upon slaves as an investment from which the owner could "reasonably expect above 16, 20 and 25 per Ct. when Rice . . . [gave] a tolerable Price . . . ."[18] But for the master to realize this remuneration the slave had to survive. Moreover, the worth of slaves conditioned or native to the country was more than strictly economic. Governor James Glen, writing to the Board of Trade in 1751, after having valued the forty thousand blacks in the province at £20 sterling per head, continued,

but this Valuation does not satisfie me, for when it is Considered that many of these are Natives of Carolina, who have no Notion of Liberty, nor no longing after any other Country, that they have been brought up among White People, And by white people have been made, at least many of them, useful Mechanicks, as Coopers, Carpenters, Masons, Smiths, Wheelwrights and other Trades and that the rest can all speak our Language, for we imported none during the War, I say when it is Considered that these are pleased with their Masters, contented with their Condition, reconciled to Servitude, seasoned to the Country, and expert at the different kinds of Labour in which they are employed, it must appear difficult if not impracticable to ascertain their instrinsick Value, I know a Gentleman who refuses five Hundred Guineas for three of his Slaves and therefore there is no guessing at the Value of strong seasoned handy Slaves by the prices of weak Raw New Negroes.[19]

Although their intrinsic value might not be ascertainable, their market price, fortunately, can be. For the attitude towards slaves as an investment, along with their pecuniary worth, was also expressed in East Florida. James Grant notified the Earl of Egmont, who balked at paying £140 sterling for a servile tradesman, that skilled slaves sold in Charleston often as high as £200 sterling and noted that he would rather pay that much for good black tradesmen than employ white indentured servants. He currently possessed, he wrote, about forty slaves, the first eight of whom had been purchased before he came from England and cost about seventy pounds sterling each. "They were all Country born but only common field Slaves. They have turned out remarkably well; for tho' I have lost three Negroes by Death I should sell those remaining for more Money than the whole cost me." He could get a hundred pounds for some of them, he went on.[20] While the assessment of

---

[18] James Glen to the Board of Trade, July 15, 1751, CO 5/373, ff. 155–57 (Public Record Office, London, England).

[19] Glen to the Board of Trade, March 1751, "An Attempt towards an Estimate of the Value of So. Carolina . . . ," received with letters dated June 24, 1751, CO 5/373, f. 137.

[20] Grant to Earl of Egmont, February 9, 1769, Grant of Ballindalloch MSS, 0771/Letterbook.

worth for skilled or seasoned slaves would be equally applicable to a select number of bondsmen in the West Indies, the general philosophy of investment surely was not. The facts of slave treatment there are *prima facie* evidence to the contrary.[21]

Some of the details of economical plantation management in South Carolina are suggested by a report on a successful venture in 1757. The owner had died, and an executor described the condition of his plantation to a relative in the West Indies:

> There are near 100 slaves great and small upon the plantation but of these scarce 40 can be deemed working hands capable to go into the field, with these hands there was made last year 1780 weight of good Indico and 240 Barrels of Rice, both of which were shipt home. If the Indico arrived and was sold at 5 sh. ster. per. lb. which by the bye is a very moderate price for good Indico, the amount will be 445 Sterling. Deduce from this 30 per Cent for freight and Insurance and there will remain about £312 neat proceeds. The Rice would not produce less than 3£ Sterling per Barrel at home which is you know 720£ amount. But then the freight which was at £4:10 & £5 per Ton, with the Insurance at 15 per Cent being deduced, the neat proceeds will not be more than about £350, so that this year the Estate has produced upwards of 600£ sterling neat from which there may be probably a Deduction for Cloaths to Negroes and overseers wages, if a fund is not raised to pay these from the Corn &c planted besides the Indico and Rice.[22]

---

[21] See *inter alia* J. Harry Bennett, Jr., "The Problem of Slave Labor Supply at the Codrington Plantations," *Journal of Negro History*, XXXVI (October 1951), 406-41; XXXVII (April 1952), 115-41; Elsa V. Goveia, *Slave Society in the British Leeward Islands at the End of the Eighteenth Century* (New Haven and London, 1965), 103-51, gives numerous examples of the thoughtless and wasteful exploitation of the slave population. Also see Frank W. Pitman, "Slavery on British West India Plantations in the Eighteenth Century," *Journal of Negro History*, XI (October 1926), 584-650; and Pitman, *The Development of the British West Indies, 1700-1763* (New Haven and London, 1917), 61-90; and Richard B. Sheridan, "Africa and the Caribbean in the Atlantic Slave Trade," *American Historical Review*, LXXVII (February 1972), 26.

Richard Dunn, *Sugar and Slaves: The Rise of the Planter Class in the English West Indies, 1624-1713* (Chapel Hill, 1972), 251, writes that in seventeenth-century Jamaica, as in eighteenth-century North America, male slaves demanded wives, and the couples reproduced. Unmarried males refused to work and went in search of women, encouraging planters to extend themselves to keep an even sexual balance. He gives examples of large Jamaica slaveholders who had a more or less proportionate number of men and women remarking that lesser planters had more difficulty. In the eighteenth century, however, when sugar production in Jamaica rapidly expanded, Dunn writes (p. 251n) that "men heavily outnumbered women in the big Jamaican slave gangs. On four of the Price estates, for example, there were 791 male and only 356 female slaves." These facts suggest support for the dichotomy between more and less intense plantation systems. Sheridan says explicitly that "slavery in the infancy of the sugar industry was considerably milder than it became in a later period of intensive culture." See Sheridan, "Africa and the Caribbean," 20. Also see Sheridan, "Commercial and Financial Organization of the British Slave Trade, 1750-1807," *Economic History Review*, Ser. 2, XI (December 1958), 249-63; and Michael Craton, "Jamaican Slavery," in Stanley L. Engerman and Eugene D. Genovese, eds., *Race and Slavery in the Western Hemisphere: Quantitative Studies* (Princeton, 1975), 241.

[22] John Murray to Sir Robert Laurie, July 12, 1757, Murray of Murraythwaite MSS, GD 219/290, ff. 48-49 (Scottish Record Office, Edinburgh, Scotland).

PATERNALISM AND SLAVERY

More than half the slave population of this plantation were not full working hands, and consequently the estate was supported by 40 percent of the labor force. Significantly, this percentage was sufficient. Most of the rest could be looked upon as a trust which would yield at some future date. This state of affairs may not have been possible in the West Indies, where many of the foodstuffs to support the slaves had to be (or in any case, were) imported. This circumstance made the keeping of slaves more expensive. On the mainland supplementary food crops frequently were a source of income as well as subsistence. In addition, they could be raised by less than full hands, that is, by older or younger men and women. So these excess slaves, far from being a drag on the plantation, were an asset—both in the years before they reached their full potential, in terms of the young, and after their prime, in the case of the old. They paid their keep and still made a contribution. Further, their contribution was very important in gauging the worth of a plantation. Accordingly, James Grant wrote a plantation adventurer, telling him that "The Gentlemen were too sanguine who informed your Lordship that 33 Negroes ought to clear £500 a year." (The optimum number for beginning a rice plantation was about thirty slaves.[23]) He went on:

I think a Planter does a great deal if he makes at the rate of eight pounds a Year of his Negroes, clear of all expences. To do that a plantation must be well established, and the Slaves must all be seasoned able working hands. There are Instances of Indigo planters doubling their Capital in a Year, but at an average Carolina planters do not make so much as I have mentioned, I mean of produce to go to Market, so as to remit the Money to Europe, for many things are consumed in a plantation which are of great utility in point of Living, tho' they cannot be converted into Cash, and I have always reckoned that an Intelligent Carolina Planter with a Capital from two to three thousand sterling (if his Negroes & Lands were to be disposed of) lives as well as any Man can do in Great Britain with an Income of £500 a Year.[24]

The picture here is of a stable planter living off the land, committed to the country, and content to live more or less within his means. This situation did not usually obtain in the Caribbean.

The patriarchal spirit inherent in the profile above, though a concomitant and not a determinant of economic management, never-

---

[23] See *inter alia* [Thomas Nairne], *A Letter from South Carolina* . . . (London, 1710), 52-53. Also Murray to Mother, March 6, 1757, Murray of Murraythwaite MSS, GD 219/287/11.

[24] Grant to Earl of Egmont, February 9, 1769, Grant of Ballindalloch MSS, 0771/Letterbook. This reference includes the quoted sentence preceding the parenthetical sentence.

theless had some humanizing aspects. Thus, Henry Laurens wrote of one slave that "he is a quiet orderly old Man, not able to do much Work and therefore is never drove to Labour, but suffer'd to go on in his own way. I observe he makes larger Crops of Rice and Corn for himself than the most able Young Negroes, which I believe is greatly owing to their Aid for they all Respect and Love him." Laurens continued: "I shall order proper Care to be taken of him, if his Life shall happen to surpass his Strength for Labour . . . and continue to make the same Annual Allowance during all the Time that he is able to perform any Work."[25] The same motivation prompted a Georgia planter to offer to sell to the same man the family of a slave he had sold previously. The "fellow," he said, "writes to his Wife frequently, and appears by his letters to be in great distress for want of her . . . ." He added that the slave would not have been sold except that "he disobliged me . . . ," but, he thought, his punishment had now been sufficient. He concluded that "A seperation [sic] of those Unhappy people, is adding distress to their unfortunate condition, for that reason I have taken the liberty to mention this matter to you; And if it is agreeable I will send them, and leave the value to your own or the Judgement of any other person after you have seen them . . . ."[26] On the other hand, it was probably just a matter of business that caused a slave dealer to report that he had obtained for a planter "a remarkable choice family. They cost no less than £290 [sterling] consist of a fellow [,] his Wife [,] two fine Grown Boys & Girl . . . . I had no directions to buy his family, but I could not get him without them."[27]

But if economical considerations were a prime ingredient of the development of a patriarchal ethos in North America, this same impulse occasionally contributed to a spirit inconsistent with our model. Peter H. Wood has related the presence or absence of a reproductive propensity to a more intense versus a less intense plantation system. As he has argued, the rapid expansion of staple-crop production in South Carolina seems to have adversely affected the situation of the slave population, which experienced a sharp decrease in its natural growth rate.[28] He has also demonstrated that in

[25] George C. Rogers, Jr., et al., eds., *The Papers of Henry Laurens* (8 vols. to date, Columbia, S. C., 1968-), IV, 148n, hereinafter cited as Rogers et al., eds., *Laurens Papers*.
[26] William Simpson to Grant, June 15, 1767, Grant of Ballindalloch MSS, 0771/243.
[27] John Graham to Grant, March 1, 1768, Grant of Ballindalloch MSS, 0771/401.
[28] Wood, *Black Majority*, 142-66. Dunn has found a roughly parallel situation in the West Indies to that Wood found in South Carolina. In the seventeenth century, although more men than women were imported, differential mortality created an equal sex ratio. But varying treatment affected the birthrate despite the sex ratio, and in the eighteenth century the condition of sexual equality in numbers was altered with an imbalance of males. In neither century did the black population reproduce itself. See Dunn, *Sugar and Slaves*, 314-25. Also see Sheridan, "Mortality and the Medical Treatment of Slaves in the British West Indies," in Engerman and Genovese, eds., *Race and Slavery*, 286-87; and Sheridan, "Africa and the Caribbean," 19-26.

the first decades of the eighteenth century the black sex ratio generally mirrored the specifications of slave traders for cargoes composed of two males for every female. There is, of course, no absolute correlation between these proportions and the ability of the slave population to reproduce itself, as Wood has admitted. Obviously, the more equal the ratio of male to female the greater the reproductive capacity, other circumstances permitting. But whereas a black population composed of 62 percent males and 38 percent females had a reproductive rate of 5.6 percent in the years prior to 1721, a growth rate which exceeded that of the white population, its natural growth decreased thereafter despite the fact that the sex ratio remained largely the same or, in some cases, improved. In St. George's Parish in 1726, for example, although the sex ratio among slaves was 129 males to 100 females, virtually the same as that of the white population, the natural increase had declined from that of the earlier period and was half that of the whites. Wood goes on to point out that the general sex ratio is not always an accurate measure of specific possibility because the sexes may be (and in this case were) disproportionate on individual plantations. But if sex ratio is not a sufficient determinant of reproductive capacity, it is undeniably an important component, and the absence of a desire to secure a more equal sexual distribution among slaves may be an important index in determining the character of a plantation system.

Some assessment of this desire may be obtained by looking at the slave trade. A document in Wilberforce House, Hull, England, giving an account of the sale of 143 slaves in South Carolina in June 1773 is useful in presenting a complete illustration of the method and rationale of sale.[29] One of the most arresting facts about the cargo, however, was that a majority of the unfortunates were female. This incident excites various considerations, such as to what extent was this a chance occurrence and how often, whether by fortune or design, did it happen. Speculations of this nature are encouraged by an enterprise of Benjamin Spencer and Company of London in 1755 and 1756. Its agents on the African coast shipped on its account in September 1755 twenty-five slaves to Antigua, composed of fifteen men, five women, three girls and two boys; in the same month ten slaves, comprising three men, two boys, two women, and three girls were dispatched for either Barbados or South Carolina, landing in the latter. The cargo with the disproportionate number of males to females went to the West Indies, while that with the more equal sex ratio ended in South Carolina. Subsequent events are more intriguing. The cargoes arrived in America

[29] "Sales of One Hundred & forty three Negroes received [from] the Snow *Robert*, Luke Mann Master from Africa on Account of Messrs. Robert Grimshaw & Co. . . ." Wilberforce House (Hull Museums, Hull, England).

just after four merchantmen, among them two French slavers, had been made prizes with the result that over eight hundred slaves had been deposited in Antigua and prices had been generally depressed. Therefore, after selling ten of the original cargo (nine had died in passage), the supercargo decided to send the rest, along with a few others from another vessel, to South Carolina, where prices were higher. This shipment of twenty-four Africans consisted of two men, eleven women, three boys, and four girls for the company's and three women and one boy for the supercargo's account. This sale did not go as well as expected because the threat of war lowered the value of Carolina produce, and the sex ratio was not helpful; as the merchant who took up the consignment mildly protested, "had they been Chiefly men instead of Women would have Answer'd better . . . ."[30] Nevertheless, these documents suggest a tendency to ship more women to the continent than to the islands or reflect an expectation that women would sell better in one place than in another. The questions become, then, how widepread was this practice or this expectation and was it significant.

Despite the fact that the usual demand for slave cargoes seems to have been almost universally in the proportion of two males to one female,[31] slave dealers were aware that disparate colonies had a differential predisposition to accept a greater or lesser percentage of one or the other gender. A London merchant instructed his correspondent in 1732 not "to meddle with any Negroes but men for Sale to the Portuguees, they seldom give Gold for any other . . . ."[32] Similarly, a Liverpool company advised a ship captain whose cargo

[30] This account is based on the Spencer-Stanhope Muniments in the Sheffield Central Library (Sheffield, England), namely, Charles Quinsac to Benjamin Spencer, November 19, 20, 30, 1755; John Guerard to Spencer, December 23, 1755, copy included with letter of February 2, 1756. Document numbers respectively are 60549/180, 60549/181, 60549/182, and 60549/150. Bills of lading for slaves shipped to Antigua, dated September 20, 1755; to Barbados, September 30, 1755; and to Charleston, November 19, 1755, numbered respectively 60550/299, 60550/359, and 60550/358. The ten slaves shipped on September 30 (60550/359) were actually directed to "Barbados or some of His Majesty's Colonies in America" and were shipped on freight with a larger cargo in the ship *Gambia*. Quinsac in his letter of November 30 (60549/182) indicated that South Carolina was foremost in his mind, though Barbados might do.

[31] See *inter alia* Herbert S. Klein, "The Portuguese Slave Trade from Angola in the Eighteenth Century," *Journal of Economic History*, XXXII (December 1972), 914n. Also see Kenneth G. Davies, *The Royal African Company* (New York, 1975), 299–300, who reports that despite the preference, an analysis of 60,000 slaves delivered to the West Indies between 1673 and 1711 indicated that three-fifths were males and two-fifths were females, or a ratio of 60 percent to 40 percent. This ratio, along with differential mortality, would be a partial explanation for the more equal sex ratio in the seventeenth-century West Indies, but it was a response to a different kind of society than was later to develop. See Sheridan, "Mortality and Medical Treatment," 287; and Sheridan, "Africa and the Caribbean," 20–26.

[32] Letter dated June 23, 1732, in H. M. Drakeford MSS 36 (County Record Office, Stafford, England).

PATERNALISM AND SLAVERY

was destined for Havana in 1803 that his vessel was legally permitted to carry four hundred slaves and added: ". . . we request that they may all be males if possible to get them, at any rate buy as few Females as in your power, because we look to a Spanish market for the disposal of your Cargo where Females are a very tedious Sale."[33] A merchant house desiring its ship to deliver slaves to the West Indies in 1761 specified that "as its most likely you'll sit at some of the windward Islands, middle size strong young Slaves will sell better than the heavy, or overgrown, avoiding too great a number of females . . . ."[34] Moreover, these cargoes were planned with contingent destinations in mind, as when the 1803 voyager was admonished not to get any slave over twenty-four years old in case he had to go to Jamaica.[35] A greater than usual disproportion of men to women seems to have confined the alternatives to either the West Indies or Latin America. Thus, an Angola vessel was ordered to collect an assortment of three males to one female which would "answer for a Spanish Contract shou'd there be one at any of the [West India] Islands . . . ."[36] A vessel trading in the River Congo was advised that if females were difficult to obtain it should purchase only males as they were certain to "make the highest Average in the West Indies, and particularly at the Havannah . . . ." Indeed, on second thought, he was ordered to "avoid females as much as possible . . . ."[37] Hence, one is brought to wonder if (and doubt that) the shipment to South Carolina in 1773 of a cargo containing more females than males was a matter of chance.[38] Even if the com-

[33] Thomas Leyland to Caesar Leyland, July 18, 1803, Leyland Papers, 387 MD 43 (Liverpool Record Office, Liverpool, England).

[34] John Maine and Company to William Hindle, Account book *Tyrell*, February 17, 1761, pp. 15–17, Raymond Richards Collection (University Library, Keele, England).

[35] Thomas Leyland to Caesar Leyland, July 18, 1803, Leyland Papers, 387 MD 32. The prohibition against slaves over twenty-four years old is explained in ship orders to John Whittle, July 2, 1798, in which Thomas Leyland wrote that "the Assembly of that Island [Jamaica] have just Imposed a further Duty of Ten pounds per head on every Slave Imported which in the opinion of Commissioners by them appointed are Twenty five years of Age or upwards." Leyland Papers, 387 MD 41.

[36] James Clemens and Company to Captain William Speers, June 3, 1767, Tuohy Papers, 380 TUO 4/2, pp. 1–3 (Liverpool Record Office). African imports into Cuba between 1790 and 1794 reflected this ratio. Slave ships landed 19,424 males and 6,535 females in Havana, for a total of 25,959, or 75 percent males and 25 percent females. See Herbert S. Klein, "North American Competition and the Characteristics of the African Slave Trade to Cuba, 1790 to 1794," *William and Mary Quarterly*, 3d Ser., XXVIII (January 1971), 98.

[37] Thomas Leyland to Captain William Young, Account book *Spitfire*, June 1795, MS/10/49 (University Library, Liverpool, England).

[38] The argument here is not that ships with such a sex ratio never went to the West Indies or elsewhere as this clearly was not the case. (See for example the Account of 508 Slaves imported in the Ship *Golden Age* from Whydah into Jamaica in December 1784 in the Liverpool Museum. Printed as Item 12 in *Liverpool History Teaching Unit*, No. 2). There were various attempts in the West Indies, especially in the latter part of the eighteenth century, to encourage reproduction among the slaves, but there was more consistency in this regard in North America.

position of the cargo was not predetermined, the destination likely would be.

In none of the regions noted by the merchants above did the slave population grow naturally, whereas by the second half of the eighteenth century in South Carolina a more settled society probably encouraged a more stable (and thereby more fertile) slave population. In any case, South Carolina planters periodically expressed a concern to create or maintain slave families. Henry Laurens, for example, indicated in 1765 that he disliked to divide families even of new Negroes,[39] and some months earlier had written to the overseer on his Mepkin plantation: ". . . I send up a stout young Woman to be a Wife to whome she shall like best amongst the single men. The rest of the Gentlemen shall be served as I have opertunity. Tell them that I do not forget their request."[40]

A similar situation existed in the Chesapeake. Russell R. Menard and Wesley Frank Craven have shown that in the seventeenth century the black population in Virginia and Maryland had an imbalance of males and females and was unable to reproduce itself, and Craven warns against too ready an assumption that seventeenth-century planters promptly recognized "the advantages of a self-perpetuating labor force."[41] Whether the advantage was immediately recognized or not, by the eighteenth century blacks in the region were beginning to show a natural increase.[42] Menard places the watershed for natural increase in Western Shore Maryland in the 1720s. This increase was not a result of a more equal sex ratio and thus, Menard asserts, did not reflect on the part of those masters who were wealthy enough to alter the sexual balance on their possessions a proper regard for black reproduction. The increase was caused, rather, by the growth of a creole population better adjusted to the environment. Allan Kulikoff suggests the beginnings of natural increase in Virginia at a few years earlier, for similar reasons.[43] But if western Maryland planters failed promptly to rec-

[39] Laurens to Elias Ball, April 1, 1765, Rogers *et al.*, eds., *Laurens Papers*, IV, 595.

[40] Laurens to Timothy Creamer, January 26, 1764, *ibid.*, 148.

[41] See Menard, "The Maryland Slave Population, 1658 to 1730: A Demographic Profile of Blacks in Four Counties," *William and Mary Quarterly*, 3d Ser., XXXII (January 1975), 29–54; Craven, *White, Red, and Black: The Seventeenth-Century Virginian* (Charlottesville, 1971), 98–103 (quotation on p. 101).

[42] Menard, "Maryland Slave Population," 42–45; and Morgan, *American Slavery, American Freedom*, 301.

[43] Kulikoff, "A 'Prolifick' People: Black Population Growth in the Chesapeake Colonies, 1700–1790," *Southern Studies*, XVI (Winter 1977), 391–428; also see Kulikoff, "The Origins of Afro-American Society in Tidewater Maryland and Virginia, 1700 to 1790," *William and Mary Quarterly*, 3d Ser., XXXV (April 1978), 226–59; and Kulikoff, "The Beginnings of the Afro-American Family in Maryland," in Aubrey C. Land, Lois G. Carr, and Edward C. Papenfuse, eds., *Law, Society, and Politics in Early Maryland* (Baltimore and London, 1977), 171–96. Herbert G. Gutman, *The Black Family in Slavery and Freedom, 1750–1925*

ognize the advantage of redressing the imbalance among their slaves, it ought not to be too much to assume that the realization came earlier in the older settlement of the Chesapeake. Certainly, Robert "King" Carter, for one, encouraged the creation of stable family units.[44] But whether for this or a different reason, by the third decade of the eighteenth century Virginia had established a reputation that encouraged the shipment of a larger percentage of women. Accordingly, in September 1721 the Royal African Company complimented the handling of a cargo by James Phipps, their agent at Cape Coast Castle, commenting that "We take Notice what you write as to the Assortmt. of Capt. Bulcock's Cargo of Slaves and approve very well your Care and prudence in enlarging the Numbr of Women[,] boys & Girls, since there is so little difference in prices at Virginia and so great on the Coast."[45]

Of course, the source of slaves had something to do with the percentage of women likely to be obtained. We have already mentioned a ship trading in the River Congo in 1795 that was ordered to push its purchase in men if it found "the usual scarcity of Females" in the region. When the British government was considering restrictions on the number of slaves to be carried per ton in British vessels during the 1780s, one official suggested that some encouragement ought to be given to carrying more women and children by altering the ratio to favor them. He noted, however, that the various regions were not equally supplied with women for sale and that even in some places where they were, those sold were unacceptable for plantation labor.[46] Herbert S. Klein avers that everywhere in the New World there was a more equal valuation of men and women than commonly assumed and that supply conditions may have been determinative.[47] The argument here, however, is that various plantation societies had differentiating characteristics that modified their demand for women in one direction or another and that their reputation in this regard was equally important.

The shift in emphasis towards a greater valuation of women in North America, then, in a period when the slave trade was still going

---

(New York, 1976), *passim*, also comments on the development of black family structure with some reference to the slave trade in the eighteenth century, though his major focus is later.

[44] Kulikoff, "A 'Prolifick' People," 399–400.

[45] London Committee to James Phipps, September 7, 1721, C 113/35, Part I, No. 142 (Public Record Office). Captain John Bulcock, of the Galley *Sarah* of London, left the Gold Coast for Virginia in April 1721 with 250 slaves and arrived at his destination June 26, 1721, with 233 slaves. Elizabeth Donnan, ed., *Documents Illustrative of the History of the Slave Trade to America* (4 vols., Washington, 1930–1935), IV, 184.

[46] James Jones to Lord Hawkesbury, June 27, July 26, 1788, Liverpool Papers, CCXXVII, Add. MSS 38416, ff. 131 and 154 respectively. See note 37 for the quotation on the scarcity of females.

[47] Klein, *Middle Passage*, 240–41.

on, is reflective of an entirely different plantation ethos than existed elsewhere. It was part and parcel of a situation and outlook that enabled North American plantations to achieve slave propagation. The components of this achievement are various and it is probable that other factors remain to be isolated. Philip D. Curtin has adduced the novel proposal that the North American phenomenon might be a concomitant of a higher proportion of preseasoned slaves arriving on the mainland, a consideration vitiated by the fact that the majority of imports into North America came directly from Africa.[48] But the different rationales of the varying plantation systems are partly an explanation; a reasoning, determined somewhat by economics, that in the one case not only provided better treatment but required a more equal balance between man and woman. This logic might not have been as strong in every period and in every region. But over the long haul it would have encouraged the importation of relatively more African women into North America than any place else in the Western Hemisphere, and the greater population growth would be partly a function of this greater human capacity. Certainly, documents of the slave trade support this view.

[48] Curtin made this suggestion in a personal communication to Herbert Klein. See Klein, "Slaves and Shipping in Eighteenth-Century Virginia," *Journal of Interdisciplinary History,* V (Winter 1975), 387*n*. Closer attention ought to be paid to this proposition, though, in view of the work of Menard and Kulikoff.

8
A Tale of Two Plantations: Slave Life at Mesopotamia in Jamaica and Mount Airy in Virginia, 1799 to 1828

Richard S. Dunn

ON January 1, 1809, John Tayloe, one of Virginia's leading planters, took a detailed census of the 384 slaves on his Mount Airy estate, listing each man, woman, and child by name, age, occupation, and monetary value. On the same day the overseer for a big Jamaican absentee planter, Joseph Foster Barham, was taking a similar census of the 322 slaves on Barham's Mesopotamia estate in which he listed each person by name, age, occupation, and physical condition. Thousands of other North American and West Indian slave inventories survive, especially in probate records, but what gives the Mount Airy and Mesopotamia lists special value is that the owners of these estates made a systematic practice of cataloguing their slave gangs annually over a long time span. The Barhams at Mesopotamia kept annual inventories from 1751 to 1832; seventy-five of these lists survive.[1] The Tayloes at Mount Airy kept annual inventories from 1808 to 1855, and forty-five of these lists survive.[2] Setting the two lists of 1809 against each other, we

Mr. Dunn is a member of the Department of History, University of Pennsylvania. He wishes to thank the American Philosophical Society for supporting his research at Oxford and London, and the National Endowment for the Humanities for supporting his research in Virginia. He drafted the article while a visiting member of the Institute for Advanced Study in Princeton, and benefited greatly from comments offered by colleagues at an Institute seminar and at a University of Pennsylvania history workshop.

[1] The Mesopotamia slave lists are filed in the Clarendon Manuscript Deposit, Barham Papers, Boxes b. 34, 35, and 36, Bodleian Library, Oxford University. The lists run in a broken series covering the years 1736, 1743-1744, 1751-1752, 1754-1769, 1771-1776, 1778, 1780-1781, 1784-1785, 1790-1806, 1808-1819, 1822-1832. From 1762 onward the lists are especially valuable because they give the age as well as the name, occupation, and condition of each slave. I wish to thank the earl of Clarendon for permitting me to use the Barham Papers, and Stanley L. Engerman both for pointing out data in this collection which I otherwise would have overlooked and for making helpful criticisms.

[2] The Mount Airy slave lists are found in four inventory books in the Tayloe Papers, Virginia Historical Society, Richmond. The earliest of these books, kept by John Tayloe III, contains slave lists for 1808-1823 and 1825-1828 (MSS 1 T2118d538). The other three, kept by his son William Henry Tayloe, contain slave lists for 1829-

can compare the structure of a Virginia plantation with a Jamaica plantation at a particular moment—just as the slave trade was closing in the United States and the British West Indies. Setting the two series of inventories against each other, we can make a running comparison between the two estates over a considerable stretch of years, and get a sense of two distinctly different slave communities in action.

Two bricks do not make a house, and it cannot be claimed that Mount Airy was representative of all Chesapeake plantations or Mesopotamia of all Caribbean estates. Still, microcosmic case studies have their utility, especially for studying such a topic as slave life, where the macrocosmic work of the cliometricians has stirred such controversy. Quantification itself is surely not at issue; the historian who wishes to interpret slave records cannot get very far without employing techniques of aggregative analysis. What is at issue is the cliometricians' habit of counterfactual hypothesis, their manipulation of synthetic figures extrapolated from mathematical models, and their certitude that by such tactics they can "correct" the "errors" of previous interpreters.[3] Quite apart from the question of the historical accuracy of the cliometricians' findings, their abstract mode of computation tends to rob men and women of individual personality, strips communities of local variety, and turns both people and places into digits in a data bank. The present essay attempts a more intimate picture of slave life. The Mesopotamia and Mount Airy inventories are so richly detailed that one can tease from them a sense of time and motion. The inventories generate statistical information about a considerable number of people—1,400 slaves owned by the Barhams and 1,100 slaves owned by the Tayloes—but the strength of the documentation lies more in its quality than in its quantity. Close examination of conditions on

---

1836 (MSS 1 T2118d13410); for 1837-1838 (MSS 1 T2118d13424); and for 1840-1847 and 1849-1855 (MSS 1 T2118a13). Probably the series began well before 1808 and continued after 1855, but other inventory books seem not to have survived. I wish to thank the Virginia Historical Society for permitting me to use the Tayloe Papers.

[3] This is not the place to detail the bulky polemical literature concerning the cliometric approach to the history of slavery. A good way to enter the debate is to compare Robert William Fogel and Stanley L. Engerman, *Time on the Cross: The Economics of American Negro Slavery*, 2 vols. (Boston, 1974), with Herbert G. Gutman's sharply critical review essay, "The World Two Cliometricians Made," *Journal of Negro History*, LX (1975), 54-227. For examples of cliometric work based more on hypothesis than on historical evidence see the essays by Richard Sutch, Jack Ericson Eblen, and Claudia Dale Goldin in Stanley L. Engerman and Eugene D. Genovese, eds., *Race and Slavery in the Western Hemisphere* (Princeton, N.J., 1975). It should be added that both the quantitative and the nonquantitative students of U.S. slavery have concentrated overwhelmingly on the years 1830-1860 because documentation for this period is much richer than for the colonial and early national periods. The Tayloe slave lists have extra value because they permit close investigation of slave life before 1830.

these two estates suggests that slave life—like any other variety of human experience—defies precise statistical formulation. Three-dimensional people emerge with variegated life histories and complex communal roles.[4]

In 1809 the slave gangs at Mesopotamia and Mount Airy were about equal in size. The 322 slaves at Mesopotamia constituted a fairly typical Jamaican plantation labor force, for the island was completely dominated by the estates of a few hundred sugar planters, and Mesopotamia was only slightly above average in strength for a Jamaican sugar estate. By Virginia standards, on the other hand, the 384 slaves at Mount Airy constituted a production force of exceptional size, for in Virginia only a few planters owned large gangs, and the great majority of slaves were distributed among middling and small farmers. Furthermore, while Mesopotamia was situated in Jamaica's richest sugar-growing district, Mount Airy lay in a tidewater region of declining agricultural importance, where the planters had switched from tobacco to wheat and corn in search of a viable product. The slaves at Mesopotamia were engaged in a labor-intensive enterprise, partly agricultural and partly industrial, that required a large, coordinated work force. The slaves at Mount Airy were doing much the same work as small free farmers elsewhere in the United States. Sugar production was the sole concern at Mesopotamia and the sole rationale for the organization of the slave force there. At Mount Airy commercial agriculture was less vitally important, and the slaves on this estate spent considerably less than half their labor in the production of cash crops.

The owners of the two estates, the Barhams and the Tayloes, had been large landholders and slaveholders in Jamaica and Virginia respectively for four generations. As big planters, they reflected in their contrasting life styles some of the basic differences between the two societies. The Tayloes lived in

[4] The Mesopotamia and Mount Airy slave lists form two of the longest series presently known. Amother long series of 27 lists from Worthy Park estate in Jamaica, spanning the years 1784-1838, is analyzed by Michael Craton and Garry Greenland, *Searching for the Invisible Man*, forthcoming. I am much indebted to Mr. Craton for letting me read his manuscript and for fruitful discussion comparing Worthy Park with Mesopotamia. For background information on Jamaica in the early 19th century see Edward Braithwaite, *The Development of Creole Society in Jamaica, 1770-1820* (Oxford, 1971), and Orlando Patterson, *The Sociology of Slavery: An Analysis of the Origins, Development and Structure of Negro Slave Society in Jamaica* (Rutherford, N.J., 1969). The closest equivalent studies for Virginia are Robert McColley, *Slavery and Jeffersonian Virginia*, 2d ed. (Urbana, Ill., 1973), and Gerald W. Mullin, *Flight and Rebellion: Slave Resistance in Eighteenth-Century Virginia* (New York, 1972). For full discussion of a Jamaican estate comparable to Mesopotamia see Michael Craton and James Walvin, *A Jamaican Plantation: A History of Worthy Park, 1670-1970* (Toronto, 1970); and for discussion of a Northern Neck Virginia slaveholder comparable to Tayloe see Louis Morton, *Robert Carter of Nomini Hall: A Virginia Planter of the Eighteenth Century* (Williamsburg, Va., 1941).

the Northern Neck, tracing their property back to Col. William Tayloe (d. 1710), who laid out a tobacco farm on the Rappahannock River, manned at the time of his death by twenty-one slaves. His son John Tayloe I (1687-1747), the chief architect of the family fortune, established an ironworks on the upper Rappahannock, opened up new farms in four Virginia counties and in Maryland, obtained a seat on the Virginia Council, and left his heirs a force of 328 Negroes, one of the grandest mid-eighteenth-century slave gangs in the Chesapeake region. In the next generation John Tayloe II (1721-1779) played the role of leisured gentleman. At Mount Airy, overlooking the Rappahannock, he built an imposing mansion and laid out a mile-long race course. Though he sat on the council, John II was better at horse breeding than at politics, and though he accumulated further property, much of it was lost during the Revolutionary War. His son John Tayloe III (1771-1828)— the owner of Mount Airy in our period—spent his youth in England at Eton and Cambridge as though no Revolution had taken place. The 1787 tax lists credit him with 11,200 acres and 290 slaves in four counties—less property than his grandfather had possessed. Still, John III remained one of the chief slaveholders in the state and by far the largest property holder in Richmond County. As his family grew (he had seven sons and five daughters to provide for), he picked up additional property in Virginia, Maryland, the District of Columbia, and Kentucky. He built the elegant Octagon House as his town residence in Washington, and founded the Tappahannock Jockey Club to promote horse breeding and racing. On his death, he left the fifth generation of Tayloes a rich legacy of twenty-three farms, three ironworks, city houses in Washington, and some 700 slaves.[5]

By contrast, the Barhams, like most big Jamaica planters, had become absentees. They traced their island holdings back to Col. Thomas Foster, who started a sugar estate in the western parish of St. Elizabeth in the 1670s, and to his son Col. John Foster (1681-1731), who opened additional sugar estates. John Foster's widow married a Jamaica physician, Dr. Henry Barham (1692-1746), and in 1736 they retired to England, well able to afford the luxury of absenteeism, since their six Jamaican properties produced a gross income of around £20,000 per year. John Foster's youngest son, Joseph Foster (1729-1789), took the name Joseph Foster Barham and inherited from

---

[5] For a biographical sketch of the Tayloe family see *Virginia Magazine of History and Biography*, XVII (1909), 369n-375n. Col. William Tayloe's estate was inventoried at his death in Richmond County Wills and Inventories, 1709-1717; John Tayloe I's will and inventory are *ibid.*, 1725-1753; John Tayloe II's will is in the Richmond County Will Book, 1767-1787—all in the Virginia State Library, Richmond. John Tayloe III's will, dated Dec. 1827, is in the Tayloe Papers, MSS 1 T2118d539-545. His property holdings for 1787-1788 have been calculated by Jackson T. Main, "The One Hundred," *William and Mary Quarterly*, 3d Ser., XI (1954), 383-384.

his stepfather the Mesopotamia estate in the fertile Westmoreland plain on the southwestern tip of the island. Though he lived as an absentee proprietor in Shropshire, Barham took unusual interest in the spiritual welfare of his slaves. He urged Moravian missionaries to come to Jamaica, where they established a station at Mesopotamia. In 1768, during the early days of this mission, eighty-four of Barham's slaves were baptized, and in 1816, when the Gothic novelist "Monk" Lewis visited Mesopotamia, he found the Moravians still at work, though only fifty slaves now belonged to their church. The owner of Mesopotamia in our period was Joseph Foster Barham II (1759-1832), of Stratford Place, Middlesex. This gentleman lived on an even handsomer scale than his Virginian counterpart John Tayloe III, was rich enough to marry an earl's daughter, and sat in Parliament. During his twenty-seven years in the House of Commons Barham worked actively for the sugar lobby, but he also voted in the 1790s to abolish the slave trade, and he published a pamphlet in 1823 favoring gradual emancipation of the slaves—on condition that the West Indian proprietors be handsomely compensated for their loss.[6]

In 1809 the Tayloes' Mount Airy estate and the Barhams' Mesopotamia estate, roughly equal in scale, were quite different in arrangement. The Tayloes' Mount Airy Department, or Rappahannock Farms, consisted of nine separate but interdependent units, each managed by its own overseer, strung a distance of thirty miles along both sides of the river. Six of these units were in Richmond County on the north bank, one was in Essex County on the south bank, and two were in King George County farther up the north bank. Perhaps forty whites lived on the whole estate; Tayloe himself had a family of eleven in this year, and most of his overseers seem to have had wives and children. The central unit, Mount Airy proper, was the home plantation where Tayloe resided for half of each year, from April to October; he took his family to Washington for the winter months. No farming was done at Mount Airy. The 106 slaves who lived there in 1809 were employed as domestic servants or craft workers. The other eight units—Old House (where the first Tayloe mansion had stood), Doctors Hall, Forkland, Mask-

[6] For information on Joseph Foster Barham see *Gentlemen's Magazine: and Historical Chronicle* (London, 1832), Pt. I, 102, Pt. II, 573, and Joseph Foster Barham, *Considerations on the Abolition of Negro Slavery* (London, 1823). The Barham holdings in Jamaica in 1739 can be traced via Edward Long's list of 428 sugar plantations for that year, Additional Manuscripts, 12, 434, 1-12, British Museum. Their holdings in 1754 can be traced via the quitrent list of Jamaican landholders, C.O. 142/31, Public Record Office. For the Moravians see M. G. Lewis, *Journal of a West India Proprietor, 1815-1817,* ed. Mona Wilson (Boston, 1929 [orig. publ. London, 1834]), 152-153. I am much indebted to Althea Silvera, West India Reference Library, Institute of Jamaica, Kingston, for additional information on the Barhams.

field, Menokin, Gwinfield, Hopyard, and Oakenbrow—were all farms, with a total of 8,000 acres. The 278 slaves who lived on these eight quarters in 1809 produced about 7,000 bushels of wheat and 14,000 bushels of corn for Tayloe to sell.[7] In the eighteenth century the Tayloes had grown tobacco on the Rappahannock, but no tobacco was cultivated in 1809.

Mesopotamia was differently organized, with all 322 slaves grouped into a single sugar production unit of 2,448 acres. Four hundred acres were planted in cane, and the rest of the land was used for cattle pens, pasture, and slave provision grounds, or was left uncultivated. Like all the big Jamaican sugar estates, Mesopotamia had its own mill for grinding the cane, a boiling house and a curing house for converting the cane juice into sugar, and a distillery for converting the sugar by-products into rum. The Mesopotamia workers produced about 250 hogsheads of sugar and 120 puncheons of rum for sale in 1809.[8] As at Mount Airy, they were divided into agricultural laborers, craft workers, and domestics, but they all lived together in a single village. To manage this work force Joseph Foster Barham II employed a small and highly transitory staff of whites. The overseer and five or six bookkeepers and artisans lived on the estate, and another two or three whites lived six miles away at the Barham storehouse in the port town of Savanna la Mar. Occasionally this staff included married men with wives and children, but most were single males, and few stayed on the Barham payroll for more than a year or two.[9] Thus the blacks at Mesopotamia outnumbered the whites by a ratio of 50:1, whereas at Mount Airy the ratio was closer to 10:1.

Today the Virginia Tayloes still live at Mount Airy in their sandstone mansion of Italianate design built in 1755. From this house, standing on a hill above the river, one can easily recapture the scene in 1809: the terraces, gardens, orangery, bowling green, and deer park, and down below the site of the old race course and the farmland where the Tayloe slaves once grew wheat and corn. But at Mesopotamia there are few tangible reminders of the Barhams' presence. Sugarcane is now cultivated at this estate by the Barham Sugar Workers Co-operative, a new experiment in which the cane workers own and operate their farm communally. The Barham Farm cane is processed at a giant modern factory several miles away. The old Mesopotamia

---

[7] No Mount Airy production figures survive for 1809, but in 1811 Tayloe's Rappahannock farms produced 8,664 bu. of wheat, of which 7,003 were sold, and the corn crop totaled 14,119 bu., with another 2,906 bu. in rent from Tayloe's tenants. See John Tayloe's Minute Book, 1811-1812, Tayloe Papers, MSS 1 T2118a10.

[8] No Mesopotamia production figures survive for 1809, but in 1817 the estate produced 254 hogsheads of sugar, of which 248 were sold, and 120 puncheons of rum, of which 113 were sold. Expense accounts, 1816-1817, Barham Papers, Box b. 33.

[9] There are lists of the white staff at Mesopotamia for the years 1789-1798 and 1816-1817, *ibid.*, Boxes b. 33, 36.

sugar works are in ruins, the old plantation house has been torn down, and only a pair of stone gate pillars and a few Barham and Moravian missionary gravestones recall the bygone activities of the slave-owning sugar magnates.[10]

What can be learned about slave life on these two estates from the Mount Airy and Mesopotamia inventories? Since space does not permit analysis of all 2,500 slaves owned by the Tayloes and Barhams during the 104 years covered by the two sets of inventories, we shall focus on the 668 slaves who lived at Mount Airy during a twenty-year span, 1809-1828, and the 548 slaves who lived at Mesopotamia during an equivalent and overlapping period, 1799-1818.[11] Tracing the individual histories of these 1,216 people, we can compare demographic trends, family structure, and labor patterns in the two slave gangs, as well as the managerial policies of the two owners.

Unfortunately, it is impossible to convey in brief compass the richly detailed biographical information in these inventories about hundreds of Mount Airy and Mesopotamia slaves. To illustrate, let us follow the careers of two people, chosen at random from the top of the alphabet—a woman named Agga at Mount Airy and a man named Augustus at Mesopotamia. In 1808 Agga was a thirty-one-year-old spinner valued at £70, which was slightly above the standard adult female price at Mount Airy. Her father had been a gardener; both parents were dead or gone by 1808. Her husband, Carpenter Harry, was forty-two in that year. Agga and Harry had four young children: ten-year-old John, who was put in a field gang for the first time in 1808 and sold in 1819; eight-year-old Michael, who entered the work force as a carpenter in 1812; three-year-old Kitty, who was sent to the Tayloes' house in Washington as a domestic in 1815; and one-year-old Caroline, who was sold in 1818. Agga had four more children in 1808, 1810, 1813, and 1818. Her two younger boys, Tom and George, became carpenters like their father and brother, and her two younger girls, Georgina and Ibby, became field hands.

---

[10] I wish to thank H. Gwynne Tayloe for generously showing me his family house at Mount Airy. The house is described by Thomas Tileston Waterman and John A. Barrows, *Domestic Colonial Architecture of Tidewater Virginia* (New York, 1968), 126-137. Mesopotamia is described by Paul F. White and Philip Wright, *Exploring Jamaica* (New York, 1969), 166-167; and the *Jamaica Daily News*, Feb. 19, 1975, has an article on the Barham Sugar Workers Co-operative.

[11] The dates chosen for Mesopotamia are a decade earlier than for Mount Airy because of an awkward three-year gap (1819-1821) in the Mesopotamia records, during which time Barham imported about 110 slaves from Springfield estate. This short gap may seem trifling, but the missing birth and death lists obscure the demographic data, and the 110 new Springfield slaves who first appear in 1822 are not identified as family groups, hence making more difficult the already formidable task of analyzing family structure. Since Mesopotamia's records are complete for the years 1799-1818, this equivalent span has been substituted.

Agga was forty-one years old when her last baby was born, and fifty-two when her husband died. In 1844, when the most detailed of the Mount Airy lists was taken, she was sixty-seven and in failing health, but still employed as a spinner. One son lived with her in 1844; two daughters and seven grandchildren lived on neighboring Rappahannock farms; the other five children and five traceable grandchildren had moved away or died.

In the Mesopotamia list Augustus is recorded as an invalid in 1809, having retired from the work force at age sixty-seven. One can trace Augustus through almost his entire long career via fifty-seven Mesopotamia inventories. He first appears on a crude list dated April 18, 1743, before the Mesopotamia managers began keeping annual records or listing the slaves by age. At this time Augustus was about three years old; his name appears toward the bottom of the boys' group with half a dozen other children who in later lists turn out to be the same age. Eight lists later, in 1756, he is promoted from the boys' to the men's group; he was then about sixteen years old and valued at £75, the price of an able adult working man at that date. In 1762, when the Mesopotamia lists become much more detailed, Augustus is described as twenty-two years old, in good health, and employed as a distiller, a post he held for the next forty years. In 1768 he was baptized into the Moravian church with the name of Peter (though he remains Augustus on the slave lists) and recovered from the smallpox without ill effect. He began to decline in health in his early forties, being characterized as "sickly" for the first time in 1781. By 1785 he was spitting blood. Nevertheless, he kept working in the still house until 1802, when he was transferred to the easier job of head watchman. In 1809, when he was retired to invalid status, he was probably two years older than his then stated age of sixty-seven, and he died at seventy-two on February 16, 1812.

As is evident from these two examples, the Mount Airy and Mesopotamia inventories are more precise and systematic documents than most population counts or census returns, including modern ones, because they were reworked annually over a long period, and because they report at least four variables per year for each slave. Thus they surmount the problem of record linkage, which is so bothersome in attempting to compare any two census lists compiled at ten-year intervals. In the Mesopotamia list for 1809 there are nine males with the name of John, and on the corresponding Mount Airy list there are twelve Johns; but each of these individuals can be distinguished from the others by maternal lineage, occupation, state of health, and especially by age. Reporting of age is seriously defective in most forms of census taking. Demographers put little credence in the stated ages of elderly people, and develop compensatory techniques to combat obvious tendencies toward age heaping—overreporting of age in even digits and in multiples of five and ten—or such more subtle problems as the under-reporting of females aged ten to nineteen and the overreporting of females aged twenty-five to thirty-four.[12]

If such defects are built into current census taking, it is obvious that age statements for slaves on most eighteenth- and nineteenth-century documents are mere guesswork. But the ages of the Mesopotamia and Mount Airy slaves can usually be accurately established. Neither set of lists shows significant age heaping or other signs of gross distortion. At Mesopotamia the Barhams kept birth registers after 1773, so that the exact ages of most slaves born on the estate during our period are recorded. The Tayloes did not keep birth registers, but the birth years for nearly half the Mount Airy slaves can be established. In compiling their lists, the census takers sometimes carelessly repeated last year's age or capriciously added ten years when a person looked old or sick. The stated ages of some elderly slaves cannot be verified and are doubtless inflated, but most other errors can be corrected. Since a great many of the slaves who were brought into the two estates arrived as children, the initial age estimates for these people are probably not wildly wrong, and the census takers endeavored to correct errors. For instance, an African girl named Matura, who arrived at Mesopotamia in 1792, was classified at first as eleven years old, but the next year her age was advanced to fourteen. Four years later, at the (corrected) age of eighteen, Matura had her first child—and as we shall see, this was early for motherhood in Mesopotamia. Thus the two series stand up under close inspection as consistent and reliable; their greatest shortcoming is that they generally identify only the mothers, not the fathers, of slave children, and thus preclude full analysis of family structure.

Table I compares demographic trends on the two estates. As is well known, over a span of two centuries the Virginia slave population experienced marked natural increase while the Jamaican slave population experienced marked natural decrease. Virginia planters imported fewer than 150,000 slaves between 1609 and 1808, and the black population of the state in 1809 was about 415,000. Jamaican planters imported something like 750,000 slaves between 1655 and 1808, yet the black population on the island in 1809 was only about 350,000.[13] As Table I demonstrates, Mount Airy and Mesopotamia reflected these demographic conditions. The most basic contrast between the two slave communities was that Tayloe's slaves increased and multiplied, whereas Barham had to keep restocking Mesopotamia with fresh purchases. The Mount Airy totals in Table I are partly conjectural,

[12] Ansley J. Coale and Paul Demeny, *Methods of Estimating Basic Demographic Measures from Incomplete Data* (New York, 1967), 19-21.

[13] For slave import and black population estimates see Philip D. Curtin, *The Atlantic Slave Trade: A Census* (Madison, Wis., 1969), 52-59, 71-74, 136-145; U.S. Bureau of the Census, *Negro Population, 1790-1915* (Washington, D.C., 1918), 45-57; George W. Roberts, *The Population of Jamaica* (Cambridge, 1957), 35-43, 65; and Braithwaite, *Development of Creole Society*, 152, 168, 207.

TABLE I
SLAVE POPULATION CHANGES OVER TWENTY YEARS

	Mount Airy, Virginia, 1809-1828			Mesopotamia, Jamaica, 1799-1818		
	Male	Female	Total	Male	Female	Total
Population at outset	219	165	384	190	174	364
Increase:						
Born	131	116	247	61	63	124
Purchased	3	0	3	28	32	60
Moved in[a]	17	17	34	0	0	0
	151	133	284	89	95	184
Decrease:						
Died	45	43	128	132	101	233
Est. died	27	13				
Sold	8	29	52	0	0	0
Est. sold	5	10				
Moved out[b]	53	31	125	0	0	0
Est. moved out	32	9				
Freed	1	0	1	2	4	6
	171	135	306	134	105	239
Population at close	199	163	362	145	164	309
Recorded Birth rate	39.83 per 1000			18.89 per 1000		
Recorded Death rate	20.64 per 1000			35.49 per 1000		

Notes: [a] From other outlying Tayloe estates.
[b] To other outlying Tayloe estates.

because John Tayloe III was constantly switching slaves from one quarter to another, or handing over farms (with slaves attached) to his sons, or opening new farms, or selling surplus slaves, without leaving adequate record. It is thus impossible to tell which of the thirty-four slaves who moved into the Rappahannock farms during this period were bought, and which were transferred from outlying Tayloe estates. Likewise, it is impossible to tell what happened to ninety-six of the 306 slaves who dropped off the Tayloe lists. But inspection of the 209 known deaths, sales, and transfers during this period reveals a persistent pattern: those Mount Airy slaves who were sold or transferred were generally in their 'teens or twenties, whereas Mount Airy slaves who died were almost always younger or older than this—under ten or over thirty.[14] The Mount Airy estimates in Table I assume that the ninety-six unidentifiable deaths, sales, and transfers follow exactly the same pattern as the 209 that are known.

The vital rates in Table I pose an interpretive problem. The demographers Ansley Coale and Paul Demeny have published 192 model life tables and nearly 5,000 stable population tables, derived mainly from twentieth-century European vital statistics and census returns, so as to estimate the full range of population parameters in various regions of the world. Historians interested in slave demography have employed these tables to project birth rates for eighteenth- and nineteenth-century slave populations at around 50 per 1,000 with death rates in the low to mid-thirties.[15] But this assumes that past slave populations experienced the same range of fertility and mortality levels as current nonenslaved populations. The demographic patterns for Mount Airy slaves do indeed seem compatible with Coale and Demeny's schedules. Tayloe kept no vital registers; hence his inventories clearly under-report infant births and deaths. Furthermore, he moved many slaves off the estate, thus depressing the fertility and obscuring the mortality of his population. But Mesopotamia's vital rates appear to be more accurate. Barham's bookkeepers kept birth and death registers that report the deaths of newborn infants. Quite possibly they failed to note many other infant births and deaths; but even if the reported vital rates are far too low, the Mesopotamia data on age composition, longevity, and rate of natural decrease make a very poor fit with all of Coale and Demeny's tables. Clearly, the demographic pattern at Mesopotamia differed radically from Mount Airy's; I believe that it also differed from all observed modern populations.

---

[14] Of the 88 Mount Airy slaves whose deaths are recorded between 1809 and 1828, 48% were children under 10, only 9% were in their 'teens or 20s, and 43% were adults over 30. Of the 121 slaves recorded as sold or transferred, 25% were children under 10, 55% were in their 'teens or 20s, and only 20% were over 30.

[15] The model tables are published in Ansley J. Coale and Paul Demeny, *Regional Model Life Tables and Stable Populations* (Princeton, N.J., 1966); their "West" family of tables—the only set to encompass non-European demographic experience—draws no data from slave populations. Jack Eblen uses these tables to postulate natural increase and high birth rates among the Cuban and Jamaican slaves. Eblen, "On the Natural Increase of Slave Populations," in Engerman and Genovese, eds., *Race and Slavery*, 244-247. However, Michael Craton's analysis of demographic evidence at Worthy Park estate, Jamaica, supports my findings.

The striking feature of the Mesopotamia vital records is the feeble birth rate. In an average year seventy-five women of child-bearing age lived on the estate, yet they produced only six recorded live births. Whether or not these figures are correct, 109 more slaves died than were born, 1799-1818, and Barham sustained his work force only by importing sixty new slaves. These new slaves came mainly in a single transaction from Cairncurran estate in Jamaica, rather than from Africa. No Mesopotamia slaves had been bought from African traders since 1792, with the result that by 1809 only 19 percent of the slave force had come straight from Africa, another 19 percent had been bought from other Jamaican plantations, and 60 percent were born at Mesopotamia.[16] Those coming from Africa had arrived at Mesopotamia in small lots of ten or a dozen; most commonly they had been teenage boys. The "seasoned" slaves acquired from neighboring plantations came in much larger groups—40 from Three Mile River estate in 1786, 61 from Southfield in 1791, 56 from Cairncurran in 1814, and about 110 from Springfield around 1820. Obviously, these people knew each other already, and they came in family groups, parents with children, the very old and the very young, and almost as many females as males. Thus while in the mid-eighteenth-century Mesopotamian slave life had been marked by ethnic diversity, with Negroes coming from various regions of West and Central Africa, speaking different languages and holding conflicting values, with the passage of time the estate became far more homogeneous, not to say inbred. Table I suggests the extremely immobile character of slave life on this estate. Negroes born at Mesopotamia almost invariably spent their entire lives there. Negroes bought from an African trader or from another planter were almost never sold again. Between 1751 and 1818 only four Mesopotamia slaves were sold, ten were manumitted, and five ran away. None were transferred to Joseph Foster Barham's other sugar estate, the Island, in neighboring St. Elizabeth parish.

Everything was different at Mount Airy. There were twice as many recorded births as deaths on this estate, few new slaves were imported, and the slight population decline between 1809 and 1828 is explained by a massive exodus of slaves—approximately 177 in twenty years. At least thirty-nine Rappahannock slaves—predominantly young males—were sent 180 miles west to the Tayloes' Cloverdale ironworks in Botetourt County beyond the Blue Ridge in 1811-1814. Another twenty-two went to Windsor Farm in King George County in 1820-1821. When John Tayloe III bought Deogg Farm in King George County in 1824, he staffed it entirely with slaves drawn from his neighboring Rappahannock farms. John III stated in his will that he disliked separating slave families, and the inventories show that he generally did sell mothers with their young children. But not always. In 1816 Forkland

[16] The origins of the remaining 2%—mainly old people who had been living on the estate since the 1740s—cannot be traced.

Cate was sold without her four-year-old son Alfred, and Gwinfield Rachel was sold without her six-year-old son Billy. A number of girls were separated from their mothers and sold at about age nine, and by the time they reached their early 'teens boys and girls were at the prime age for transfer to a new quarter or for sale off the estate. For example, a boy named John, after spending the first decade of his life with his mother at Doctors Hall, was put in the field gang at Forkland at age eleven, was then switched to Old House at age fifteen, and was sold at age eighteen. Almost all of these Negroes must have been native Virginians. Back in the early eighteenth century the Tayloes had established their slave force with Africans, but John III had no need for the slave trade and small need for slave purchases. Only three of the 668 slaves on the Mount Airy lists for 1809-1828 are identified as "newly purchased." Starting as he did in the late 1770s with about 200 Negroes, Tayloe could, through the process of natural increase alone, come close to building up his force to the point where, fifty years later, he had 700 slaves to distribute among his seven sons.[17]

Examination of the shifting population on one Rappahannock farm, the Hopyard quarter, shows how Tayloe's system worked. This farm had a slave force of forty-six in 1809, of whom twenty-six were classified as working field hands. During the next twenty years thirty slaves were born at Hopyard, three grown slaves moved in, and thirty-nine dropped off the lists. Five of these thirty-nine were moved to other Rappahannock farms, four were moved to Cloverdale, nine died, and eleven were sold—ten of them in a single year, 1816, when two women, Kesiah and Patty, were sold together with eight of their young children. Ten of those who disappeared from the Hopyard lists cannot be traced, but judging by their ages, five of them probably died and five were probably transferred out of the Rappahannock district. Thus Hopyard—a comparatively small population unit—produced some twenty-five slaves for sale or transfer in twenty years. By 1828 the Hopyard gang had dropped to forty, but since twenty-seven of these were classified as working hands, the farm was as strongly staffed as in 1809. Only

[17] My figures on the total size of John Tayloe's slave force are only approximate. I have tried to trace his holdings from 1787 onward through the annual county personal property tax lists in the Va. State Lib., but this is hard to do, since he and his sons held slaves in many counties and the tax assessors frequently confused their slaves with slaves held by planters named Taylor. The process followed by John Tayloe III is easier to trace in the records of a contemporary South Carolina cotton planter, Peter Gaillard, who inherited his property in 1784 and retired from business in 1825. Gaillard started with 134 slaves, bought 125 more (mainly in the 1800s in order to set up his three elder sons), and sold 35. During his 40 years of plantership his slave force doubled through natural increase: 456 births as against 237 deaths. Thus by 1825 he was able to give 433 slaves to his eight children and still have 10 left to attend him in his old age. See Peter Gaillard's Account and Memorandum Book, Gaillard Papers, South Carolina Historical Society, Charleston.

nineteen of the forty-six slaves who were there in 1809 were still there twenty years later. But some families remained largely intact. Four men and four women, very probably married couples, and nineteen of the four women's children and grandchildren accounted for most of the Hopyard population in 1828.

At Mount Airy, as at Mesopotamia, the age profile of the slave population fluctuated considerably from year to year. In Table II age distribution within the two slave gangs has been averaged out over a twenty-year period to show basic differences between the proportions of males and females, young and old, on the two estates. Several obvious and important differences emerge. Mount Airy had a much higher proportion of children—naturally enough, since the recorded birth rate was greater. Mount Airy had a much younger population, with the median age at twenty, as against twenty-seven in Mesopotamia. Mount Airy had a higher proportion of males, including young adult males of prime working age (20-34). Mesopotamia had a higher proportion of females in every age group above fourteen, most particularly young women of prime child-bearing age (20-29). This makes the low Mesopotamia birth rate especially puzzling. Mesopotamia also had a higher proportion of relatively old people, with nearly 20 percent of the population beyond the age of forty-five, and a higher percentage than at Mount Airy beyond the age of sixty.

At Mesopotamia, as was generally the case on West Indian sugar estates, the females proved tougher than the males and better able to survive the trauma of slavery. In the eighteenth century, when the Barhams kept restocking from African slave traders, they maintained a pronounced male

TABLE II
AGE DISTRIBUTION OVER TWENTY YEARS

	Mount Airy 1809-1828		Mesopotamia 1799-1818	
	Percent of Total Pop.		Percent of Total Pop.	
Age	Males	Females	Males	Females
0-14	22.1	17.9	14.7	14.4
15-19	6.1	4.0	4.6	5.0
20-24	5.5	3.8	3.8	4.0
25-29	3.9	3.4	4.5	4.1
30-34	4.1	2.8	3.1	3.4
35-39	3.2	2.9	5.7	4.3
40-44	3.3	2.5	5.0	3.7
45-	7.5	7.0	8.9	10.8
Totals	55.7	44.3	50.3	49.7

majority—as high as 148 males for every 100 females in 1772, for example. But between 1799 and 1818 this male preponderance disappeared, and by 1818 the females were in a decided majority (88/100). The significance of this shift in the sex ratio becomes clearer if we focus attention on those Mesopotamia slaves of prime working age, between seventeen and forty years old, who were categorized as "able bodied" rather than sickly, weak, or diseased. The number of healthy slaves of prime age at Mesopotamia was always strikingly small—ranging from one-fourth to one-third of the total slave force. During the 1799-1818 span 52 percent of the healthy prime-aged slaves at Mesopotamia were women. Thus much of the heavy labor at Mesopotamia had to be performed by females who—especially since they produced few children—occupied a role radically different from that taken by women in most western societies.

At Mount Airy the women remained a minority. Part of the explanation is demographic, for, as Table I shows, there was a surplus of male births between 1809 and 1828. But the sex ratio was also powerfully affected by the Tayloes' transfer of slaves from Mount Airy. Clearly, the Tayloes valued women well below men, as shown by the inventory of 1809, where the females are priced at an average of £53 6s. 2d. apiece as against £64 2s. 2d. for the males.[18] About two-thirds of the slaves transferred from the Rappahannock district to other family farms were males—thus helping to build a solid majority of men on the Tayloe work gangs. Enough females stayed at Mount Airy to perform domestic tasks and to assure a healthy rate of natural increase. The rest were sold.

About three-quarters of the Mount Airy slaves sold between 1809 and 1828 were female. Of the twenty-nine female sales for which we have definite record, four were small children, all sold with their mothers. Sixteen were girls aged nine to seventeen; of these, eleven were sold separately from their kin and none had babies of their own. The remaining nine were mature women, of whom seven were mothers, sold in combination with thirteen of their young children. From a practical standpoint it was doubtless sensible to sell slave girls at the preadolescent/adolescent stage of life, when they commanded a price that amply repaid the cost of their upbringing and when they were old enough to work for a new owner, yet still too young for motherhood. There is no evidence whatsoever that the Tayloes practiced slave breeding in the sense that they mated Negroes forcibly, frequently, or promiscuously in order to sell the surplus progeny. But plainly the Tayloes did prefer male to female workers, and they maintained an artificially unbalanced sex ratio in their Mount Airy slave gang.

---

[18] Likewise at Mesopotamia the female slaves in scattered lists from 1786 to 1814 were priced at an average of £71.17.9 apiece as against £84.0.3 for the males.

The conditions described thus far did nothing to bolster family life among the slaves. The character of household and family structure at Mesopotamia and Mount Airy is difficult to discover, since the white men who compiled the slave inventories took small interest in such matters and seldom even identified husbands or wives. But it is certain that many slaves on both estates maintained conjugal family units. For family structure at Mesopotamia the best evidence comes in 1814, when the fifty-six newly purchased slaves from the Cairncurran estate are listed in family groups rather than by the usual division into males and females.[19] The first sixteen entries from this list are reproduced below to show the character of record keeping at Mesopotamia:

Name	Age	Occupation	Condition	Value (£)
Smart	37	Head Driver	Prime	200
Lettice	35	Field	Prime	145
Peggie	12	Field	Healthy	100
Job	5	Small Gang	Healthy	50
Camilla	33	Field	Able	160
Bob	3		Yaws	30
Leicester	5		Yaws	40
John Savey	9		Healthy	60
Exeter	55	Head Mason	Aged	140
Sally	37	House Cook	Prime	160
Bessie	13	House Wench	Healthy	130
Richard	11	Small Gang	Healthy	85
Joe	7	Small Gang	Healthy	70
Mary	5		Healthy	55
Jean	3		Healthy	40
Ann	1½		Healthy	30

These sixteen persons evidently belonged to three families: Smart and Lettice with two children; Camilla with three children and no mate; Exeter and Sally with six children. To be sure, Smart and Exeter are not specifically identified as husbands or fathers, but unless we are to suppose that the record keeper put down names at random—which is highly implausible—the list indicates that these men headed their respective families. A conspicuous feature of the list is that very young children worked in the field in the Small Gang, also known as the hogmeat or grass gang.

Camilla, without a mate on arrival at Mesopotamia, gave birth to her

[19] The list of Cairncurran slaves, dated Jan. 29, 1814, is in the Barham Papers, Box b. 34.

next child fourteen months later. Her small boys, Bob and Leicester, suffered from a common West Indian infectious disease, the yaws, characterized by skin eruptions and bone lesions. John Savey may have been a mulatto, named after his white father, which would explain why he was unemployed although a healthy nine year old. However, he was priced low for a mulatto, was never recorded as mulatto in the Mesopotamia lists, and soon went into the field gang in which mulattoes almost never worked. Exeter appears rather old for his wife and brood of young children, but he was evidently a robust person, still alive at age seventy-three when the last Mesopotamia list was made in 1832. Sally, a prolific breeder by Jamaican standards, had two more children after arrival at Mesopotamia. Viewed collectively, forty-five of the fifty-six Cairncurran slaves arrived in family groups.

How many of these Cairncurran families lived in nucleated households at Mesopotamia is unknown. Evidence from other contemporaneous Jamaican estates argues that rather few did so. Mating was often casual, parents frequently lived in separate establishments, and households containing children were more often headed by women than by men. In the Jamaican slave kinship network, the maternal bond was the key element, for a grown son tended to live with his mother until her death or stayed as close as possible in the house next door.[20] Demographic conditions also powerfully affected family life. In the case of the Cairncurran slaves, by 1832 when the last Mesopotamia inventory was taken, eighteen years after their arrival, twenty-two people from this group of fifty-six had died, and nineteen of the survivors were weak, diseased, or invalided. Of the twelve family groups in 1814, eight had lost one or both parents by 1832, and four had lost one or several children. The one stalwart exception was Exeter's and Sally's family of eight, all still alive, the father nearly blind and the mother caring for him, while their six grown children labored in the cane fields.

The other great impediment to family life among the Mesopotamia slaves was the white overseers' and bookkeepers' sexual exploitation of the black women. Fourteen of the 124 babies born at Mesopotamia between 1799 and 1818 were mulattoes or quadroons—11 percent of the total births. This ratio may not seem high, but it must be remembered that only about six white men lived on the estate at any one time as against some ninety black men between the ages of seventeen and fifty. A white man living at Mesopotamia was twice as likely as a black man to sire a slave baby, a

[20] See B. W. Higman, "Household Structure and Fertility on Jamaican Slave Plantations: A Nineteenth-century Example," *Population Studies*, XXVII (1973), 527-550. Compare Charles B. Dew's analysis of slave family life at a Virginia ironworks, 1811-1813, in "David Ross and the Oxford Iron Works: A Study of Industrial Slavery in the Early Nineteenth-Century South," *WMQ*, 3d Ser., XXXI (1974), 189-224.

finding which provides some idea of the frequency of interracial sexual intercourse. The whites preferred their slave mistresses to be young: of the nine women who bore mulatto or quadroon infants during this twenty-year span, one was only fifteen years old and four others were under twenty. The whites also preferred light-complexioned women, such as Mulatto Ann, who worked in the overseer's house from age ten on, and bore two quadroon children when she was sixteen and eighteen. Neither Ann nor any of the other women used by the whites at Mesopotamia was manumitted. A twenty-three-year-old field slave named Judy, the mother of two Negro children, bore a mulatto boy named Archibald; she remained in the field gang and bore six more Negro children. Her son Archibald, being a mulatto, was placed in the overseer's house at age six and trained as a carpenter at age sixteen, but he must have been an unhappy person for he became a chronic runaway and died of the yaws when he was twenty-seven. Another sad story is the case of Batty, a field slave who caught the eye of the overseer, Patrick Knight, when she was a young mother in her twenties. Batty had two daughters by Knight, and in 1803, when she was twenty-nine, he manumitted both girls, took Batty out of the field gang, and installed her in his house. For the next thirteen years Batty lived with Knight and bore him two more children. Her sixth baby was a Negro child, and at about the age of forty she contracted the Coco Bays, a disease akin to leprosy. So Batty was sent back to the Mesopotamia work force, though she was sick, to toil her declining years as a washerwoman and grass cutter.

In Virginia mulattoes and quadroons were not identified as such in the records; thus there is no direct evidence of miscegenation at Mount Airy. But the evidence of disrupted family life is abundant. In 1816, when the Tayloes sold Hopyard Kesiah and Hopyard Patty with eight of their children, two of Patty's teenage children stayed at Hopyard and no male mates went with Kesiah and Patty. Similarly, in 1820, when Bob and his wife Betty were sent from Oakenbrow quarter to Windsor farm with their three youngest children, their older three teenage children stayed at Oakenbrow. A fuller sense of the disjointed nature of family life at Mount Airy comes from inspection of the inventory for 1835, a few years beyond our period. This is an exceptionally interesting list because it identifies both parents of all the small children on the estate and thereby records many of the slave marriages.[21] The following extract shows the character of this 1835 list:

---

[21] The inventory for 1835 is in the first inventory book kept by William Henry Tayloe, John III's third son, who inherited Mount Airy and four adjoining farms in 1828 (Tayloe Papers, MSS 1 T2118d13410, 43-46). William Henry Tayloe's inventories for 1844 and 1845 are likewise useful, since they identify the parents of almost all the adult slaves living at Mount Airy in those years. *Ibid.*, MSS 1 T2118a13.

Car[penter]. Bill and his wife Esthers .... wife Winney Jr.
Children
    Winney 8
    Anne 7
    Juliet 5
    James 3
    Charlotte .... Inf[ant]

Tom and his wife Winneys
    William .... out in field
    Grace 7
    Chapman 4

    Urias 7
    Paul 5
    China 3
    Prince .... Inf[ant]

Marilla's—Husband Decd.
    Rose 8

Jacob and his wife Mary
    Kate .... 10 in Ala[bama]

Altogether, thirty-five Mount Airy slave families can be reconstructed by combining this inventory of 1835 with the earlier lists.

The five Mount Airy families shown in the illustration above are not listed completely. Esther had a ten-year-old daughter working at the Tayloe house in Washington; Mary had a fourteen-year-old son working in a field gang; and Marilla had five grown children, three of whom lived on the same quarter with her in 1835. Except for Winney's twelve-year-old son William, this list excludes working children, but when Mount Airy boys and girls did enter the work force they sometimes were sent far away—like ten-year-old Kate in the example above, who went to Alabama in 1836. When this list was drawn up, the Tayloes were in the process of moving Negroes from several of their Virginia estates to Alabama in order to open new cotton plantations there. By 1836 William Henry Tayloe had transferred forty-seven of his Mount Airy slaves—mainly young unmarried men and women—to Alabama, and by 1855 he had more slaves there than in Virginia.

The Mount Airy kinship network was pervasive in 1835; almost every slave had several blood relatives living on the estate. Yet of the thirty-five identifiable slave families, fifteen were lacking one or both parents, and in most families the children were widely dispersed. Furthermore, in the twenty families where husband and wife were both still living, only five couples regularly lived together. The others worked at separate quarters, often many miles apart, which doubtless helped to depress the Mount Airy fertility rate. Carpenter Bill in the illustration above was a polygamist who lived at Mount Airy while his two wives, Esther and Winney, lived at Landsdown with parallel sets of young children. Another polygamist in 1835 was a man named Oliver who was owned by a neighboring planter. Oliver had two wives and eight children at Doctors Hall quarter. Three other outside males were married to Mount Airy women; one of them was a free black named David, by whom Forkland Criss bore a family of ten slave children. As at Mesopotamia, most young Mount Airy children seem to have lived with their mothers but not with their fathers. And since the Tayloes kept more male than female slaves, a great many Mount Airy men had no marriage partners unless they mated with women from neighboring estates. According to the 1835 list, almost all the eligible Mount Airy women had husbands, while only a third of the men in their twenties and thirties had local wives—a fact that must have contributed powerfully to masculine feelings of inadequacy and frustration.[22]

We turn now to the employment pattern on the two estates. Table III compares the distribution of jobs at Mount Airy and Mesopotamia in 1809. According to these figures, Barham employed his slave force much more fully than did Tayloe. At Mesopotamia four-fifths of the Negroes were allotted tasks, and many more women and children worked than at Mount Airy.

Tayloe designated as working hands only those slaves mature and strong enough to do a full "share" or a half "share" of labor apiece—in practice, all able-bodied Negroes over the age of ten[23]—but the younger children also must have had light tasks to keep them occupied, if not productive. At Mesopotamia boys and girls were put to work at age six, generally in the hogmeat or grass gang, to gather grass and straw from the fields to feed the livestock. At age ten they usually graduated to harder assignments, such as hoeing the young cane and carrying dung in the third field gang. At fifteen or sixteen they might move up to the second field gang and spend the next few years weeding the cane and cleaning the pastures. At about twenty they would be ready for the Great Gang and its backbreaking toil of digging the cane holes and cutting the ripe cane. Slaves who started out as field laborers were rarely switched to craft or domestic jobs. Only workers in their prime did the heavy field labor. At Mesopotamia men who were past forty and sickly became jobbers or watchmen; old or ailing women became field cooks, nurses, or washerwomen. Very likely some of the older Mount Airy slaves listed as nonworkers performed similar marginal tasks.

On both plantations a majority of the laborers in 1809 were field hands, but there were a good many skilled and favored job holders. At Mesopotamia a small managerial elite—the drivers of the field gangs, the chief craftsmen, and Quasheba the African female doctor—received special rations of rum each week. The Mesopotamia craft workers were all male; almost all of them

[22] For a considerably more positive picture of the slave family in 18th-century Maryland see Allan Kulikoff, "The Beginnings of the Afro-American Family in Maryland," in Aubrey C. Land *et al.*, eds., *Law, Society, and Politics in Early Maryland: Essays in Honor of Morris Leon Radoff* (Baltimore, 1976).

[23] The overseer on each Rappahannock farm got one or more "shares" of the crop, encouraging him to extract maximum output from the workers, who in 1809 produced an average about 50 bu. of wheat and 100 bu. of corn apiece.

TABLE III
LABOR PATTERNS AT MOUNT AIRY AND MESOPOTAMIA IN 1809

	Mount Airy			Mesopotamia		
	Males	Females	%	Males	Females	%
A. *Workers*						
Drivers	0	0	0.00	4	1	1.55
Craftworkers	33	14	12.24	25	0	7.76
Domestics	18	17	9.11	8	15	7.14
Field Cooks	0	0	0.00	0	6	1.86
Fieldworkers	86	58	37.50	45	92	42.55
Jobbers	7	0	1.82	7	0	2.17
Transport	5	0	1.30	6	0	1.86
Stockkeepers	0	0	0.00	14	2	4.97
Watchmen	0	0	0.00	19	0	5.90
Nurses	0	0	0.00	0	11	3.42
Total	149	89	61.98	128	127	79.19
B. *Nonworkers*						
Too young	61	63	32.29	20	16	11.18
Too old	2	5	1.82	4	6	3.11
Too sick	7	8	3.91	11	10	6.52
Total	70	76	38.02	35	32	20.81
C. Totals	219	165	100.00	163	159	100.00

had learned their jobs as boy apprentices and pursued the same routine for years. The fact that nearly half of them were over the age of forty in 1809 indicates—as one might guess—that craft workers survived longer than field laborers. No fewer than ten houseboys, maid servants, and cooks waited on the overseer and the bookkeepers, and four slaves assisted the Moravian missionaries at their chapel. At Mount Airy the proportion of skilled and semi-skilled workers was considerably higher—34 percent of those employed, as compared with 21 percent at Mesopotamia.[24] Because the Tayloes lived at Mount Airy, they had twenty-six domestics, and the overseers at the outlying farms had one or two house servants each. As at Mesopotamia, the Mount Airy artisans and domestics entered their jobs as children and, once established, were far less likely to be transferred or sold than were the field laborers.

[24] This figure at Mount Airy is unusually high. Fogel and Engerman, in their analysis of southern U.S. slave occupations around 1850, claim that 26.3% were in skilled or semi-skilled jobs (*Time on the Cross*, I, 38-40), but Gutman argues that the true figure was 15% or lower ("The World Two Cliometricians Made," *Jour. of Negro Hist.*, LX [1975], 111-128).

The labor pattern at Mount Airy was designed to achieve almost total self-sufficiency. The field workers raised corn and pork—the staple slave foods in Virginia—and tended their vegetable gardens in off hours. Using cotton grown and ginned on the estate, and wool sheared from local sheep, the seventeen spinners and weavers made coarse cloth for slave apparel and fine cloth for household linen. The four shoemakers tanned and dressed leather from Mount Airy cattle to make coarse shoes for the slaves and custom shoes for the Tayloes, as well as harness for the horses, mules, and oxen.[25] The nine smiths and joiners built and repaired wagons, ploughs, and hoes, and shod horses, while the twenty-two carpenters, masons, and jobbers moved about the estate erecting and repairing buildings. Tayloe's wagoner carted goods from one unit to another, and his schooner, *The Federalist*, manned by four slave sailors, carried his cash crop of Rappahannock wheat, flour, corn, and oats to Baltimore or Alexandria.

By contrast, the Mesopotamia labor pattern was by no means designed for self-sufficiency. In order to keep his slaves alive and working, Barham shipped food, tools, and clothing from Britain. On the mountain land bordering the estate the slaves cultivated crops of cocco roots and plantains—an equivalent to Mount Airy's cornmeal—but quite often these provision crops were ruined by tropical storms or drought. In 1815 and again in 1816 the Mesopotamia overseer bought a ton of cocco roots as emergency rations to prevent starvation. For protein the slaves depended largely on meager allotments of salt herring shipped from England—150 barrels in 1817, or half a barrel per slave per year.[26] Livestock were plentiful on this estate—448 cattle in 1809, for example—but they were used to produce manure for fertilizer or as draft animals, and only three steers were slaughtered annually for the slaves as a special Christmas treat. No cloth was made at Mesopotamia; instead, Barham bought about two thousand yards of coarse oznaburgh linen annually, or seven yards per slave, together with thread, scissors, and needles, so that his Negroes could make their own clothes. There were no shoemakers—unneeded since slaves wore no shoes—and no local ironworks to supply ironmongery and nails. The three Mesopotamia blacksmiths did not manufacture farm implements as at Mount Airy, so that every tool and piece of machinery had to be imported. Not even the labor force was self-sufficient. Whenever major repairs were needed at the sugar factory, managers hired white masons and coppersmiths rather than trust the work to their slave craftsmen. And during the two years 1816-1817 they paid £930—the cost of a dozen new slaves—in order to hire extra Negro laborers to hole

---

[25] The Mount Airy artisans also did custom work for the Tayloes' white neighbors. See Tayloe Account Book, 1789-1828, Tayloe Papers, MSS 1 T2118d357.

[26] The Mesopotamia expense accounts for 1816-1817 are in the Barham Papers, Box b. 33.

the cane fields, clean the pastures, and plant provisions, since the regular work gangs could not handle all the necessary field tasks.

Another big difference between the employment patterns of the two estates is that the female slaves did much more of the basic labor at Mesopotamia. Table III shows that two-thirds of the agricultural laborers on that estate were women and girls. Even on the Mesopotamia Great Gang, where the hardest work was done, there were thirty-one women and twenty-two men in 1809. Females did much of the heavy labor at Mount Airy also, but more of them worked in craft or domestic jobs, and nearly half were excused from employment. Motherhood was no excuse, however. Thirty-eight Mount Airy women had one or more living children under the age of six in 1809, and all but two of them had full-time jobs. Sally, a twenty-nine-year-old field hand at Doctors Hall, had five young children in 1809 and gave birth to eight more by 1826, while continuing to do her "share" of farm labor. Motherhood was more honored at Mesopotamia, because the birth rate was so alarmingly low. Here the overseer made a practice of moving pregnant members of the Great Gang to the second field gang, where the work was lighter. Matura, the mother of five children, was taken off the Great Gang permanently in 1809 so that she could look after her youngsters. At Mesopotamia all mothers of newborn infants were paid a bonus of £1 in cash "for raising their children," and nursing mothers received a quart of oatmeal and a pint of sugar each week.[27]

Not surprisingly, the Mesopotamia management favored those mulatto and quadroon slaves who had been sired by the white staff. Of the seventeen mulattoes and one quadroon living on the estate in 1809, six were house servants, two were carpenters, one attended the Moravian missionaries, and the rest were too young for employment. Mulattoes at Mesopotamia generally began work when they were nine years old, three years later than the Negroes, and they never labored in the fields. At the opposite end of the spectrum, the management discriminated particularly against native Africans. Of the sixty-two persons at Mesopotamia in 1809 who had come via the slave trade, only Quasheba the doctor was recognized as an important figure. Over 80 percent were relegated to gang labor in the fields.[28]

But how meaningful were these occupational titles? Did the field hands really spend all their time in the fields or the blacksmiths at the forge? Fortunately, the Tayloe plantation records include three work logs dated just before and after 1809, showing the actual tasks performed each day of the

---

[27] Mesopotamia food allotments, June 1802, *ibid.*, Box b. 36; expense accounts, 1816-1817, *ibid.*, Box b. 33.

[28] Of these Africans, 43% were field workers in 1809, and another 39% were jobbers, watchmen, nurses, or invalids who had formerly worked in the field gangs. Craton and Walvin find much the same pattern for mulattoes and Africans at Worthy Park (*A Jamaican Plantation*, 138-140).

year by the Mount Airy craft workers, and each week of the year by the Rappahannock farm gangs.[29] From these logs it is evident that the Tayloe slaves did indeed have distinct occupations. The 144 field workers labored almost exclusively on agricultural tasks, spending more time on the corn crop than on any other job—some twenty weeks during the course of the year. They spent ten weeks on the wheat crop and three weeks on the oats crop. The artisans joined in the grain harvest for two weeks in June and July but otherwise worked exclusively at their crafts. Work assignments were variegated, with new tasks assigned every two or three days. Throughout the year the slaves were kept busy six days a week. Apart from Sundays, they had nine days of vacation: Easter Monday, Whit Monday, two free Saturdays in May and July, and a five-day Christmas break from December 25 to 29.

Though the Mount Airy artisans and field hands had separate and specialized functions, their work rhythms were closely synchronized. During the coldest six weeks from mid-December to the end of January, when the previous season's crops had all been harvested and the winter wheat was in the ground, the field workers and jobbers shucked and beat corn, cut and hauled timber to the saw mill for fence rails and posts, and cut ice from the Rappahannock creek. The carpenters meanwhile operated the saw mill; while the smiths and joiners repaired ploughs, wagons, and harnesses for spring ploughing. In February the field hands worked with the carpenters and joiners in putting up fencing, with the jobbers in clearing and manuring the fields for ploughing, and with the sailors in loading the previous year's corn on *The Federalist* for shipment to Alexandria. In March the field gangs planted oats, in April corn, and in May cotton and peas, while the smiths and joiners repaired their tools. In mid-April, just before the Tayloe family arrived for the summer from Washington, eighty-five laborers from five of the Rappahannock farms came to Mount Airy to dress up the mansion lawn. In late May the field workers began to weed the corn, while the smiths, joiners, and carpenters were making and mending rakes, cradles, and scythes for the coming harvest. In mid-June the wagoner went to Kinsale, a nearby town, to fetch five barrels of whiskey for the harvest. The wheat crop at Old House, Doctors Hall, and Forkland was harvested in one frenzied week during mid- or late June; forty-five extra hands were pressed into service—the smiths, masons, joiners, carpenters, and jobbers cut and cradled the grain, while the spinners, shoemakers, and weavers raked and bound it into sheaves. John Tayloe III personally supervised operations; in June 18, 1801, he com-

---

[29] John Tayloe III's Minute Book for Jan. 1-Dec. 7, 1805, Tayloe Papers, MSS 1 T2118a8, records craft work daily and field work weekly. His Minute Book for 1811-1812 does the same for Jan. 9-Sept. 4, 1811, and Feb. 10-Dec. 31, 1812, *ibid.*, MSS 1 T2118a10. His Spinning Minute Book itemizes the amount of cotton and wool spun at Mount Airy every week for Jan. 1806-Dec. 1807, *ibid.*, MSS 1 T2118a9.

plained to a correspondent of being "just from my harvest field and fatigued to death."[30]

In July the seasonal pressure continued, as all hands joined for a week to cut and rake the oats crop at Old House. During the next weeks the field gangs worked mainly in the corn fields, hilling and hoeing the plants. In August they cut the hay and threshed the wheat. The jobbers helped with the hay, the carpenters made grain barrels, and the smiths and joiners worked as usual on ploughs and wagons. In September the wagoner and sailors helped the field hands to load Tayloe's schooner with wheat for the Baltimore market. Now it was time to gather the corn leaves as fodder, and to start the long process of seeding the winter wheat. With the fall racing season approaching, the smiths set to work to shoe Tayloe's racehorses. Shortly after the races, the Tayloes departed for Washington, and the craft workers could now make necessary repairs on the mansion house—as in 1805, when the carpenters, joiners, masons, and jobbers worked for a month reshingling the mansion roof under the supervision of a hired white builder. In October the spinners joined the field hands at picking cotton, and from mid-November to mid-December, with the wheat fields finally seeded, the field hands harvested the corn and hauled the stalks to the saw mill. Just before and after Christmas, the slackest work period of the year, Tayloe's masons, carpenters, and jobbers were sent to repair the Richmond County courthouse.

For the Mesopotamia labor force no equivalent work logs have survived. But a field labor book, dated 1796-1797, for Newton plantation in Barbados—a sugar estate of about the same strength as Mesopotamia—records the daily tasks of field and craft workers throughout the year.[31] The work pattern recorded at Newton did not necessarily hold true for Mesopotamia, since Barbadian and Jamaican planting practices differed significantly. Nonetheless, these Barbados work logs are certainly of some help. For one thing, they suggest that craft workers on a Caribbean sugar estate were less clearly differentiated from field workers than they were on a big Virginian estate like Mount Airy. The coopers and masons at Newton estate spent a full two months each year working with the field hands on the sugar harvest, and once the cane was processed, the specialists in the sugar factory labored in the fields for the next six months. Field laborers were ranked by ability in both systems, categorized in Virginia as full "shares" or half "shares," and were sorted in the Caribbean into three gangs, with the first gang (Great Gang) always assigned the hardest jobs. As at Mount Airy, work assignments were

---

[30] John Tayloe III to John Rose, John Tayloe Letter Book, 1801, *ibid.*, MSS 1 T2118d170.

[31] Newton Plantation Field Labor Book, May 5, 1796-Apr. 26, 1797, Newton Papers, MS 523/110, University of London Library. A second field labor book for the adjoining Barbados plantation of Seawalls, running from Jan. 1 to Sept. 4, 1798, is *ibid.*, MS 523/122.

changed every few days, and it would be a mistake to suppose that slaves on a sugar estate spent all their time planting and cutting cane. The Newton logs show that the first and second gangs spent the equivalent of six months per year in sugar production, three months raising guinea corn for cattle and slave food, two months repairing the cattle pens, and another month at such miscellaneous tasks as cultivating yams, potatoes, and peas.[32] On one day each year all plantation work was stopped, and every man, woman, and capable child was given cloth, a needle, and thread, and set to work stitching together his or her set of clothes for the following year.

At Mount Airy the Tayloes stretched seasonal employment into year-round employment by letting their slaves work at a leisurely pace. Three masons took fifteen working days to build a cottage chimney; thirty working hands at Old House took eight weeks to sow 258 bushels of wheat. Only at harvest time did people work under extreme pressure for several weeks. But in the Caribbean the sugar harvest lasted for four months or more, from January or February to May or June. And while at Mount Airy the heaviest labor was done by horses, mules, and oxen, at Newton—or Mesopotamia—the slaves did the work of draft animals. The Newton first-gang slaves spent nearly one week in every month at the brutal task of cane holing by hand. Crop time was the period of prime pressure, since the various stages in the sugar-making process were so closely synchronized. At Newton the fifty strongest members of the first gang cut the cane and ten members of the second gang loaded it into carts and took it to the mill where another ten workers from the first and second gangs ground it, while fourteen workers processed the cane juice at the boiling house, and three still-house workers converted molasses from the boiling house into rum. This was not a continuous four-month process; every few days the cutting gang was shifted to lighter tasks in order to recruit strength. But the work clearly took its toll. At Newton, in relatively slack periods, only 3 or 4 percent of the field workers reported sick, but after a week of holing or cutting cane the number rose to 9 or 10 percent. Even the holidays were fewer at Newton than at Mount Airy—only four days off per year: Good Friday, a free day in mid-October, and two days at Christmas.

We are now in a better position to examine the demographic contrast between Mesopotamia and Mount Airy, which is the most mysterious and also the most crucial aspect of our inquiry. Why did the Mesopotamia slave population suffer such pronounced natural decrease while the Mount Airy slave population enjoyed pronounced natural increase? Why in particular did the Mesopotamia women produce so few children, only half as many as the

---

[32] There were 155 workers on average in the three Newton plantation field gangs in 1796-1797, compared with Mesopotamia's 137 field workers in 1809.

TABLE IV
MOTHERHOOD AT MOUNT AIRY AND MESOPOTAMIA

	Mount Airy, 1809-1828		Mesopotamia, 1799-1818	
	Number	%	Number	%
A. Childless Women	50	32.67	100	50.00
Mothers	103	67.32	100	50.00
	153		200	
B. Size of Completed Families:				
1 child		4.55		36.84
2-3 children		9.09		26.32
4-6 children		40.91		26.32
7-9 children		36.36		8.77
10-13 children		9.09		1.75
		100.00		100.00
Average no. live births per mother		6.36		3.07
Average age of mother at first live birth		19.32		20.47
C. Infant and Childhood Mortality:				
Percent died during first year		6.94		10.09
Percent died aged 1-10		15.28		16.06
Percent surviving past 10th birthday		77.78		73.85

Mount Airy women? Table IV compares the females aged seventeen and over who lived on these two estates during our twenty-year span. There are 200 potential mothers at Mesopotamia and 153 at Mount Airy. The smaller number at Mount Airy reflects the fact that many females on that estate were sold or transferred before they reached child-bearing age. Of the 200 Mesopotamia women, exactly half appear to have borne no live children. The other 100 are identified in the estate records as mothers—including seventeen women who bore some or all of their children on other Jamaican plantations before coming to Mesopotamia. At Mount Airy two-thirds of the women can be identified as mothers. This percentage is undoubtedly too low, for the Mount Airy records fail to pick up either the mothers of children who were grown by 1809 or the young future mothers who had children after they were transferred off the estate in their late 'teens or twenties. Of the 153 Mount Airy women in Table IV, only seven can definitely be categorized as sterile. In my opinion, close to 90 percent of the Mount Airy women bore children, compared with about 55 percent at Mesopotamia.

The other striking difference is that the Mesopotamia women who did bear children had much smaller families. As Table IV shows, only 37 percent of the Mesopotamia mothers had four or more children, whereas 86 percent of the Mount Airy mothers had families this large. The average Mesopotamia mother had less than half as many children as her Virginia counterpart. She entered into her first successful pregnancy a year later. If she bore more than one child, the births were spaced five months farther apart than at Mount Airy, and she had her last baby three years earlier. These averages conceal much significant variation. If we focus for a moment on those women who lived through the entire thirty-year period of possible reproduction (ages fifteen to forty-four), we find enormous range in age at first birth, child spacing, and age at last birth. The youngest mothers on both estates were only fourteen years old, while two Mesopotamia women had their first babies at age thirty-two. About 20 percent of the Mount Airy mothers gave birth before they were seventeen, twice the percentage at Mesopotamia, which suggests a possible difference in the age of menarche. At the other end of the cycle, nearly half of the Mount Airy mothers bore children into their forties—again twice the percentage at Mesopotamia. The presence of the white managerial staff was clearly a factor in the Mesopotamia fertility schedule, since more than half of the babies produced by mothers under the age of seventeen on this estate were mulattoes. Minny, a seamstress at the overseer's house, was the most prolific Mesopotamia mother; she bore eight sons and two daughters in twenty-five years, and three of these children were mulatto. Minny was forty-five when her last boy (a Negro) was born in December 1815, and to honor the occasion the baby was named Joseph Foster Barham. But Minny was a rarity in a community of small families.

According to Table IV, relatively few infants died on either estate, reflecting the incompleteness of infant birth and death records, especially at Mount Airy. The figures for children over age one are more reliable, especially for Mesopotamia with its immobile population. Unless the Mesopotamia infant records are completely misleading, newborn babies and young children died at about the same rate on the two estates, thus ruling out infant mortality as a factor in explaining the demographic difference between the two plantations. The key issue is clearly fertility—the low fertility of the Mesopotamia slave women.

There is no simple explanation for the low fertility at Mesopotamia. At least half a dozen separate factors contributed to the problem. In the first place, women lived less long than at Mount Airy, and experienced fewer years at risk of pregnancy. Half of the childless Mesopotamia women and a third of the mothers died in their twenties or thirties, during the prime childbearing years. At Mount Airy only about a tenth of the women died at this age. Yet longevity is certainly not the only factor in our equation, nor probably the most important one, for the chief point to be made here is that the Mount Airy slaves had quite long life expectancy, and even the Mesopotamia slaves had longer life expectancy than many populations with stronger birth rates. At age seventeen a Mount Airy female could expect to live thirty-nine more years, a Mesopotamia female thirty-one years. Over 80 percent of the Mount Airy women lived long enough to reach menopause. Even at Mesopotamia the childless women averaged twenty-one years at risk of pregnancy without producing a live birth, and 40 percent of them lived on the estate throughout the years of possible reproduction. Likewise, most Mesopotamia mothers of one or two children lived on for many years after their babies were born. Nearly 60 percent reached the age of menopause, and the average mother bore her children within a space of seven years while experiencing twenty-three years at risk of pregnancy. While the longevity of the Mount Airy women thus helps to explain their large families, life expectancy at Mesopotamia was not sufficiently restricted to account by itself for the low birth rate in this community.

A second factor is the sex and age ratio, which in Caribbean slave populations is frequently seen as the chief reason for low fertility. As Table II has shown, however, neither the sex nor the age structure at Mesopotamia was unfavorable. The sexes were always much better balanced on this estate than at Mount Airy, where the young men consistently outnumbered the young women by nearly two to one. There was always a larger number of women of child-bearing age at Mesopotamia than at Mount Airy. Indeed, the age structure of the Mesopotamia population was seemingly very favorable to high fertility. Between 1799 and 1818 the proportion of females aged fifteen to forty-four averaged 24.5 percent of the total Barham slave force. At Mount Airy, because so many girls were sold or transferred, the proportion of potentially fertile females was significantly smaller: between 1809 and 1828 it averaged 19.4 percent. Thus if imbalance between the sexes and a low proportion of young women depress the birth rate, one should look for problems at the Tayloe estate rather than at Mesopotamia.

A third factor is the presence at Mesopotamia of African-born women with African child-rearing habits. It has been argued that Caribbean slave populations had low birth rates because so many of the women were Africans. Fertility rates in West Africa are thought to have been generally low; and when African women were shipped to America they reproduced less actively than creole slave women, being habituated to a long nursing period with a resultant wide spacing between births.[33] At Mesopotamia, however,

[33] This line of argument is advanced by Michael Craton, "Jamaican Slave Mortality: Fresh Light from Worthy Park, Longville and the Tharp Estates," *Journal of Caribbean History*, III (1971), 1-27, and by Russell R. Menard, "The Maryland Slave Population, 1658-1730: A Demographic Profile of Blacks in Four Counties," *WMQ*, 3d Ser., XXXII (1975), 29-54.

the differences between African-born and creole women were not pronounced. Of forty-two African women during the 1799-1818 span, twenty-three (55 percent) were childless, and the nineteen African mothers averaged 2.42 children each. They bore their children at no wider intervals than the creole women, so the probability is that all the mothers followed the same nursing practices. The Mesopotamia women collectively did space their children more widely than at Mount Airy, with an average interval between live births of three years and three months, compared with two years and ten months at Mount Airy.[34] But there is little evidence that this was because they nursed their infants longer. If we compare the most fecund mothers on both estates—those who bore seven or more children—we find no difference whatsoever in child spacing. The Mesopotamia and Mount Airy mothers in this category both averaged two years and six months between births. In fact, these women generally gave birth every two years, with an occasional shorter gap when the nursing period was broken off by the death of the last-born infant, and an occasional longer gap perhaps caused by a miscarriage or stillbirth. Mothers of small families, especially at Mesopotamia, had extremely irregular birth intervals, ranging up to ten years, and such eccentric spacing is better explained by sexual abstinence, intermittent fecundity, miscarriages, and abortions than by nursing habits.

A fourth factor is the debilitating work regimen imposed upon the Caribbean slaves, which presumably robbed them of vitality and dulled their sexual instincts. As we have seen, the labor pattern was more punishing at Mesopotamia, and women of child-bearing years performed much more heavy work than at Mount Airy. Unfortunately, there is no way of demonstrating a correlation between this debilitating labor and slave infertility. Analysis of job distribution at Mesopotamia during the 1799-1818 span shows that mothers and childless women held the same range of jobs in almost exactly the same proportions. The heaviest labor was done by the Great Gang, and on an average 34 percent of the Mesopotamia women over seventeen were assigned to this gang. Of the childless women in Table IV, 36 percent worked in the Great Gang; of the mothers with one to three children, 32 percent worked in the Great Gang; of the mothers with four or more children, 35 percent worked in the Great Gang. Similarly, at the other end of the spectrum, the lightest work was done by the domestic servants who attended the white staff; 6 percent of the Mesopotamia women were assigned to these jobs—5 percent of the childless women, 7 percent of the mothers with one to three children, and 6 percent of the mothers with four or more

---

[34] Herbert S. Klein and Stanley L. Engerman, who stress the difference between U.S. and British West Indian nursing practices in an unpublished paper, "The Demographic Study of the American Slave Population," find a U.S. child-spacing interval of only 2.5 years and a British West Indian rate of 3.5-4.0 years.

children. These figures seem to show that work assignments had no effect whatsoever on the birth rate. Yet it remains difficult to believe that years of rugged labor in the cane fields did not reduce slave procreation, increase the chance of miscarriage, and lower slave fertility.³⁵

A fifth factor, possibly the most important one, is inadequate nutrition, which can depress fertility by impairing female reproductive development. Biologists have established that the timing of the adolescent growth spurt and menarche, the maintenance of regular menstrual function, the recovery of reproductive ability after childbirth, and the timing of menopause are all directly affected by the female's state of nourishment. A severely undernourished woman, with inadequate fat storage in her body, will achieve menarche belatedly, will experience irregular menstrual cycles or none at all, and will enter menopause early. If she manages to conceive and bear a child, pregnancy and lactation will draw thousands of calories from her meager energy stores, so that she will be slow to recover reproductive ability for another pregnancy.³⁶ Certainly the Mesopotamia women betray symptoms of impaired reproductive development: delayed first births, long and irregular birth intervals, and early final births.

The estate death records contain some evidence of dietary deficiency, as well as bad sanitation, which further undermined the health of the Barham slaves. By the manager's own reckoning, 11 percent of the slaves died "bloated" or from "dropsy." These vague terms covered a wide range of bodily swellings, undoubtedly caused in many cases by protein and vitamin deficiencies. Another 9 percent died of "flux" or various forms of dysentery, promoted by unsanitary living conditions. This was a larger number than the 8 percent who died from malaria and other tropical fevers. Another 11 percent of the deaths were attributed to such African diseases as hookworm, guinea worm, yaws, leprosy, and elephantiasis, which were much more common in Jamaica than in Virginia. Among the remaining chief causes of death were—in rank order—tuberculosis, pleurisy, smallpox, epilepsy, and venereal disease. Nearly 3 percent of the slaves were said to have died in plantation accidents, but only two men in eighty years committed suicide, and

---

³⁵ It is perhaps significant that 40% of the Mesopotamian live births were concentrated in the four months October-January and only 25% in the four months April-July. A number of the pregnancies that should have come to term in April-July may have resulted in fetal deaths because of the mothers' field labor during the sugar harvest season.

³⁶ Rose E. Frisch, "Demographic Implications of the Biological Determinants of Female Fecundity," *Social Biology*, XXII (1975), 17-22.

³⁷ A comparable set of death records for Worthy Park estate in Jamaica has been analyzed by Michael Craton, with findings very similar to mine. I am indebted to Mr. Craton for the use of his unpublished essay, "Death, Disease and Medicine on Jamaican Slave Plantations: The Example of Worthy Park, 1792-1838."

only five women died during childbirth. Obviously, these mortality statistics cannot be taken too seriously, especially since 17 percent died of "old age" and another 9 percent died as "invalids." In any case, the correlation between ill health and infertility is inexact. A woman named Esther suffered continuously from "weakness" and yaws during her life at Mesopotamia, but she bore six children. Luna was incapable of employment because of her bad sores, but she gave birth to seven children. Ophelia contracted venereal disease when she was thirty and was retired from the field gang, but she had two more children and lived another twenty years. And Sabina began to experience epileptic seizures at twenty-eight but bore three more children and died at sixty-three.

Still another factor that could have contributed to low fertility at Mesopotamia is the aggressive role of the white overseers and bookkeepers, who requisitioned the sexual services of a good many young slaves. These women, if they wished to prolong their status as concubines, may have aborted their mulatto offspring in order to keep physically attractive. The large number of mulatto births at Mesopotamia, however, makes this proposition unlikely. Other Mesopotamia women may have practiced sexual abstinence or committed abortion because they could not stand the prospect of bringing babies into a world of enslavement. Several of the Mesopotamia children who died during infancy were "overlaid" by their mothers. Were such events accidental or intentional?

Finally, we should not forget the Mesopotamia men. It is surely a mistake to focus exclusively on the female slaves, for the males must have had much to do with the low birth rate. If longevity was a factor, the men had shorter life expectancy than the women, and many Mesopotamia females must have stopped bearing children when their mates died. If the work regimen was a factor, it was the men who were forced into the most exhausting jobs. If nutrition was a factor, the men suffered more than the women from chronic debility. Frequently weak or sick, flogged and maimed far more often than the females when they resisted or malingered or ran away, humiliated by an arrogant cadre of white masters who took their women, the Mesopotamia men had lost all powers of leadership and independence—and this loss in psychic power may well have drained their sexual potency.

Thus, despite the wealth of statistical information about slave life at Mesopotamia and Mount Airy, many questions cannot be answered and many issues cannot be settled. Ultimately the observer who wonders why Jamaican slaves behaved differently from Virginian slaves is reduced to armchair psychology. But at least we can document telling differences—some of them rather surprising—between the two communities. At Mesopotamia the slave population was cooped up; at Mount Airy it was in constant movement. Mesopotamia families were more inbred than Mount Airy families, and less disrupted by sales and transfers. Mesopotamia women were tougher than their men and dominated the work force, whereas Mount Airy women were reckoned to be more marginal than their men and were frequently sold. Vocational opportunities were narrower at Mesopotamia than at Mount Airy, the work load was harder, the food and clothing were less adequate, the disease environment was more threatening. In consequence, Mesopotamia slaves died earlier than at Mount Airy and produced fewer children.

The owner of Mesopotamia, Joseph Foster Barham, was a conscientious master by the Jamaican standards of the day. Few other sugar magnates paid missionaries £70 per year to instruct their slaves in the Christian religion, or published arguments for the abolition of West Indian slavery,[38] or required their agents to keep meticulous track of the blacks who inhabited their estates. During the eighty years of record keeping the Mesopotamia slaves never rebelled; very few of them ran away permanently; and during a smallpox epidemic the white managers showed their humanity by inoculating even the aged invalids in order to keep them alive. Nonetheless, the Mesopotamia records give us a deeply depressing picture of slave life on this estate, a picture reaffirming the universal opinion of modern scholars that Caribbean slavery was one of the most brutally dehumanizing systems ever devised.

John Tayloe III likewise emerges from his letter books and account books as a thoroughly benevolent and well-intentioned master. Working and living conditions on his Rappahannock estate must have been exceptionally relaxed, for Tayloe was no profit-maximizing entrepreneur. His well-worn fields produced modest yields, his work force was far larger than necessary, and the rhythm of the place evoked leisured gentility rather than business efficiency. And yet this paternalistic planter manipulated and exploited his slaves to a high degree. Thus my portrait of slave life at Mount Airy is ambiguous, and just how it fits into the general debate about the character of slavery in the United States is rather hard to say. As Stanley Elkins has recently pointed out, a long scholarly tradition that stressed white brutality and white damage has been superseded by a new emphasis on black achieve-

---

[38] While Barham was Wilberforce's ally in the British abolition movement, his tract, *Considerations on the Abolition of Negro Slavery*, was studded with pejorative remarks about the Negro as a person and a worker. In 1806 he offered Wilberforce a solution to the West Indian labor problem: substitute Chinese coolies for Negro slaves. Robert Isaac Wilberforce, *The Life of William Wilberforce*, III (London, 1839), 272.

ment and black resistance.³⁹ The Mount Airy evidence can be used to buttress either side of this debate, but on the whole I believe that it better supports the older tradition. While Tayloe was not a brutal master, his slave regimen was, it seems to me, designed to thwart black achievement and to defuse black resistance. His system offered extremely little scope for the dynamic economic, cultural, and social slave achievements currently celebrated by Robert Fogel, Stanley Engerman, Eugene Genovese, and Herbert Gutman.⁴⁰ But however noxious Tayloe's system, if one had to be a slave, Mount Airy was a better place to live than Mesopotamia.

³⁹ Stanley M. Elkins, "The Slavery Debate," *Commentary*, LX (Dec. 1975), 40-54.

⁴⁰ I am thinking here of Fogel and Engerman's *Time on the Cross*, Eugene D. Genovese's *Roll, Jordan, Roll: The World the Slaves Made* (New York, 1974), and Herbert G. Gutman's *The Black Family in Slavery and Freedom, 1750-1925* (New York, 1976).

9
Slaves and Slave Masters on Eighteenth-Century St. John
Karen Fog Olwig

Historical anthropology, with its in-depth analysis of the daily life of people living in the past, provides a special perspective on the development of culture. It has been recognized that this is particularly the case in the "colonial situation" where the confrontation of radically different cultures has led to the construction and reconstruction of new cultural forms. As noted by Bernard Cohn, with regard to this process:

> Both white rulers and indigenous people were constantly involved in representing to each other what they were doing. Whites everywhere came into other peoples' worlds with models and logics, means of representation, forms of knowledge and action, with which they adapted to the construction of new environments, peopled by new 'others'. By the same token these 'others' had to restructure their worlds to encompass the fact of white domination and their own powerlessness (1981:218–219).

The following case study of the confrontation of European and African cultures on 18th century St. John, one of the Danish West Indian islands, shows, however, that on the level of daily material life, the process of "restructuring" was not necessarily dominated by the Whites.[1] By focusing on the way in which European perceptions of Blacks and Whites were molded by the colonial experience, it can be seen that the development of an Afro-Caribbean culture on the part of the slaves presented a challenge to White supremacy which forced the Whites to restructure their own culture. The paper thus raises basic questions about the validity of assumptions based primarily on European experience and ideas or modes of analysis rooted in European society and economy. On the level of daily life peripheral "primitive" peoples did much to undermine European civilization, creating countercultures and economies which were not, it turns out, so "powerless" as suggested by Cohn.[2]

The European Perception of the Africans

The European pattern of utilizing and perceiving the Africans had long been well established when St. John was settled by the Danes in 1718. During the 17th century, most of the West Indian islands had been colonized by Europeans for the purpose of establishing sugar plantations using slave labor. With the virtual extinction of the aboriginal Indian population in the Caribbean during the 16th century, this labor power was imported from Africa. By 1700, a type of colonial society therefore had emerged in most of the Caribbean, consisting of two main groups of people: a small European upper class owning and supervising plantations, and a large African lower class performing the hard, physical labor needed to operate the estates. The great difference in physical appearance and cultural background between the Europeans and the Africans was used by the upper class as a means of keeping the two groups separate. Denigration of the Blacks, furthermore, served to maintain the superior status of the Europeans in the socio-economic hierarchy of the plantation society (for general discussions of this see van der Boogard 1982; Smith 1982).

The colonization of St. John occurred according to the already established pattern. The entire island was divided into estates which were converted to sugar or cotton plantations with the help of European capital. They were managed by European owners or overseers with a labor force of African slaves. Descriptive sources from 18th century colonial society in the Danish West Indies provide revealing information about White attitudes towards the Black slaves. The Danish civil servant and plantation owner Reimert Haagensen, writing in 1758, describes the Blacks as "made to be thralls, in that they don't conceive of anything but constant labor and do it therefore with pleasure, yes singing and in high spirits, as if they were the happiest creatures ... their black skin witnesses their evilness and their predestined thralldom, so it is reasonable that they ought not have any freedom" (1758a:51)[3] Haagensen also claimed that the slaves did not understand how to work in a systematic fashion when they started on the plantations, and that this was a sign that they lived as savages in Africa (1758b:309). John Lorentz Carstens, who was born of a Danish planter family in St. Thomas, and who himself ran plantations on St. Thomas and St. John until he moved to Denmark late in the 1730's, also regarded the slaves as wild, heathen savages:

> ... there are so many idols as there are kinds of people among them, and the manners and occupations that they can be recognized by are just as different; ... some are more savage, murderous and devillish than others and deal with witchcraft and devil-fantasy (1981 [1740's]:83).

Carstens was not alone in this view of the slaves. According to Oldendorp, who spent a year in the Danish West Indies in order to write an account of Moravian missionizing among the slaves, it was typical for Whites to regard all Blacks as nothing but devillish creatures (1777:529).

The slaves' dress also branded them as sub-human. Carstens described it in this way:

> The little clothing that these plantation slave men and women use is nothing but a piece of cloth which they carry around their waist in order to hide their shame, otherwise they walk stark naked all over. There is one nation among them, which has a black and greasy piece of linen, which they call a pansa, tied around the waist, which covers only their shame; the slave women have a short and skimpy piece, loosely hanging in front of their shame, and they only do this if the master threatens them with punishment if they do not, since they have otherwise no modesty. These people, when they will thank or honor somebody, show their compliment or thanksgiving by presenting their shame and nakedness (1981:90).

Carstens compares the slaves' appearance and habits with those of animals. He describes, for example, Negroes whose "procreative limbs both in thickness and size look like the genitals of the dumb beasts and like them they are supposed to be located one inch higher than on others'" (ibid.:94). These animalistic characteristics also hold true for their home life, he believes. According to Carstens, the thin mats on which the slaves sleep at night are spread out "close to one another on the earthen floors, and they lie on them among one another as beasts" (ibid.:90). As far as their sexual relations are concerned, Carstens states that "most of the slaves mix with each other where they come, just as dumb beasts" (ibid.:94). In a similar vein, Hans West, school master on St. Croix during the 1790's notes that "these early maturing creatures hardly reach the age of 10–12 years before they begin to mix with the other sex at random" (1790:43).

European perceptions of slaves as reflected in these descriptions hardly constitute merely spontaneous reactions towards non-European strangers. The European view of the Blacks as animals seems instead to be a means of rationalizing European exploitation of the slaves' labor power in the fashion of beasts of burden. The fact that slaves were regarded as animals, in other words, was part of the ideology of suppression which Europeans created to deal with Blacks. This accords well with descriptions of the way in which the plantations were run in the Danish West Indies during the 18th century. They all document an inhuman work day for the slaves, often in excess of 12 hours six days a week and consisting of only hard, physical toil. They were "driven" to this toil with whip and stick, as if they were beasts.

The West Indian Colonial Situation: The Development of an Afro-Caribbean Culture

Closer scrutiny of the plantation society which emerged on St. John during the 18th century reveals that the view of slaves as animals, whose brute strength was to be exploited as efficiently as possible, did not entirely reflect the reality of life on the island. A journal from Carolina plantation from the years 1766–67 shows, for example, that a fair proportion of the slaves escaped work on the plantation. Of the approximately 90 adult slaves, between seven and 21 were ill; 10–11 were manqueroons, i.e. disabled who had no labor power to offer the plantation at all, and anywhere from one to eight were maroons, i.e. run-aways from the plantation. This clearly shows that the classification and utilization of slaves as beasts was not fully realized in daily life. Unlike animals, slaves were able to report to the sick house of the plantation and, while there is no doubt that many of them were actually ill, it was not always possible to distinguish the ill from those who took themselves a free day in the sick house. West claims that the slaves reported to their beds when it suited them and that their masters had to spend fortunes having them examined by doctors (1790:45). The many manqueroons also presented a problem. If the slaves had been regarded merely as animals, they would have been destroyed when they ceased to be of economic value to the plantation. The fact that more than 10 % of adult slaves were disabled proves that the slave owners, despite everything, recognized the slaves as human beings and let them live even when they were unproductive and an economic burden. The maroons also illustrate the slaves' humanity. Marronage from the plantations was a clear sign of the wish for self-determination which the slaves displayed towards the plantations and the inability of the plantation owners to cope with it. It became a contributing factor in the undermining of European control (SAA 62).

It was the plantation economy itself and the slaves' culture which most effectively subverted the slaves' inhuman status and put to shame the European view of slaves as animals. It was the duty and responsibility of the planters to make sure that the slaves received food so that they maintained their value as labor. Under ideal conditions, such as on Barbados, virtually all land was devoted to commercial crops, and all slave labor was devoted to export production (Mintz and Hall 1960). Reproduction of the labor force was based on imported food stuffs and fresh labor stock. On St. John, where there was a great deal of mountainous land not suitable for sugar cultivation, the planters chose, however, to let their slaves cultivate their own provision crops on small plots on the unused land. In this way the planters avoided having to import foods from Europe and North America, which were expensive and only in irregular supply. The common practice was to release slaves from plantation work for two hours at noon and all Sunday in order to cultivate their own subsistence crops. Outside the season of sugar harvest, which lasted for about half a year, slaves also had Saturday afternoons to work in their provision grounds. The slaves, originating from areas in Africa with horticulture, had an excellent knowledge of the cultivation of tropical root crops, vegetables and fruit, and gradually developed a subsistence economy which, to a great extent, made them independent of the planters. Indeed many slaves were capable of cultivating a surplus, as well as raising pigs and fowl for sale. This led to an extensive exchange of goods, not just informally on St. John, but also, through a more formal system of trade, with the neighbouring island of St. Thomas where, unlike on St. John, there was a market (see Olwig 1981a, 1981b, 1983).

The planters and the administrative authorities chose to ignore this trade as far as possible. Nonetheless there were times when they had to face the fact that there was an extensive economic system about which they had little knowledge, and which they did not control. When the Danish administrator on St. John, for example, asked the slave Maria why she wanted to go to St. Thomas, he was told that she "wished to travel in order to collect her money which she had outstanding". Surprised, the administrator asked: "What kind of debts does she as a slave woman have outstanding?" and was told that it was money for bread which she had sold on credit on St. Thomas (LA 72: October 5, 1804). The planters usually prohibited their slaves from travelling to St. Thomas because they feared they would not return. It was therefore necessary for most slaves to trade through boat captains who plied the waters between St. John and St. Thomas. Even though the planters were not happy about this trade, they usually had to accept it, since they did not want to invest in importations of food stuffs and other necessities which slaves were able to obtain themselves through their economic system. It was increasingly apparent, however, that this slave economy presented not just a strategy of survival, but had become a means of maintaning traditions concerning horticulture and trade that most slaves knew from Africa. The planters, in other words, had opened the door not just to a subsistence economy that would save them money, but also a cultural system that could not be controlled, once it was reconstituted (cf. Mintz 1974).

In addition to bringing with them knowledge of methods of cultivation and trade (which flourished in the West Indies), the African slave also arrived with well-defined ideas about the family and its importance. Even though the Africans came from different tribes, each with its own system of kinship, they all shared certain basic values as far as the family was concerned. As stated by the Jamaican sociologist Orlando Patterson, "Every West African transported to the New World would have valued highly the principle of kinship as a basis for social organization, the love and respect of kinsmen, the welding force, the permanence and the communal nature of marriage, and the solidarity of the extended household" (1982:139). The forced exodus from Africa, where individual slaves were removed from their home without kinsmen or spouse, meant that a new family had to be created in the West Indies. To a certain degree this had to be adapted to the conditions of slavery and the plantation system. The principle of kinship became the nucleus of this family form.

The individualized Africans who arrived in the West Indies apparently created fictive kinship with one another in order to fill the vacuum in which they found themselves. The Africans thus formed ties with those who came from the same tribe in Africa or had even just travelled on the same boat to the New World. They also sought to establish families through conjugal ties, as far as this was possible under a plantation regime where the owners could sell individual slaves freely and where a staggering mortality rate made permanent relations difficult. There are indications that the conjugal tie was more important for the kinless African slaves than for the Creole slaves who were born in the West Indies and surrounded by kinsmen. As the slaves had children and thereby kin, they developed a family system which centered on these. A slave child always belonged to the mother's plantation, and those slaves who were born and reared on their mother's estate therefore were encircled by maternal kin. If the father belonged to the same estate, his kinsmen also were present. In many instances, however, this was not the case, because with the widening of the matrilateral kin relations within the plantation, slaves often began to choose partners on other plantations. This occurred particularly on small plantations where the number of possible partners was restricted, and each person was related to a growing number of resident slaves. The slave John Williams thus described a common situation when he explained to the administrator that, whereas his father did not belong to his plantation Brownsbay, his mother and eight brothers did, and that he was, besides, "related to almost the entire stock of the plantation" (LA 42: June 25, 1845).

Most adult slaves lived in separate households in small huts, but, as court records reveal, there was lively exchange between households with relatives giving each other foodstuffs and helping one another care for the small children. In many ways they therefore functioned as one big household, despite the fact that they lived separately. If a slave had problems of any sort, it was usually a relative, who was nearest, who helped. In the same way, close relatives protected each other against slaves on other plantations, if disagreements should arise. This does not mean that the extra-plantation husband-wife relationship or the paternal relatives were of little significance. The fact that they belonged to other plantations meant that they could step in when extraordinary situations developed, and the normal plantation system broke down. Maroon slaves, for example, often sought help from relatives outside their own plantation. The great importance that slaves attached to having close relatives on the plantations did mean, however, that they, as far as the every day aspects of life were concerned, were less dependent on their spouses and had a relationship with them that was more based on affection than on practical need. Slaves therefore regarded the permanent family as being first and foremost that network of matrilateral kinsmen with whom they lived on the plantation, and with whom they freely exchanged help and subsistence goods. The slaves, in this way, maintained "the African principle of kinship as a basis for social organization" and recreated a sort of extended household among those relatives living on the plantation. The husband-wife relationship, on the other hand, acquired a less practical, more affective meaning. (For a more detailed discussion of the slaves' family system see Olwig 1981a, 1981b).

It is apparent that plantations benefited from the presence of an Afro-Caribbean culture with the extensive production of subsistence goods, exchange of goods and trade of produce, as well as the strong community of kindred who supported and helped the weaker slaves. It was, in fact, this culture which formed the real basis of the slaves' continued existence and thus the reproduction of the slave labor force. It can be argued, therefore, that the slaves' culture was merely an adaptation to the plantation regime. In this way, it enabled even the weakest plantations to carry on production, even though they were not able to pay for, let alone support their labor force. Sidney Mintz is much closer to the heart of the matter, however, when he characterizes the Afro-Caribbean culture which developed in the West Indies during slavery as a mode of response as well as resistance to the plantation system (1974:132–133).

The Confrontation of European and Afro-Caribbean Culture

The very same Afro-Caribbean culture which made European plantation production possible also gave the slaves a measure of autonomy and independence. And the slaves' self-contained activities placed increasing pressure on the Europeans' position of domination and superiority. The legal system reflected the problem of controlling slaves who had become more and more autonomous. Danish Law from 1683 formed the official basis of the legal code in the Danish West Indies, and in 1756, a special English version of the law was published for use on the islands where only a minority among the white population knew Danish. This law did not contain any provisions concerning ownership of human beings, and regulations concerning the treatment of slaves could only be based on legislation concerning private property. The colonial judges were thus presented with the rather difficult task of interpreting the Danish law according to local conditions. When slaves were abducted from the Danish islands to be sold on the neighbouring islands, for example, the guilty were judged according to a paragraph dealing with flagrant theft of horses or large cattle in the field (Olsen 1983:306–7). Two slave codes were drawn up in the 1700's in order to make provisions for the special conditions of slavery in the West Indies. Furthermore, a number of local ordinances and placards were published. The primary purpose of these was to "protect the slave owners' property right against—paradoxically—the property itself, which had the quality of being able to oppose the owner" (ibid.: 307).

While the slave law from 1733 defined the slave as the owners' private property without any rights in themselves, except for the right not to be murdered, this law went on to describe the duties which slaves were to suffer if they failed to perform to their owners' satisfaction. By doing so, the law deviated from its basic conception of the slave as nothing but private property. It would be unthinkable and absurd to imagine a law that listed the duties of the other elements in the planters' private property, such as the sugar mill or the domestic animals, and then spoke of the punishment these things were to be accorded if they did not behave. The slave law goes on to prohibit various forms of behavior on the part of slaves, for example, the usage of "Negro instruments" at parties and funerals, or witchcraft. These prohibitions sought to erase the Afro-Caribbean culture; they can also be regarded as indirectly acknowledging the fact that slaves, in fact, possessed customs and traditions which raised them above their objectified status.

In the slave code from 1755, slave rights are mentioned for the first time. This law stipulated that slaves were to receive a weekly ration of food and regular supplies of clothing; the law granted them the right to attend church and marry, and made baptism of children obligatory. This law can be interpreted as a total abandonment of an earlier effort to dehumanize slaves and to ignore their claims to certain rights. The law can also, however, be interpreted as a sign of an increasing anxiety on the part of the colonial authorities over the seemingly more and more uncontrollable slaves, and, thus, as an attempt to gain control of their social and economic activities. If the planters had fed the slaves, for example, the economic activities of the slaves could have been checked and brought under control. Also, if stable marriages and Christianity could be introduced among the slaves, they might establish family relations, which were, in the European view, more orderly and therefore easier to grasp and regulate. This interpretation of the 1755 law is supported by the fact that the slave law also contained a number of regulations which sought to acquire some control over the slaves' culture. It was to be illegal for the slaves to sell their produce, whether on the plantations or at the market, unless they were registered and issued permits as sellers by the authorities. The law also suggested the appointment of inspectors who were to maintain order at the markets. As far as marriages were concerned, the slaves' right to marry was limited by the fact that their owners were to consent to the choice of partner (Hall 1977). The slave law therefore was rather an indirect recognition of the slaves' increasing autonomy, and an attempt to gain some control over them.

The plantation owners and the local colonial administration were not ready, at this time, to grant slaves even these limited rights, but insisted, rather, on maintaining the sharp distinction between an objectified class of slaves and a cultured European class of humans. The slave law from 1755, which had been drawn up in Denmark, was never made public in the West Indies, and in its stead a series of local placards and ordinances were issued for the purpose of regulating the slaves' free time and freedom of movement. In 1759 a regulative was published which limited the slaves' visits to other plantations, even when the visits were to spouses. Movements of slaves in the towns also were circumscribed by certain rules: slaves, for example, were not allowed to ride in town streets. In 1765, a proclamation restricted the slaves' opportunity to attend meetings in church during the week by prohibiting church going after 8 p.m., when they were not required to work in the plantations. The planters initially attempted to prevent the Moravian missionaries from preaching among the slaves by whipping them or refusing to provide boat passage to the missionaries (Oldendorp 1777:528). A placard from 1774 prohibited "wakes", i.e. gatherings kept by slaves in connection with births and funerals (Hall 1977).

The Development of a White "Mode of Response"

The White planter class attempted to hold its ground against the slaves with the help of not just these written regulations, but, perhaps more importantly, through the application of a long series of unwritten rules that distanced them as much as possible from the slaves. Furthermore, they sought to lead a way of life that was as far removed as possible from that of the slaves. The sources thus reveal that the White West Indian culture appeared just as exotic to metropolitan observers as did the Black culture.

In other words, it is reasonable to assume that this exotic European society was a mode of response to the slaves' culture. While the slaves' daily life was characterized by drudgery from morning to night in keeping with their status as a sort of work animal, the Whites' life was characterized by round the clock laziness. According to Carstens, Whites were "very indolent, to the point that they do not have to move their hand to move anything from one place to another; yes, they are so lazy that a slave with a wisp must at all times follow them to their secret room" (he means that Whites were so lazy that they didn't even wipe themselves). Men spent some time supervising plantation business and women did a little sewing, but these tasks were not very demanding. As Carstens notes,

> besides all sleeping late in the morning, all Whites at this time of the day appear as if they were dead. Immediately afterwards they drink tea, after which both sexes spend the rest of their time until late into the night playing cards ... especially the women-folk; the men play billiard in the pubs; this is the steady and common work they perform (1981:70–71).

This luxury life naturally depended on a great deal of service on the part of slaves, and White homes had a large number of house slaves:

> The lord or master has his male slaves, the lady or mistress has her female slaves, the children also have their own according to their sex, so that the most superior can have in their houses 16–24 such servants or attendants surrounding them, but a common citizen only 4–6; this is according to the honor and rank, and the size of the household. All these house slaves must serve and wait, wherever their master or mistress either sit, lie, walk, stand, or where they travel back and forth (ibid.: 79–80).

Hans West notes that

> the Creole ladies are so taken by the vanity of having their table surrounded by serving Negroes and Negresses that they can hardly see a beautiful Negress without demanding to have her, just as connoisseurs of horses, when they meet a horse of good creation.

Apparently the house slaves took advantage of the large number of servants used in the house, so that they refused to perform duties that were not part of their normal routine. West thus claims that it was difficult, if not impossible for a master to make a slave that "cooked, sweep the floor, or the one that swept, wash a piece of clothing, or even make the seamstress mend" (1790:45).

Just as the Whites in contradistinction to the Blacks were living in the greatest of comfort, they also reserved a special kind of clothing to themselves. The Whites were dressed following European fashions, and on St.

Croix they even wore wigs when going to the towns (Oldendorp 1777:258–9). Most slaves, on the other hand, were scantily clad, being supplied with no or very little clothing by their owners. Even so, there were regulations concerning which type of clothing they were allowed to wear. According to Carstens, slaves were prohibited from wearing shoes or proper hats, and their clothes had to be made out of "either crude linen, stripes or checkered, or poor cotton, as no slaves can wear silk, damask or jewels, except what their lady or lord can give them" (Carstens 1981:79). The few slaves who might accumulate the means to wear elegant clothes and in this way pose a threat to the superior appearance of Whites, were hereby prohibited from doing so. There were, however, no restrictions on the right of the slave owners to decorate their slaves, their private property, if they fancied doing so. This was common in the case of house slaves, who, as noted, often were chosen on the basis of their looks. Modesty alone, of course, dictated that all slaves working in the house be clothed (Haagensen 1758a:57–58).

Even though the Whites were surrounded by slaves in almost all aspects of their lives, it was taboo to socialize with slaves. Open fraternizing resulted in the White being ostracized from White society. In a court case concerning unruliness at a party on a plantation on St. John, the host, planter Durloo, claimed that a certain Lieutenant Stürup, who had gone to the party uninvited, was to blame:

> The witness had earlier seen Stürup dance and jump in his Negro houses with both free Negroes and slaves, which was the reason why the witness did not want to invite him to his table nor the ladies want to dance with him (LA 67: November 28, 1778).

It was also inconceivable that Whites ate with Blacks. Oldendorp, in his account of the Moravians' missionary activities in the Danish West Indies, stated that even though the homes of converted slaves were clean and proper, considerations of status prevented Whites from sharing a meal with them in their houses (1777:377).

It was naturally also an unwritten rule that Whites did not marry Blacks (ibid.: 265). Nevertheless it was the order of the day for White men to have sexual relations with slave women. One of the reasons for this was that there was an excess of White men in the West Indies, working within the colonial administration and the military or as overseers of plantations. Many of the married men who had their families with them on the island also used slave women sexually. For White men such relationships were not problematic so long as the slave women were used only as sexual objects. Black women were regarded as being promiscuous and more than eager to offer themselves to their masters. Oldendorp admits, however, that even if the women wished to resist sexual ouvertures from their masters, they might as well give in, having no legally recognized right to deny them (ibid.: 268).

Whereas a slave woman was regarded and treated as a kind of sexual object that had to be at the Whites' disposal, the Whites held a diametrically opposed view of the White woman, in that she must be virtuous and a guardian of the family's honor. West describes the life of White women on St. Croix as severely circumscribed and dull:

> Women lead very restricted lives and have little fun.—Their only parades take place, when they let themselves be invited to Tea-treat or Teawater. They meet at 5 o'clock in the stiffest manner, enjoy a cup of teawater as strong as lye, eat different kinds of jam and baked goods, doing this almost without saying a single word to each other; finally, after an hour's time, they take their leave, thus putting an end to this pretty merriment. The men, on the other hand, eat regularly at each others' homes and feel thereby and because of their business less of a lack of the free and genuine fun of life.

West adds that the very limited social life of the women

> makes them on all occasions so withdrawn that one hardly sees them smile. And just as one truthfully can say that their behavior is unusually modest, I also cannot deny that this modesty develops into a stiffness and an indescribable boredom in all their behavior. Rarely do they speak a single word, much less contribute to the enjoyment of life and the party (West 1790:80–81).

The White women kept their hands and teeth as white as possible and did not tolerate any uncleanliness or disorder in their clothing (West 1790:81). This would have required them to stay inactive and indoors most of the time.

It seems that the White women on St. John tended to lead a similarly passive life style, where physical activity was held at a minimum. In a court case we hear of a woman, "madam Wood", who was completely immobile and thus not able to act as a witness in court. The reason was her "corpulence which made it virtually impossible for her to even attend church, so that when this occurred she had to be carried in a hammoc by 8–10 Negroes" (LA 65: August 24, 1761). The records also indicate that White women were modest and prudish. In another court case we thus learn that when the free Black woman Theodora Finch used "indecent" language in front of the White planter's wife Mrs. McCullen, this caused the latter to be thrown into a hysterical fit (LA 59: December 16, 1830).

For White women to have sexual involvement with Blacks was an almost unspeakable offence. In the rare instances when this did happen, the women were punished severely. When, for example, Anna DeKooning, daughter of a plantation owner, gave birth out of wedlock to a child of "color a Negro" the court on St. John banished her from the Danish West Indies and denied her "all the capital she possesses ... for her committed promiscuity and frivolous connection ... with a heathen and black slave." She received this harsh punishment "as a deterrent to others and as just punishment for herself" (LA 64: February 14 and June 19, 1758).[4]

The great number of sexual unions which took place between White men and Black women in the long run posed a serious threat to the rigid division of the population into two separate and entirely different classes. These unions regularly resulted in the birth of mulatto children. Officially they were defined as slaves and belonged to the mother's owner. In reality however, many mulatto children were freed by their fathers, and it was also quite common that a slave woman was freed after she had born a number of children to a White man. Gradually a third class of free colored therefore emerged which broke down the cleavage between the White masters and the Black slaves. On St. John many of the freed settled as small farmers, fishermen, sailors and traders, attempting to find occupational niches that were not already filled by slaves or Whites. Many chose to move to the city on St. Thomas, where there were better economic opportunities. Some of the free colored adopted the Whites' life style in order to set themselves apart from Black slaves, and in so doing created much uneasiness among Whites, who saw their prerogatives threatened by half-castes who could not be contained by slave regulations. West, for example, describes with scorn how these free colored regarded themselves as gentlemen and ladies and greeted each other in this tone: "How do you do Sir. I hope you well Mam (Madam)" (1790:48). Apparently some copied the dress and mannerisms of the Whites in what West took to be an attempt to pose as members of the upper class. He describes colored women in this way:

> with gravity these arrogant queens with slow steps shoot their proud bodies forward through the street, decorated with straw hats or fine English Castor hats, long gold ear rings, gold necklaces, several rings on all their yellow fingers, and armbands; dressed in magnificent clothing of expensive English or East Indian print, the finest muslin, or clear cambric with long trains, which, flopping along the street, announce their standing (ibid.:48).

This class of free colored grew considerably in size as well as economic importance during the 18th century and became an increasingly influential force that contributed to undermining the White system of suppression.

Conclusion

During the 1700's two sharply distinguished social, economic and cultural groups were confronted with one another in the colonial situation of St. John. On the one hand, there were the Europeans who colonized the island and imported Africans to the island by force for use as slaves in a plantation system established solely for the benefit of Europeans. They brought with them a set of values which stressed the supremacy of their own civilization and the sub-human nature of all non-Europeans. Their perception of the Africans who were to be used in their colonization efforts was molded not just by common European notions of primitives, but also by the very exigencies of the colonial enterprise. On the other hand, the Africans brought with them knowledge and ideas about a radically different culture. They did not readily fall into the role, staked out for them by the Europeans, as objectified slaves eager to serve, but rather, drawing on cultural resources brought with them, they built an Afro-Caribbean culture in an attempt to create their own autonomy and an independent existence. The Europeans, threatened by this assertion of cultural and economic self-reliance, reacted by elaborating a style of life which was defined in opposition to that of the slaves. While the latter's existence was dominated by hard, physical work, that of the Europeans was characterized by virtually a total lack of any kind of productive activity. While the slaves had to be dressed most scantily and in crude fabrics, the Europeans were enveloped in elaborate and elegant draperies that limited their mobility and symbolized their inactive lives. While the White woman had to lead a secluded, virtuous, strictly monogamous life, the Black woman had to be prepared to gratify her master's sexual demands, however excessive. The White man's free access to the Black woman was in contrast to the absolute taboo placed on any sort of sexual or even social congress between Black men and White women. In the long run White culture was self-destructive, because it spawned a White lifestyle in which any expenditure of hard, physical work, energy and, it seems, even creative thinking was regarded as degrading. The physically inactive life in luxury could not adapt to the new social and economic order that emerged when the plantation regime based on slavery ended.

This case study of the confrontation of European and African culture in the West Indies illustrates the importance of examining European and non-European cultures on their own terms. Afro-Caribbean culture cannot simply be reduced to a matter of adaptation or resistance to the colonial system; its African heritage had an integrity of its own which was reconstituted and further developed in the West Indies. On the other hand, the

228   Karen Fog Olwig

European perception of the West Indies and the way in which White culture evolved in the area cannot be understood primarily in terms of inherited European patterns of thoughts and behavior. These patterns, rather, mask the real confrontation going on at the level of daily life in the West Indies. The very inadequacy of European ideologies, in fact, favored the development of new cultural forms by blinding plantation society to the real processes going on. European culture in the West Indies, and the European view of the African in the West Indies, manifested, to a great extent, a mode of response to the strong Afro-Caribbean culture that was created in the New World. This is apparent in the colonial situation where the real meeting between Western and "primitive" worlds took place.

NOTES

1. I would like to thank Michael Whyte and Michael Harbsmeier for reading and commenting on the article. It is a revised and translated version of a paper presented at the 10th Meeting of Nordic Ethnographers, October 1982, Copenhagen.
2. The case material is drawn from published and unpublished accounts of colonial St. John, but references are also made to the other two islands in the Danish West Indian colony, St. Croix and St. Thomas. Four main descriptive sources are used: 1) *En Almindelig Beskrivelse om alle de Danske, Americanske eller West-Indiske Ey-Lande* written in the 1740's, published in 1981. Though there is some uncertainty about the author, the book is attributed to Johan Lorenz Carstens. Carstens was born in the Danish West Indies and owned plantations on St. Thomas and St. John. He moved in 1739 with his family to Denmark, where he settled on a large estate which he purchased. The manuscript therefore is written in Denmark. 2) *Beskrivelse over Eylandet St. Croix i America i Westindien* from 1758 written by Reimert Haagensen. He was born in Denmark, but moved to St. Thomas in 1739, where he worked as a book keeper for the West Indian Company, which administered the Danish West Indies until 1755. Haagensen also owned cotton plantations. 3) *Geschichte der Mission der evangelischen Brüder auf den caraibischen Inseln S. Thomas, S. Croix und S. Jan* written by C. G. A. Oldendorp in 1777. Oldendorp had been sent to the Danish colony in 1767–68 in order to write an account of the Moravian missionary work among the slaves, and the book is based on this trip. It contains a description of the missionary activities and of the plantation society and the slaves' African background. 4) *Beretning om det danske Eiland St. Croix* from 1790 by Hans West, who functioned as school head master on St. Croix in the 1790's. He attempted to establish an institute of education for well-to-do children in the Danish colony, but had to close the school due to poor enrollment. Apparently the wealthy preferred to send their children abroad for education. West then stayed on as a public notary. (See Bro-Jørgensen 1966: 257; Vibæk 1966: 72, 125, 216, 228; Nielsen 1981 [1970]). In general, Oldendorp's account is probably the most sympathetic to the slaves, though it is very guarded in order not to offend the Danish authorities, on whose goodwill the Moravian missionaries depended. Carstens's description is very frank and seems to benefit from having been written in Denmark, removed in time and space from the Danish West Indies. Haagensen's book is influenced by the business interests of the author, whereas West's book seems to reflect some

dissatisfaction with the low level of culture and education which the author encountered among the Whites as well as the slaves.
3. All sources, whether they appeared originally in Danish or German, have been translated into English. It has not been possible, unfortunately, to maintain the old-fashioned quality and the strong language of the texts.
4. A Commission had been set down in Denmark to decide on the issue of sexual relationships between Whites and Blacks. It concluded that relationships between White men and Black women should be treated as harshly as those between White women and Black men. This decision, however, seems to have been ignored (Olsen 1983).

REFERENCES

VAN DER BOOGART, ERNST. 1982. Colour Prejudice and the Yardstick of Civility: the Initial Dutch Confrontation with Black Africans, 1590-1635. In *Racism and Colonialism*, R. Ross, ed. Leiden: Martinus Nijhoff Publishers.

BRO-JØRGENSEN, J. O. 1966. *Dansk Vestindien indtil 1755*, vol. 1 *Vore Gamle Tropekolonier*, J. Brøndsted, ed. Denmark: Fremad.

CARSTENS, J. L. 1981. [1740's] *En Almindelig Beskrivelse om Alle de Danske, Americanske eller West-Indiske Ey-Lande. Dansk Vestindien for 250 år siden*. København: Dansk Vestindisk Forlag.

COHN, BERNARD S. 1980. History and Anthropology: The State of Play. *Comparative Studies in Society and History*, 22: 198-221.

HAAGENSEN, REIMERT. 1758a. *Beskrivelse over Eylandet St. Croix i America i Westindien*. København: Lillies Enke.

– 1758b. Om de på Øen St. Croix værende sorte Hedninge eller Slaver. *Oeconomisk Journal*, Februarii-Maaned.

HALL, NEVILLE. 1977. Slave Laws of the Danish Virgin Islands in the Later Eighteenth Century. In Comparative Perspectives on Slavery in New World Plantation Societies. V Rubin and A. Tuden, eds. *Annals of the New York Academy of Sciences*, vol. 292.

MINTZ, SIDNEY W. 1974. *Caribbean Transformations*. Chicago: Aldine.

MINTZ, SIDNEY, and DOUGLAS HALL. 1960. The Origins of the Jamaican Internal Marketing System. *Yale University Publications in Anthropology*, 57: 1-26.

NIELSEN, HERLUF. 1981. [1970] *Inledning til En Almindelig Beskrivelse om Alle de Danske, Americanske eller West-Indiske Ey-Lande*. København: Dansk Vestindisk Forlag, pp. 9-14.

OLDENDORP, C. G. A. 1777. *Geschichte der Mission der evangelischen Brüder auf den caraibischen Inseln St. Thomas, S. Croix und S. Jan*. Barby: Christian Friedrich Laux, 2 vols.

OLSEN, POUL ERIK. 1983. Danske Lov på de vestindiske øer. In *Danske og Norske Lov i 300 år*. København: Jurist- og Økonomiforbundets Forlag, pp. 289-321.

OLWIG, KAREN FOG. 1981a. Finding a Place for the Slave Family: Historical Anthropological Perspectives. *Folk*, 23: 345-358.

– 1981b. Women, 'Matrifocality' and Systems of Exchange: An Ethnohistorical Study of the Afro-American family on St. John, Danish West Indies. *Ethnohistory*, 28(1): 59-78.

– 1983. Sorte Danskere. *Skalk* 3: 18-27.

PATTERSON, ORLANDO. 1982. Persistence, Continuity, and Change in the Jamaican Working-Class Family. *Journal of Family History*, 7(2): 135-161.

230   Karen Fog Olwig

SMITH, RAYMOND T. 1982. Race and Class in the Post-Emancipation Caribbean. In *Racism and Colonialism*. R. Ross, ed. Leiden: Martinus Nijhoff Publishers.

VIBÆK, JENS. 1966. *Dansk Vestindien 1755—1848*, vol. 2 *Vore Gamle Tropekolonier*. J. Brøndsted. ed. Denmark: Fremad.

WEST, HANS. 1790. *Beretning om det danske Eiland St. Croix i Vestindien fra Junii Maaned 1789 til Junii Maaneds Udgang 1790*. Kiøbenhavn.

Archival Sources

Schlimmelmann Archiv Ahrensburg (Familienarchiv). Landesarchiv, Schleswig, Germany.

62. Carolina Plantagenjournal 1766–67, Westindiske Besitzungen (Carolina, La Grange, Princesse) 1767–1863.

Vestindiske Lokalarkiver, Landfogeden på St. Jan. Rigsarkivet, Copenhagen, Denmark.

42. Politiretsprotokol, 1841—1851.
59. Politiretssager, 1829–1836.
64. Landretsprotokol, 1752–1759.
65. Landretsprotokol, 1760–1771.
67. Landretsprotokol, 1777–1782.
72. Landretsprotokol, 1802–1807.

10
Toussaint Louverture and the Slaves of the Bréda Plantations
David Geggus

The Bréda plantations in Saint Domingue's northern plain are famous in Haitian history because of their connection with the revolutionary leader, Toussaint Louverture. It was on the estate of Haut-du-Cap, just outside of modern Cap Haitien, that Toussaint was born, apparently around 1745. There, as a young man, he supposedly benefitted from the benevolent attention of the manager, Bayon de Libertat, who ran the plantation for the absentee owner and who later became attorney for both Haut-du-Cap and its sister plantation in the adjacent parish of Plaine-du-Nord.[1] Both estates were situated in what would become the heartland of the great slave revolt of 1791 which, according to some, Toussaint secretly organised.

Laid out around the village of the same name and adjoining the main colonial highway, the Haut-du-Cap plantation was the only sugar estate in the parish of Cap Français Its slaves lived within forty minutes' walk of the city of Le Cap and mingled daily with the population of Haut-du-Cap *bourg*. The district's thin, stony soil limited the estate's production, even in a good year, to little more than 100,000 lbs. of low-grade refined sugar *(commun* or *blanc)*. Its owners, therefore, tended to neglect it in favour of their other plantation, which commonly made over five times as much revenue.[2] This was situated some three miles away in the middle of the lush plain that was checkered with over two hundred other such plantations.

Gabriel Debien has already provided a fascinating portrait of the plantations' slaves during the two years leading up to the 1791 revolt.[3] The Bréda estates were then racked by hunger, overwork, high mortality and numerous runaways. The purpose of this paper is to look at the plantations in the period immedi-

Toussaunt Louverture and the slaves

ately preceding, and in particular to analyse the structure of their workforces and place them in comparative perspective. First, however, it is proposed to examine critically the little that is known about Toussaint Louverture's connection with the Bréda estates.

Toussaint Louverture and the Bréda Plantations

In 1799, when Toussaint as Governor of Saint Domingue was approaching the height of his power, he received an ingratiating letter from the comte de Noé, one of the owners of the Bréda estates.[4] The plantations had been sequestrated by the colonial government[5] and the absentee count wanted their revenue sent to him in Europe. Nostalgically, he recalled the good treatment Toussaint had received 'in the past' and he expressed confidence in his attachment to the Bréda family. The confidence proved to be justified; the plantations' revenue was thereafter passed on annually.[6] As is well known, Toussaint's relations with Bayon de Libertat were also notably amicable after his rise to power. The ability to do favours for one's poverty-stricken former masters was doubtless a source of subtle pleasure, and Toussaint was anyway a charitable man. Nonetheless, it seems clear that Toussaint's experience of slavery at Haut-du-Cap had been exceptionally benign.

Toussaint's favoured treatment as a slave may have been due to his talents as a horse-doctor, which, according to General Kerverseau, earned him promotion from stable-lad to coachman.[7] He might also, as his son later claimed, have been descended from an enslaved African 'prince' to whom the plantation owner had accorded special status.[8] Toussaint himself, when famous, said he had been freed by Bayon de Libertat some twenty years before (therefore around 1777), though Bayon in his old age implied Toussaint was still a slave when the Revolution began.[9] Until recently, historians concluded that Toussaint probably enjoyed the unofficial 'freedom of the savanna' and that he remained living on the Haut-du-Cap estate, perhaps as a coachman or in some supervisory capacity, up until the great slave revolt of 1791.

However, in 1977, Gabriel Debien, Jean Fouchard and Marie-Antoinette Menier published documents from the French archives which showed that as early as 1776 Toussaint was in fact a freed man, that he himself owned at least one slave and that he later rented a small property on the other side of the plain in the parish of Grande-Rivière.[10] He had thus ceased to be even a privileged slave long before the revolution. Having been formally manumitted, he belonged to the class of *'gens de couleur libres.'*

Toussaint was of course a common name in Saint Domingue and, as will be seen, the Bréda plantations had more than one slave of that name. Nevertheless, there seems no reason to doubt that the Toussaint Bréda mentioned in the 1770s documents was indeed the same man who adopted the surname Louverture in the summer of 1793. Yet there remains a substantial problem. Why does no other source, printed or manuscript, identify Toussaint as a free man? Why did so many contemporaries think that he had been a slave up until the Revolution? Early in the Revolution he is never referred to as *nègre libre*, the usual term for a free black. He appears sometimes as "Toussaint à Bréda," a designation more usual for slaves, or more ambiguously as "Toussaint Bréda," or "le nègre Toussaint." Many contemporary sources imply that he still lived on the Haut du Cap estate.[11]

It may be that there is no necessary conflict between the documentary evidence and tradition. While one would not expect a *bona fide* free black, who had been legally manumitted and who owned property, to be still resident on his former master's plantation, Toussaint may have been exceptional and combined salaried employment with independent entrepreneurship. Perhaps his close association with the Bréda estates obscured the fact that he was free. Or maybe by 1789 he was so well known in the northern plain that some never bothered to use the term *nègre libre* when referring to him.

One possibility is that, after being freed, Toussaint lived close by the Bréda estate on a small property that belonged to his son-in-law. Philippe Jasmin Désir was a free black, a little older than Toussaint, yet who seems to have married one of his illegitimate daughters. He was the owner of the land and slaves in Grande-Rivière which Toussaint leased in 1779-81.[12] As, under the law, his first and last 'European' names were not officially recognised, he may have been the same 'nègre Jassemin' mentioned in notarial papers as owner of a property almost adjoining the Bréda plantation in the steep hills above Haut-du-Cap.[13] But, again, Jasmin was not an uncommon name.

Toussaunt Louverture and the slaves

However, even if Toussaint did not actually live on the estate as a free man, there remains the problem of why he is not mentioned in any of the extant papers of the Bréda plantations. One would expect at least his manumission to have been discussed in the letters of Bayon de Libertat and the comte de Bréda. Antoine François Bayon de Libertat was attorney for the two Bréda plantations and resident at Haut-du-Cap from August 1772 to July 1789. He was not, as most authorities claim, merely the manager of Haut-du-Cap and for only a short period of time. His correspondence survives for the period 1772-85, and that of his successor for 1789-91.[14] This raises the question of exactly when Toussaint was freed.

According to Debien, Menier and Fouchard, he was manumitted in 1776. They cite a marriage register entry of September 1777, which concerns an African slave formerly belonging to 'Toussaint Bréda'. The document bears a margin note referring to a legal manumission in 1776.[15] The wording is admittedly ambiguous, but it seems to me obvious it was the African who was freed in 1776 and not Toussaint. The priest was legally obliged to verify the status of the bridegroom, but not that of his former owner. Besides, Toussaint could scarcely have been freed, bought a slave and manumitted him all in the space of a year. The 'twenty years' of freedom that Toussaint spoke of in 1797 should not be taken too literally.

On the other hand, if it was indeed Bayon de Libertat who freed him as Toussaint said, this presumably must have occurred after 1772, since the comte de Bréda did not recognise Bayon as his attorney until the end of that year. Moreover, during the summer and autumn of 1773, Bayon was temporarily replaced by a rival, which would narrow even further the period during which he could have freed Toussaint from the count's service. Yet, although Bayon's letters mention many slaves by name and some requests for manumission, they contain no reference to Toussaint. A complaint of March 1775 that the plantation had no horse doctor might suggest that he was by then already free (or that his later reputation as *docteur-feuilles* was ill-founded). Contrarily, a reference in October 1775 to 'Barbe, chambrière, votre affranchi' gives the impression that the estate had only one freed slave, who was a woman. However, there seems no question that Bayon could have surreptitiously freed Toussaint, since the comte de Bréda's nephew lived in the colony close to

Haut-du-Cap. Must one presume that a vital letter is missing from the existing correspondence?

Should such a letter not be missing, I think the best interpretation of the evidence we have runs as follows. Toussaint was freed by Bayon de Libertat some time in the early 1770s, but he was by then no longer a slave of the comte de Bréda. He must have been purchased by Bayon before he became the count's attorney in 1772. We know that Bayon was already attorney for three nearby sugar plantations, and he claimed to know the Bréda estates extremely well.[16] Bayon had lived in Saint Domingue since 1749, and it was in fact the comte de Bréda's previous attorney who named him as his successor. Hence he could easily have known Toussaint as a boy before becoming formally connected with the Bréda plantations. We also know that when Bayon moved to Haut-du-Cap he brought most of his domestic slaves with him, and that these included several coachmen.[17]

Moreover, from 1777 onwards, Bayon began slowly building up his own sugar estate ten miles away in the parish of Limbé. Though he remained living in the *grand' case* at Haut-du-Cap, this must have entailed periodic absences. Toussaint, now free and still living close by, may at such times have acted in an informal supervisory capacity. This would explain why some contemporaries wrongly thought he was the estate's slave-driver.[18] As others have said that Toussaint was in charge of the livestock on Bayon's own plantation,[19] he might also have spent time in Limbé, either before or after Bayon's dismissal by the Bréda heirs in July 1789. Since there is documentary evidence that at the start of the slave revolt the Bayon family was rescued by a slave from their Limbé plantation,[20] the well-known story that Toussaint shepherded his master's wife to safety may well refer to this event rather than to anything that happened at Haut-du-Cap.

In 1785, on the other hand, when the only surviving lists of the plantations' slaves were drawn up, Bayon de Libertat was still acting as the estates' attorney, and Toussaint may have been working with him as his *homme de confiance*, or simply farming a plot of land on the margin of the estate. No longer part of the plantation's property, he is naturally not listed in the inventory.

Toussaunt Louverture and the slaves

It is these *états de nègres,* drawn up in April 1785,[21] that we propose to analyse in this paper. While it is not certain that the slaves listed were actually under the supervision of Toussaint Louverture, one can be sure that they were people whom Toussaint knew well. Many of the eldest of them must have been his boyhood companions.

The Slaves in 1785

Both the Bréda plantations were unusual in that their workforces were heavily African. In the 1780s, locally born creoles constituted nearly two-thirds of the slave population of the sugar plantations in northern Saint Domingue, and among adult slaves Africans and creoles were equally balanced.[22] On the Bréda estate at Plaine-du-Nord, however, creoles accounted for only 41% of the workforce, and merely 28% of the adults. At Haut-du-Cap, moreover, Africans made up 64% of the slaves and fully three-quarters of the adults. These, then, were communities where the voice of Africa was still strong.

Apart from this low degree of creolisation, the Bréda plantation at Plaine-du-Nord was remarkably typical of the *sucreries* of the northern plain on the eve of the Haitian Revolution. It thus provides a useful point of comparison with its sister estate at Haut-du-Cap, which was in many ways exceptional.

With only 154 slaves,[23] Haut-du-Cap was a relatively small sugar estate. Though these were far more common in Saint Domingue than one might imagine, (given the reputation of *la perle des Antilles*), the Plaine-du-Nord workforce of 210 slaves was much closer to the norm both in size and structure. Just over half its slaves (55%) were young adults aged 15 to 39; 27% were older, with 7% over 59, and 18% were children. Males slightly outnumbered females (106:100).

At Haut-du-Cap, the demographic profile was less favourable. Sexual imbalance was much more extreme (152:100), and especially among the young adults (157:100). Even if the African men did not practise polygamy, one in three of the young males would ordinarily have been without a partner. It is true that children accounted for 29% of the workforce, but this had nothing to do with its fertility. Two-fifths of the children were recently purchased Africans. Less than half of the slaves (47%) were in the prime working age-group of 15-39, but fewer than 5% had passed 60 years of age. Life-expectancy, then, seems to have been shorter here.

David P. Geggus

Bayon preferred to purchase Africans before they reached adulthood, as they gave less trouble, he said, and suffered less mortality in the 'seasoning' period.[24] He seems to have been right. The two workforces had been increased by 27 new Africans in the previous year (most of them in the last four months) and none had died, though one, an epileptic, was returned to the merchant who sold him. In other respects, however, vital rates on the two plantations contrasted sharply. At Haut-du-Cap, over the previous fourteen months, the slaves exhibited birth and death rates equivalent to 1.7% and 5.8% per annum. With such a rate of natural decrease it was necessary to purchase six or seven slaves per year simply to keep the workforce up to strength. The Plaine-du-Nord slaves, however, experienced a birth rate of 3.6% p.a. and death rate of 2.6% p.a. The workforce was thus growing naturally, but this was no doubt an exceptional year.

The fact that these data include the birth and death of a child who died at one month old would suggest that they might not suffer from underregistration of births. The underregistration of births resulting from infant mortality is one of the main pitfalls in the study of slave demography, especially when the sources are, as here, terse annual summaries. Bayon himself did not include the child in his listing, merely a footnote, but it appears nonetheless that the rates given may be complete. Infant mortality can therefore be put tentatively at around 30% for Haut-du-Cap and about 10% for the two plantations combined. The average age at death in both workforces was 37. Stated causes of death included dirt-eating and dropsy (2 cases each), smallpox, yellow fever (maladie de Siam), abscess on the chest and an accident. Then aged about 40, Toussaint would already have seen about half of his contemporaries die.[25]

Of the two slave communities, Plaine-du-Nord was the healthier by a substantial margin. If we disregard the slaves, aged over 59 who are listed merely as 'infirme,' but include those who are younger but described as old or weak, we find that 11% (24) of one workforce but 18% of the other workforce suffered some sort of physical disability. The average for sugar estates in the region was 9%. In inventories such as this, only the permanently sick were counted, not those temporarily in the plantation hospital. In both cases, the locally-born creoles were healthier than the Africans, as was usual, and the men were healthier than the women, which was uncommon but not

Toussaunt Louvertire and the slaves

rare. Even so, the women and Africans at Plaine-du-Nord were healthier than the men and creoles at Haut-du-Cap. Place of work, rather than gender or origin, was therefore the key factor here.

Paradoxically, Haut-du-Cap actually enjoyed a more salubrious location than low-lying Plaine-du-Nord, but its slave-quarters were more dilapidated and its resident manager negligent and cruel.[26] Perhaps more important, the slaves' gardens in the hills behind were difficult to reach and on poor soil. Throughout his correspondence, Bayon calls the Haut-du-Cap slaves lazy, because they neglected their gardens and preferred to gather fodder grass and wood which they could exchange in the nearby village for rum, sweetmeats and bread. The Plaine-du-Nord slaves, by contrast, were 'gorgés de vivres' (stuffed with provisions).[27]

Fully half the 52 sick were described in vague terms such as ill. weak or (prematurely) old. Three were listed as dying *(valétudinaire);* another three as having pains. Though such categorisation tells us little, it is intriguing to find the term *fatras* (feeble, trash) employed. An uncommon word, it appears on no other known plantation inventory, but provides a curious link with Toussaint Louverture, whose childhood nickname is said to have been 'fatras bâton.'[28] Among the more recognisable ailments were three cases of chest disease, one of epilepsy *(mal caduc),* three of elephantiasis or phlebitis *(jambes enflés/gros pieds),* four of ulcers, one of gravel, one of weak sight, one of leprosy, and surprisingly, only one of hernia. One woman had also lost an arm, probably when feeding cane into the mill.

Turning to the Africans who bulked so large on these plantations, we find no fewer than 25 different 'nations' listed, about 20 on each estate. French colonists seem to have been far more interested than the British in their slaves' origins. However unsatisfactory are the ethnic labels they used, they deserve to be taken seriously.[29] Whatever one thinks of the well-known generalisations about ethnic personality-traits, or European ignorance of Africa, the planters' terminology certainly reflected solid physical facts, (sex ratio, average height), and was the joint product of African knowledge as well as European observation – hence the use here of the Fon terms 'Nago' and 'Tapa' to describe the Yoruba and the Nupe.

Those known as Congos formed the largest group of both Bréda estates, as on almost every other Saint Domingue plan-

David P. Geggus

tation by this date. While the designation 'Congo' at least serves to distinguish Bantu from non-Bantu slaves, it is unhappily the most general of the terms employed. It seems to have been applied to almost all blacks coming from south of the Cameroons. A large proportion, however, was no doubt Bakongo from the Congo basin,[30] while certain detailed plantation inventories suggest that their northern neighbours, the Mayombe, made up about a sixth.[31] Here, the Congos together with the Mondonga, also Bantu, constitute 32% of the Africans, in fact 1 in 5 of all the slaves on the two plantations.

Like most sugar planters, Bayon did not like Congos. They fell sick, neglected their gardens, and ran away. He bought them only because after 1774, he had no choice. Nineteen out of twenty slaveships, he claimed, carried Congos. For a while, his attitude grew more favourable. The women were willing workers, gregarious and had lots of children. But by 1784 he was swearing he would buy no more. Sooner or later, it seemed, they gave way to despair and refused to work.[32]

With at least twenty ethnic groups living together in the slave quarters, the pressures towards acquiring a common creole tongue were evidently great, and the chances of an African language surviving were slim. Toussaint Louverture, we are told,[33] nevertheless managed to master perfectly the language of his Arada parents, which was presumably Fon. Particular interest therefore attaches to the number of Arada and other Dahomeyan slaves living on these plantations, especially as they are credited with the dominant influence in the shaping of Haitian culture. Taking the Arada, Fönds and Foeda together with their Adia (Aja) neighbours, we find that they account for only 10% of the Africans. Certainly, they had been more numerous in the previous decade when, in November 1774, Bayon had purchased 40 Arada in one lot. Several, like Baptiste, the old gate-keeper at Haut-du-Cap, must have been there since before Toussaint was born. In view of Toussaint's parentage, it is doubtless significant that the Arada were Bayon's preferred ethnic group. 'Ils n'ont aucun volonté,' he noted. They submitted silently to plantation discipline and worked well. They were more expensive than Congos.[34]

As usual, the second most prominent grouping were the Nago or Yoruba (13%), followed by Bambara (10%) from the Upper Senegal. Probably from the same savanna region were the dozen Mali (5%), who were perhaps Mandinka speakers. Slaves from

Toussaint Louverture and the slaves

these far-inland regions were invariably mostly male. The Mesurades and Miserables (8%) came from the coast around present-day Monrovia and its hinterland. Known as Canga in other parts of Saint Domingue, they have yet to be identified with a known ethnic group. Also from the 'Grain Coast' were the Aquia (Akwa), Quicy (Kissi) and Sosso, never very numerous in the West Indies. The Ara were an Ibo sub-group. All the Ibo together constituted 7% of the Africans. As only the women were considered good agricultural workers, Ibos tended to be avoided by sugar planters and to be more common on coffee plantations, where selection was less rigorous.

Only a tiny handful of these Africans are listed under what appears to be their original name. At Haut-du-Cap we find only an old Bambara named Maha, and another, much younger, called Thomanis. At Plaine-du-Nord, Quoicou, an Aja, and Couby, an Arada, apparently bore West African day-names (i.e. Quaco and Cubba). Doussou was an Ara, aged 50, and Quimby, a Timbou, from the highlands of Futa Jallon. Another Timbou was called Cotocolly, but this may have been a nickname, since it is the name of an ethnic group from northern Dahomey. Similarly, the three women (one Ibo, two Congo) called Zaire were probably named by the plantation manager. The other Africans bore a mixture of saints' names, military-type nicknames and names derived from classical antiquity. The names of all the creoles, on the other hand, were Christian; they had been baptised. Each of the plantations had a Toussaint. One was a Yoruba field slave, the other a creole boilerman.

The boilermen made up the largest group of 'elite' slaves on the two plantations. Plaine-du-Nord had a dozen, three of them being master boilermen. At Haut-du-Cap, which made much less sugar, there were two masters and seven others. All but one of these boilermen were Africans. Important though the post was for a plantation, it was probably not perceived as a high status position. Almost everywhere, creoles tended to avoid work in the boiling house's stifling heat. Other slaves employed in the two Bréda factories were two disabled Africans in charge of heating the refineries and five old or sick Africans who, at Plaine-du-Nord, 'capped' with wet clay the terra-cotta moulds of sugar.

Next most numerous among the specialists, with 8 on each plantation, were the drivers of the ox-carts and their assistants.

David P. Geggus

Nearly half were creoles, including all 4 master carters. The 3 *commandeurs,* or slave-drivers, were also creoles, and of the domestics (2 cooks, 2 laundrywomen, 2 servants), only one was African, a Congo. With almost no exceptions, the only Africans to work in the *grandes cases* of Saint Domingue were Congos. They were preferred apparently because of their temperament and their ability to speak a 'pure' creole (close to French?).[35] Whether this was an aspect of the Bantu 'adaptability' of which Roger Bastide wrote, or was due to a prior acquaintance in Africa with Portuguese, deserves investigation.[36] The relative paucity of domestics, and the marked preference for creoles in all these posts, were standard features of sugar plantation society in late colonial Saint Domingue.

Among the artisans (5 coopers, 1 mason, 1 carpenter), creoles also predominated, though the 3 stockmen and 3 nurses were mainly Africans. The 4 millers, who also looked after the mules that turned the mills, were equally balanced between the two. The dozen supernumerary posts (gatekeepers, watchmen, hedgecutters) were filled almost entirely by aged Africans. All in all, specialists accounted for just over 20% of both workforces, an average proportion.

With half of the men counted as specialists, the majority of the field slaves were women, which was usually the case. The main field gang was the driving force of a plantation; its size and make-up largely determined its productive capacity. Healthy field slaves aged 15 to 50 constituted 42% of the slaves at Plaine-du-Nord, a percentage slightly better than the norm. At Haut-du-Cap, however, (excluding the 17 year old creole listed as a frequent runaway) such slaves represented merely 25% of the workforce, the lowest proportion we have found anywhere.

While it is not known exactly how tasks were allotted in the canefields, the fact that women performed the bulk of unskilled manual labour on sugar plantations was doubtless an important factor behind their universally low levels of fertility on such estates. It is true that fertility levels of slave populations may have been understated by historians because of lack of data on infant mortality; the fertility index (children 0-4 per 1,000 women 15-44) is especially liable to be distorted by abnormally high numbers of infant deaths. On the other hand, the continual importation of adolescents and young adults gave populations where the slave trade operated relative advantages

Toussaunt Louverture and the slaves

with respect to fertility,[37] as did the lack of the economic and social constraints operating in free society. When examining slave fertility in a comparative context, the effect of these two biases is to some extent reduced one by the other.

Fertility seems to have been extremely low in Saint Domingue's northern plain. While one might expect an 18th century European or modern black African population to have a fertility index of between 700 and 900 (children 0-4 per 1,000 women 15-44), we have found a mean index for 19 northern plain *sucreries* of only 278.[38] The Bréda estates were fairly close to the norm. At Plaine-du-Nord the fertility index was 300; at Haut-du-Cap, 250.[39]

Overall, the main determinant of slave fertility appears to have been age-structure, followed by work load, creolisation, and disease. Unfortunately, since it is not known how much cane was planted on these two estates, one cannot say whether slaves were worked harder on one rather than the other. At Plaine-du-Nord, output per fieldhand was about 50% higher than at Haut-du-Cap, but this may have been due to the latter's inferior soil and millwork. As for creolisation: African women tended to be more frequently childless and bear fewer children than did creoles, at least on sugar plantations. The Bréda estates were no exception here. Although only a minority of the mothers are identified on the lists, it is clear that in every age-group creoles were far more likely to have children than were Africans.

Fertility would doubtless have been lower on the two estates but for their peculiarly favourable age-structure. Slave women in the northern plain seem to have reached their peak fertility in their late twenties. Those aged 25-34 usually constituted about a third of those in the fertile age-band, but at Haut-du-Cap they made up almost a half of the fertile women, and at Plaine-du-Nord just over a half. As the proportion of Africans in the fertile age-band was almost the same for the two estates, it may be this difference in age-structure, along with differing degrees of morbidity, that explains their differing levels of fertility.

However, there were of course other factors which influenced fertility that may have been significant at the level of individual plantations. It is alleged, for example, that proximity to the town of Cap Français encouraged prostitution or promiscuity among female slaves and that this diminished fertility.[40] There is

David P. Geggus

reason to believe that on certain estates this was true.[41] It may therefore be relevant that the *sucrerie* at Haut-du-Cap was one of the closet of all to Cap Français. as well as being much frequented by visitors from Haut-du-Cap *bourg*. After troops were billeted there during the American War, Bayon complained that three-quarters of the women had venereal disease.[42] Another variable that may have affected fertility was the character of the plantation manager. The manager of Haut-du-Cap was apparently brutal and negligent, and he was dismissed in 1790 by Bayon de Libertat's successor, who called him a 'tyrant.'[43]

Towards a Revolution?

It has sometimes been claimed that the conditions of slave life in Saint Domingue were worsening towards the end of the *ancien regime*. Soil exhaustion and spiralling slave imports supposedly caused food shortages and a heavier workload as planters sought to maintain profits. According to this view, slave discontent was already mounting when the colony was swept into the maelstrom of the French Revolution.

The Bréda papers offer some evidence to support such an interpretation, but they also give cause for scepticism, and are anyway too fragmentary to give a clear picture of any trends that may underlie this period.

Slave imports in the 1780s certainly never attained the levels of 1776-78 when Bayon was buying up to forty slaves in a year. The output of both plantations did increase between the two periods, and notably at Haut-du-Cap. However, as both workforces were on average about one-fifth larger in the 1780s than before the American War, only at Haut-du-Cap can one speak of increasing output per capita, and perhaps therefore of increasing workload. The surviving plantation accounts suggest that sugar production probably peaked in 1784-85 (a post-war glut?) and then dropped after the drought of 1786, falling further the following year when new canes were drowned by heavy rain. Output seems to have never fully recovered, but high prices in 1789-91 brought partial compensation. The drought of 1790 was exceptionally severe, and the correspondence published by Debien shows it caused great hardship for the slaves.[44] Withered crops had to be replanted in hard-baked earth; food was scarce, disease rife, and many slaves died. The two workforces' rate of natural decrease (2.9%) was double what it had

Toussaunt Louverture and the slaves

been in 1784-85. However, this was due to an abrupt fall in the unusually high birth rate at Plaine-du-Nord. Death rates on the two plantations (5.3% and 3.0%) were overall no higher than in 1784-85.

At the beginning of 1790 the combined workforces were a little smaller than five years previously. In the intervening period, the slaves of Haut-du-Cap seem to have fared remarkably well while the Plaine-du-Nord workforce had declined by 6.6%, perhaps because of its greater proportion of aged slaves. During the drought of 1790, however, it was the Haut-du-Cap slaves who suffered most. Moreover, although Villevaleix, the ambitious new attorney, purchased 30 new Africans that year,[45] he sent few or none to Haut-du-Cap, which was put up for sale. These new purchases brought the strength of the two workforces to a new peak, slightly above that of early 1785. Nevertheless, one cannot say that the proportion of unassimilated Africans in the slave population was notably greater on the eve of the slave revolt than it had been six, or sixteen, years before.

On the other hand, since almost all the new arrivals were males in their twenties, they did considerably increase the already large percentage of young men living in the Bréda slave quarters. And in a year of extreme drought, of course, they put further strain on the plantations' food supply. Droughts were not uncommon in Saint Domingue, but when Bayon de Libertat managed the estates, he had avoided buying new slaves when they could not be adequately fed.[46] The drought of 1790, furthermore, was the worst in living memory.[47]

If conditions were changing in other respects, it is difficult to see a clear pattern. By the late 1780s, overt cruelty was probably becoming less common everywhere in Saint Domingue, as sensibilities changed. In 1773, the attorney who briefly replaced Bayon de Libertat had terrorised the workforce of Haut-du-Cap because of a poisoning scare, torturing several slaves and causing two to commit suicide.[48] By contrast, Valsemey, the harsh and negligent manager employed by Bayon, was dismissed in July 1790, after a slave he had punished died and nine others ran away. These slaves returned as soon as Valsemey was replaced but runaways were to remain a recurrent problem. The new manager found that Haut-du-Cap had acted as a magnet for fugitives from other estates, too, who hid in the slave quarters and the hilly part of the plantation. He clamped down on them, arresting 27 in two months. He also tried to restrict the easy access to the plantation of the inhabitants of Haut-du-Cap village, which he considered abusive. Frequent references to 'good order' in Villevaleix's correspondence suggest increased care combined with tighter surveillance (though this was the impression every attorney wished to give).

For the slaves, of course, such 'improving' management was a mixed blessing. In 1791, Villevaleix began a project to replace some of the slaves' dilapidated huts with a wooden barrack-type structure. Apart from any unwelcome regimentation this may have implied, it meant cutting trees in places Bayon had considered inaccessible. The workforce had therefore to haul down lumber from the mountains at the same time as it was digging a new system of drainage ditches, planting new canes and bringing in the crop.

Yet in August, just a week before the revolt broke out in the plain, the slaves of Haut-du-Cap apparently rallied round Toussaint and the slave-driver Bruno to extinguish a fire that had been started in several cane-pieces. On their own initiative, it was said, the manager being absent, they then cut and processed the cane before it could spoil. When the uprising began, if we are to believe the same anonymous source,[49] who was probably Bayon de Libertat, the slaves of Haut-du-Cap did not join in but remained on the estate. They may not have had much choice, of course, since Haut-du-Cap soon became a military camp. However, two years later, after the burning of Le Cap and the abolition of slavery, the same blacks, at Toussaint's urging, offered help to the former attorney, suggesting he resume management of the plantation, 'saying that since I had lost everything, they would see to the needs of my family.'[50]

NOTES

1. Not to be confused with *la plaine du Nord*, the northern plain itself.
2. Incomplete accounts and some commercial correspondence for the period 1784-91 are in Archives départmentales de la 'Loire-Inférieure'. Nantes, E 691. Despite its water-logged location, the Plaine-du-Nord estate could produce up to 400,000 lbs. of good quality refined sugar ($1^{er}/3^{me}$ qualité).
3. G. Debien, *Etudes Antillaises*, (Paris, 1956), 143-173.
4. Public Record Office, London, CO 137/50, letter of 8 April 1799.
5. Toussaint himself appears to have leased the Haut-du-Cap estate: G. Debien, 'Les biens de Toussaint Louverture', *'Rev. de la soc. haitienne d'hist.*, no. 139 (juin, 1983), 70.
6. V. Schoelcher, *Vie de Toussaint-Louverture*, (Paris, 1889), 392

Toussaunt Louverture and the slaves

7. Archives Nationales, Paris, (hereafter: 'A.N.'), CC9B 23, report of 6 sept. 1801.
8. Bibliothèque Nationale, Paris, Manuscrits, Nouv. Acq. Fr. 6864, ff. 37-48. Apart from this story, the document says almost nothing about Toussaint's life before the Revolution.
9. See the documents printed in Schoelcher, *Vie*, 388-392.
10. G. Debien, 'M.-A.' Menier, J. Fouchard, 'Toussaint Louverture avant 1789: Légendes et Réalités,' *Conjonction, Revue Franco-Haitienne*, no. 134 (juin-juillet, 1977), 67-80.
11. E.g. Archivo General de Simancas, Guerra Moderna 7157, Garcia to Acuña, 23 de nov. 1793, enclosure no. 2; above, note 9. We do know that in 1779 Toussaint still lived in the region of Haut-du-Cap but not exactly where: above, note 10.
12. See note 10. In P. Pluchon's admirable biography he wrongly appears as Toussaint's father-in-law, and also light-skinned: *Toussaint Louverture: de l'esclavage au pouvoir*, (Paris, 1979), 18, 21.
13. A.N. Section d'Outre-Mer, Notariat, reg. 856, deed of sale by Rostaing, 27 avril 1783.
14. Bayon's letters are in A. N., 18 AP 3. They are an important and totally neglected source for the impact of the American War of Independence on Saint Domingue. For the correspondence of Villevaleix, Bayon's successor, see above, note 2. A small amount is also in private hands: Debien, *Etudes*, 144-5.
15. The document is reproduced in the article cited above, note 10.
16. A. N., 18 AP 3, letter of 21 dec. 1772.
17. Ibid., and letter of 3 nov. 1776. Note that this reconstruction also accords with the tradition preserved by Schoelcher *(Vie 88)* that Bayon 'attached Toussaint to his personal service and later made him his coachman'
18. Bibliothèque Municipale, Auxerre, Ms. 331, f. 181. In view of Debien's statement that the Haut-du-Cap slave list did not include a *commandeur* or driver (below, note 21), one might be tempted to guess that Toussaint regularly supervised work in the fields, as such an omission would be extraordinary. However, a slave-driver is in fact listed, a creole named Bruneau. He was also mentioned by Bayon de Libertat: above, note 9.
19. Memoir written around 1800, cited in Pluchon, *Toussaint*, 18.
20. Public Record Office, London, WO 1/58, 11: Bibliothèque Municipale, Nantes, Ms. 1809, f. 188. However, the rescuer in these stories is a slave and it is the whole family that is rescued.
21. A.N., 18 AP 3. Short analyses of these lists have been published by G. Debien in *Revue de la Faculté d'Ethnologie* (Port au Prince) nos. 10 and 11 (1965, 1966). However, our reading of the documents, which are difficult to decipher, differs substantially from the versions printed in these articles, and also from the summary of them published in G. Debien, *Les esclaves aux Antilles Francaises*, (Basse Terre/Fort de France) 1974, 64-65.
22. A study of twenty other *sucreries* in the region can be found in D. Geggus, 'Les esclaves de la plaine du Nord à la veille de la Revolution Française,' *Revue de la société haitienne d'histoire*, nos. 135, 136,

144 (June and Sept. 1982, Sept. 1984). A conclusion incorporating the Bréda data is published in No. 149 (Dec. 1985).
23. The statement in R. Korngold, *Citizen Toussaint*, (London, 1945), 51, that it had over 1,000 slaves is somewhat exaggerated.
24. A.N., 18 AP 3, letter of 8 avril 1773.
25. One must remember that the continued importation of adolescents and young adults created artificially youthful populations. Such populations are not therefore comparable with populations where the slave trade did not operate.
26. Debien, *Etudes*, 167-9.
27. A.N., 18 AP 3, letters of 27 mars 1775, 28 dec. 1778, 18 avril 1784, 10 avril 1785.
28. A.N. (Cols.), CC9B/23.
29. The point is argued in D. Geggus, 'The slaves of British-occupied Saint Domingue: an analysis of the workforces of 200 absentee plantations, 1796-97,' *Caribbean Studies, 18* (April, 1978), 5-42; and D. Littlefield, *Rice and Slaves: Ethnicity and the Slave Trade in South Carolina*, (Baton Rouge, 1981), chs. 1 and 2.
30. M.L.E. Moreau de Saint-Méry, *Description . . . de Saint-Domingue*, (Paris, 1958), vol. 1, 52-53.
31. A.N., Section d'Outre-Mer, Notariat de Saint-Domingue, registres 293 and 294, plantations Larousselier and Gagnard; A.N., 107 AP 124, dossier 4, Gallifet inventory 1783.
32. A.N., 18 AP 3, letters of 14 juin and 6 oct. 1775, 8 fev. and 9 juin 1777, 18 avril 1784, 10 avril 1785.
33. See above, note 8.
34. A.N. 18 AP 3, letters of 14 juin 1775, 4 fev. 1784. In the 1770s, male and female Aradas sold for 1800 and 1700 *livres* respectively, Congos for 1700 and 1600 *livres*.
35. Moreau de Saint-Méry *Description*, vol. 1, 52-53.
36. R. Bastide, *African Civilisations in the New World*, (New York, 1971), 106. In South Carolina, too, Congos along with Ibos were considered the best linguists: Littlefield, *Rice*, 151, 158-9. Ibo is a 'semi-Bantu' language. Evidence from both Saint Domingue and South Carolina suggests that at least some Congos spoke Portuguese: D. Geggus, 'On the eve of the Haitian Revolution,' in G. Heuman, ed., *Out of the House of Bondage; Runways, Resistance and Marronage, in Africa and the New World*, London, 1986; Littlefield, *Rice*, 132.
37. However, at the same time, the presence of young women who had spent less than five years in a colony also artificially depresses the fertility index.
38. See above, note 22.
39. Correcting for the bias mentioned in note 37, the two indices are respectively 301 and 309.
40. A.N., 107 AP 128, dossier 1, letter of 15 dec. 1785.
41. Geggus, 'Plaine du Nord . . . partie IV.'
42. A.N., 18 AP 3, letters of 7 juillet 1783, 18 avril 1784.
43. Debien, *Etudes*, 167-8.
44. Debien, *Etudes*, 163-173. Though the attorney, as Debien suggests, may have exaggerated the effects of the drought to hide his own in-

Toussaunt Louverture and the slaves

competence, many other sources testify to its unusual severity. See below, note 47

45. The accounts for 1790 list only 24 new additions, but it is evident that Villevaleix forgot to include six children he purchased in May and mentioned in his correspondence.
46. See above, note 24. Moreover, none appear to have been bought during the last five years that Bayon was attorney.
47. Odelucq, attorney of the famous Galliffet plantations, observed it was worse than the drought of 1776: A.N., 107 AP 128, dossier 1, letter of 8 avril 1790. At Haut-du-Cap, the fountain dried up for the first time ever: Debien, *Etudes* 164. Cf. Arch. dept., Nantes, E 691, letter of 27 jan. 1791; Geggus, 'Plaine du Nord ... partie II' 16, and on the Galliffet plantations, 'Plaine du Nord ... partie III' 33-34.
48. A.N., 18 AP 3, letters of Aug. 1773.
49. Letter written to the Paris *Moniteur* in 1799, reprinted in Schoelcher, *Vie* 389.
50. Ibid.

Ethnic Origins: Bréda Plantation, Haut du Cap

	Men	Women	Children	Total
Creole	13	14	26	53
Mulatre		2		2
Unknown			1	1
Congo	16	13	2	31
Mondongue	1	1		2
Bambara	12	1	2	15
Mali	6			6
Aquia		2		2
Quicy	1			1
Mesurade		1	7	8
Miserable			4	4
Thiamba	2			2
Adia	1			1
Foeda	3			3
Arada	2	1		3
Cotocoli	1	1		2
Barba		1		1
Tapa	1			1
Nago	3	5		8
Ara	1	1		2
Dimba	1	4		5
Idra			1	1
	64	45	45	154

David P. Geggus

Ethnic Origins: Bréda Plantation, Plaine du Nord

	Men	Women	Children	Total
Créole	24	24	35	83
Grif			3	3
Congo	8	25		33
Mondogue		5		5
Bambara	4	3		7
Mali	6			6
Timbou	3			3
Mesurade	3	2		5
Miserable	1			1
Soso	2			2
Minha	3			3
Thiamba		1		1
Adia	3	2		5
Foeda		1		1
Fonds		1		1
Arada	6	2		8
Somba	1			1
Barba	3	1		4
Tapa	1	2		3
Nago	13	8		21
Ara	2			2
Ibo	7	5		12
	90	82	38	210

Age Structure Plaine du Nord

	Male	Female
0-4	10	8
5-9	2	8
10-14	6	4
15-19	8	1
20-29	32	43
30-39	17	14
40-49	15	5
50-59	10	12
60-69	5	6
70-79	2	1
80-89	1	
	108	102

Age Structure Haut du Cap

	Male	Female
0-4	5	2
5-9	9	6
10-14	15	8
15-19	10	7
20-19	24	12
30-39	10	9
40-49	5	4
50-59	12	9
60-69	2	4
70-79	1	
	93	61

11
Freedom and Oppression of Slaves in the Eighteenth-Century Caribbean

Arthur L. Stinchcombe

Sociologists have had great trouble developing a sociology of freedom and of its opposite, slavery. Orlando Patterson started with the sociology of slavery (1967) and developed freedom as its opposite (1991). I follow Patterson in starting my investigations in the Caribbean at the height of slave society[1] in the late eighteenth century, before "amelioration" or "emancipation."

But I do not follow Patterson's mature work (1991). He shows how the history of the idea of freedom was shaped by the social and normative experience of its opposite, slavery. I treat freedom or liberty as the high end of a continuous empirical variable in the eighteenth-century Caribbean, a variable whose low end is slavery in the ideal-typical sense. In particular, I study how the restriction of the possibilities among which slaves could choose was greater in some slave islands than in others, and less among slaves serving some functions for their masters that required slave loyalty, enthusiasm, or discretion.

I define freedom as a set of liberties. As the argument develops, it will be clear that many of the decisions slaves in fact took freely were not protected by law. John R. Commons's (1924:92–100;11–46) definition of liberties enables me to conceive of slaves' freedom as a variable made up of the liberties they in fact enjoyed, whether or not they were defended in the law. Because of the way restrictions on slave liberties were defined, low slave freedom means high liberty of the slave owner to do as he or she likes with the slave.

By a liberty, Commons means a decision that someone can take even if the consequences damage or help others, so the decision may mean a loss to one other, but a gain to a third person. For example, Spanish law provided that if slaves of different owners married, one or the other owner had to sell

---
*Direct correspondence to Arthur L. Stinchcombe, Sociology, Northwestern University, Evanston IL 60208. Terry Boswell, Rogers Brubaker, Christopher Jencks, Mindie Lazarus-Black, John Markoff, and Charles Tilly commented on previous versions of this paper. Concentrating the historical information into tables, thus making room for the argument, was suggested by Carol A. Heimer. Much of the historical information appears in an appendix available from the author, which tells how the tables were generated. Roughly one-fourth of the research and writing was supported by Heimer's summer salary, one-fourth by a Guggenheim Foundation fellowship, and half by paid leave from the College of Arts and Sciences of Northwestern University. The support of the Willima and Marion Haas Fund for the map is gratefully acknowledged.

[1] Goveia (1980 [1965]:vii), defined this term in a way slightly different from mine, but the main island she studied, Antigua, was one of the most "slave society" islands in the late eighteenth century by the definition I am using here.

his or her slave so that the marriage could exist. This created a slave legal liberty, but in fact there was often no practical possibility for such unions to be created (e.g., owners tended to import males) and no real way for slaves to call on the law. The legal fact shows that more legal liberties existed at the time in Spanish colonies than in British colonies, but the practical fact meant that on sugar plantations in Cuba (a Spanish colony) slaves were too dominated to marry, while in Trinidad (a Spanish colony that became British with Spanish law in 1800) many could marry.

The reason the social structural answer to the practical existence of liberties disagrees with the legal answer depends on Commons's observation that a liberty creates an *exposure* of others to the different consequences of different choices by the free person. When the person exposed to the consequences likes them (or can contract to get the right decision for a price), all goes well. But when a practical slave liberty damages masters, laws may be tightened, informal sanctions within the practical liberties of slave owners may be brought to bear, or male African slaves only may be imported.

Freedom, then, is here defined as a latent, usually unobservable, conceptual variable describing the sum of practical liberties of a slave life, decisions that slaves could in fact take, rather than the sum of legally defended slave liberties (which were very minimal indeed). Every practical existence of a slave liberty was a limitation on the liberty of the slave owner to do as he or she liked with the slave. We can look for the indicators of freedom defined in this way, its legal and social causes, interpret the motives of slaves to seek more freedom by means other than laws, and perhaps ultimately reconstruct the life experience of the difference between legal slavery and legal freedom when slaves were manumitted or emancipated. The definition then is a sum of practically available liberties, including, in particular, the social capacity to get others to suffer the consequences of the practical freedom of slaves to decide.

BACKGROUND

The conception of slavery as a dichotomous legal status represented in laws, and contrasted to a status of freedom, is of course irrelevant to my purpose of describing and explaining variations among slave islands and among slaves within islands. But most comparative work on slavery (e.g., Tannenbaum 1946; Klein 1967; Goveia [1965] 1980; or Patterson 1991) treats the *elements* of slavery as dichotomies in the law, a list of various things that are permitted or forbidden to slaves that are permitted or forbidden to free men (or to proletarians, serfs, subfeudatories, freeholders, nobles, free women, or other contrast groups). The dichotomy between slave and free is then constituted by its subdichotomies of legal freedoms or constraints, and slave systems can then be compared by the components that constitute their overall contrast between slave and free.

As Patterson pointed out in his *magnum opus* (1991), such a pattern of contrasts between slaves (of various kinds) and free people (of various kinds) constitutes a particular society's definition of freedom; it is what all nonslaves hold in common. A particularly crucial aspect of freedom so conceived is the right to call upon courts or other authorities, more or less separate from one's owner or superior, to defend one's rights, or to defend one's practical freedom by using other more or less legal liberties, such as emigration, rebellion, or the right to duel.

The dispute between Tannenbaum (1946) and Klein (1967) on the one hand and Moreno Fraginals ([1964] 1976) on the other, for example, provides a contrast between such conceptions of how to analyze slavery, legal dichotomies versus daily practice. The central question Moreno Fraginals asks is how far variations in legal rights between Spanish and English colonies (here especially Cuba and Virginia) influenced the realization of slavery found on the Cuban *ingenios* (in Spanish the word for plantations refers to the sugar mill rather than the planted fields; in both Spanish and English colonies, sugar plantations had both) so as to make it different from that in Virginia. Moreno Fraginals argued that the probability of concrete oppression is better predicted by the demands of sugar plantations and the drive for cheap labor through legal and illegal coercion of slaves than by differences in legal freedoms or in the possibility of appeal to the church authorities. In contrast, Williams ([1944] 1964:6–7) argued that free labor was cheaper if already there because it was more productive—but it would be more expensive to move, and this would imply that coercion of the slaves was inefficient once they were in the Caribbean. Broadly, my approach agrees with that of Moreno Fraginals as against Klein and the others, that economic dominance of sugar and planter power determine the oppressiveness of slavery. But I argue that this gives a different prediction for late eighteenth-century Cuba as a whole than he found on the nineteenth-century *ingenios* (these had less than half of Cuban slaves in the nineteenth century, and even less in the late eighteenth century).

My conception of the degree of freedom of slaves is *the inverse of the probability of coercive limitation in daily life* of many rights *less often interfered with* among free men. Here differences in the probability of coercive limitation of rights in everyday life define the difference between slave and free, and those probability differences can be large or small in different societies. Free women of the eighteenth century, by this definition, were less free than free men, but freer than slave men or women. That is, free women could make more decisions about daily life without coercive interference than could slaves.

One main argument here is that the degree to which law and political authority ferreted out incipient slave liberties or patches of freedom and relentlessly invented laws to suppress them, was itself shaped by the determinants of planter power. Thus, in some sense, the effectiveness of the law in depriving slaves of liberty on a daily basis was another effect of the same cause as frequent intervention by the planter in slave daily life within that law. This is because slave law was a dynamic achievement of planter power, just as the concrete elimination of choice and appeal to the courts or police for slaves was. Sugar plantations of long duration on more self-governing islands caused both legal and daily life dynamics and so slaves were less free on sugar islands.

To be more explicit about the implications of this for the empirical argument, the dependent variable explained in Tables 1 and 2 that follow is not explicitly measured. As explained in more detail in the theory here and in an appendix on the Constitution of the Data, the dependent variable is an estimate of how many possible liberties of slaves the government of an island devotes itself to limiting. (The appendix is not shown here, but is available on request from the author—hereafter it is referred to as "Author's Appendix.") The basic thing being protected by such a government is the liberty of the planter to do whatever he or she wants with a slave. The ingenuity devoted to making governmental provisions against all liberties of slaves that might interfere with such liberties of planters is the implicit explicandum of the tables.

A melange of observations went into my judgment of how much island colonial governments limited slaves' freedom (which, to anticipate, is the unobserved dependent variable in Table 2): interference with slave marriage, not allowing slaves provision grounds, making manumission (the release of a slave from bondage by the owner) difficult, forbidding slaves to sell things, requiring slaves off the plantation to carry a permission from the master, forbidding missionaries from converting slaves, regulating singing, prohibiting magic, prohibiting sleeping in houses of nonslave relatives, prohibiting the naming of slave children after their White father—in a deep slave society the list goes on and on. But I could only form an impressionistic diagnosis of how active the government was in devoting itself to such regulation. Data on any one of the items on this operationalization of the degree to which an island had a slave society government was quite likely to be missing for half or more of the islands.

Deliberate institutional action to restrict choices varied among island governments. For example, the Spanish colonial governments of Santo Domingo (the Domincan Republic) and Puerto Rico and the Dutch government of Curaçao spent almost no effort to make sure slaves had no choices, no liberties, and made some effort to restrict the liberties of planters. The government of Barbados, in contrast, did little else but make sure that slaves could choose almost nothing, and that planters could therefore choose all aspects of slaves' lives. I will call Barbados in the eighteenth century "more of a slave society" than Puerto Rico, Santo Domingo, and Curaçao, because although all had slaves, the latter did not spend much governmental effort making sure they could not choose anything. Thus the first dimension I use to discuss the "freedom" variable *among slaves* (ranging from slave to free) is the degree to which their island government devoted itself to their "unfreedom." The best measure of freedoms granted by individual masters—whether or not those masters granted legal manumission (termination of bondage)—is not at all a good measure of this inter-island variation. (For reasons, see the Author's Appendix.)

But even in Barbados, some slave owners gave some of their slaves their freedom. The enthusiasm of slaves for being free rather than slave, even though there were many restrictions on the possibilities of the "free colored" Barbadians,[2] shows that they thought free was better than slave. And there is every sort of evidence that slaves in the categories most often freed (e.g., domestic servants, soldiers, skilled workers, mistresses) were treated more like free people, even before they were freed. The owner, as well as the society, could restrict or expand the possibilities among which slaves could choose, and of course the owner's free discretion of which slaves should have more liberties was rarely legally specified or recorded. And the way they expanded them in daily life in Barbados, or failed to restrict them in Curaçao, formed regular social patterns that can be explained.

This informal system of slave owners providing some slaves rights to choose, even ultimately sometimes formal legal freedom, is thus another form of variation from slave to free among formal slaves. In the worst case, an owner's slave mistress in Spanish colonies could be tortured. A wonderful description of an awful case in Trinidad after 1800 that had British authorities applying Spanish law, is Naipaul's "Apply the Torture" ([1969] 1984: 182–221). But the whole context of the case shows that this slave, a White man's mistress, had a great deal of effective freedom. In fact, owners' mistresses and their children tended to end up free rather than tortured, and moderately often ended up rich (e.g., Cauna 1987:53–56). The extreme case for labor was for the slave to work under the whip to the limits of endurance in holing for the young sugar plant. In fact, some slaves in Spanish colonies (in the eighteenth century the main ones were Trinidad, Cuba, Santo Domingo, and Puerto Rico) were paid wages (Boin and Serrule Ramúe [1979] 1985:61, 63). The right to inherit freedom from one's slave mother and part of the estate from one's planter father was surely a step toward freedom for a person born a slave, and the right to spend one's own wages is generally taken as a test of free labor. So the worst case under the law was sometimes not the average case.

Holt's (1992) study of the emancipation of slaves in Jamaica has an intellectual strategy comparable to mine here. He showed that the political process of emancipation restricted the choices of ex-slaves so they would work for planters "freely" for low wages. In part, this required taking away the rights to houses they had built and subsistence plots for which they had broken ground in their "free time" as slaves, so that they would have to earn them back by working on the plantation. But there was no agitation among Blacks to recreate slavery so they could claim houses and subsistence plots. That is, Holt studied one island as it changed from a moderately oppressive slave society to a "free society ruled by a planter legislature"; Holt's main point was that this latter society was not very free.

I have chosen instead to study cross-island variation between about 1750 and 1790. The intensity of government concern to preserve slaves' lack of freedom varied from one island to another. In many ways, Blacks on Dutch Curaçao or Aruba were freer under slavery in the mid-nineteenth century than were the emancipated slaves in Jamaica that Holt studied. A commercial aristocracy did

---

[2] The term "colored" was generally used in the sources for this paper to designate persons of mixed African and European ancestry. Particularly when referring to slaves, "colored" meant "of mixed ancestry." However, the term "free colored" was used to designate any former slave freed during a period when there was still slavery, whether the slave was of mixed or exclusively African ancestry. As I explain in the argument connected to Table 3, there was a correlation between color and freedom, with those having European ancestry being more likely to be manumitted—but that correlation was much less than 1.0. I cannot be more precise in distinguishing ancestry and condition of servitude than the sources I use. It is a separate question, not analyzed here, why Whites made little social distinction by color among free people with African ancestry.

not need gang labor in the fields—it needed agents to help them run their businesses and their homes. They did not devote their government to depriving slaves of freedom, nor did they devote their daily personal dealings with their slaves to restricting slaves' choices.

By specifying the causes of variations among islands in the degree to which planter power could create a slave society, I describe one main force that restricted the choices of slaves. By specifying when masters found it to their advantage to leave some liberties in the hands of individual slaves, sometimes manumitting them (freeing them from bondage)—that is, by specifying when slave owners, though legally permitted, did not push slave lack of freedom to its utmost—I describe informal transactions that increased the possible choices available to slaves and explain the variation within islands in the degree of freedom that slaves experienced.

I argue that this tack toward understanding variations in freedom *within* islands, as well as *between* them, helps avoid defining freedom, or slavery, by its essence. Defining things by their essences is always troublesome in an explanatory science. So defining slavery by its uttermost extremity, by the fact, for example, that rape of slaves by their owners usually could not be punished, does not explain why mistresses of White men were disproportionately colored, were sometimes given their freedom, and sometimes inherited part of their lover's estate.

Nor does the extremity of hard work under the lash in holing a field to plant sugar cane explain why skilled workers on plantations or dock workers in towns were more often given their freedom, more often made contracts for their services with people other than their owners, and more often rented houses from urban landlords and bought their own food than did slaves on sugar plantations. Worst case scenarios tell us whether we are in a slave society, perhaps, but they do not tell us about the expansion and contraction of the space of choice in the lives of individual slaves. Such scenarios may be good guides to the macrosociology of freedom, to whether there has been governmental care on a given island to make sure that slave owners are not forbidden to push their liberties with slaves to the extreme, but they are not a good guide to the informal part of the sociology of slavery and freedom.

I define the degree to which an island was a "slave society" as the degree to which the island government devoted itself exclusively to making the liberties of the planters in their property unlimited, and had the powers to do a good job of that. This, then, is the first determinant of how oppressed a slave was. I argue that the main determinants of the degree to which an island limited slave liberties (to maximize owners' liberties to use slaves as they pleased) were: the degree to which an island was a sugar island (sugar planters were the largest and most demanding users of slave labor); the degree of social and political organization of planters (better organized planters could better build the island society around oppressive sugar slavery); and the political place of the planters in island government and of the island government in the empire (the more powers local planter government had, the less limited it was in building a slave society).

Conceiving of planter institutional power as the institutionalization of a planter liberty over his or her property means that the higher slave owner power is, the more the owner can treat a slave any way he or she pleases.[3] We now are inclined to moral judgment of the slave system by what was the worst that could happen to the slaves, and rightly so. But that was not the way a slave had to look at it in order to try to live a decent life within that system. In particular, planter owners could supervise slaves closely in gang labor in the fields and make no promises, or they could negotiate contracts with their slaves, or even set them free—and which the owner chose mattered a lot to the slave.

The very thing that made slave owners powerful—the existence of a slave society—made what they wanted to do the main determinant of what happened to the slave. If we study who it was that the planters set free as the extreme manifestation of owner liberty, we find systematic and powerful patterns in how much the "deals" slave owners made with their slaves resembled those they made with free people, including

---

[3] Women treated their slaves (mostly domestic slaves) much differently than men treated their slaves (mostly field hands).

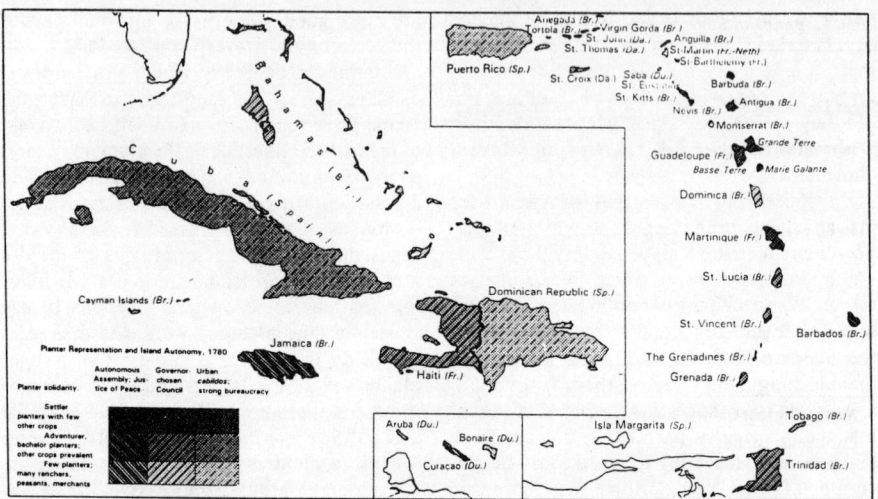

Figure 1. The Caribbean Islands in the Late Eighteenth Century: Planter Representation and Island Autonomy

Note: Dominance of sugar planters is indicated by the shading, and empire on the island toward the end of the century by abbreviations. Trinidad's government was Spanish, its planters mostly French, and by 1800 it became British with Spanish law still valid; I have chosen to call it British on the map and have shaded it according to the planter dominance in the English period. Isla Margarita did not have a separate island government. The map was prepared by Prof. John C. Hudson, Northwestern University.

"deals" that set the slaves free, that extinguished the relevant slave owners' liberties. Slave societies probably varied in their manumission rates, but (as explained in the Author's Appendix) available data make manumission rates very difficult to determine. But the will of an individual slave owner on an island with a given intensity of slave society determined which slaves on that island would be freer; differential manumission rates of groups within islands can indicate that.

The core bourgeois liberty is the freedom to alienate property, to truck, barter, and exchange. The distinctive thing about slaves as a type of property is that one can alienate them to themselves, can give them the liberties to decide what to do. Hence manumission, the individual granting of freedom, is a sensitive tracer of which slaves were most treated as free people.

My argument is that the central determinant of within-island variation among owners in treating slaves more like they were free was the owner wanting the slave to be a responsible agent in unsupervised services or work, work involving care or enthusiasm or risk to the worker, or work requiring loyalty that could be easily betrayed. Thus, when the slave owner wanted trustworthy agency by slaves, he or she treated them as if they were free, as if they had rights, and in the extreme gave them rights.

EXPLAINING INTER-ISLAND VARIATION IN SLAVERY: SUGAR CANE, PLANTER DOMINANCE, AND ISLAND AUTONOMY

In Tables 1 and 2, I apply the core variables of my analysis of the degree to which an island was a slave society to most of the Caribbean islands. The units of analysis in Table 1 are "islands," in the geographical and economic sense, unified bounded economic and geographical systems in the eighteenth century. (See Figure 1.) As explained in the Author's Appendix, what are conventionally called islands are in fact usually "clouds" of islands unified by geography, economy, and history. The units of analysis in Table 2 are political units treated as separate by their imperial governments. They were often unified on a larger scale than the islands of Table 1,

SLAVERY IN THE EIGHTEENTH-CENTURY CARIBBEAN

Table 1. Period of Sugar Frontier by Degree of Dominance of Sugar, for Caribbean Islands

Sugar Dominance[b]	Period of Frontier[a]		
	Before 1750	1750–1800	After 1800
80 percent or more	Barbados Antigua Martinique	St. Croix Guadeloupe	Tobago
50 to 80 percent	St. Kitts	Jamaica[c] Haiti Grenada Tortola	Trinidad
50 percent or less	Nevis	St. Vincent St. Lucia	Cuba Puerto Rico Santo Domingo

Notes: Highest Planter Economic and Social Dominance is in upper left corner.

Caribbean Islands where sugar cultivation was never important are: British in 1800s—Caymans, Bahamas, Dominica, Montserrat, Anegada, Barbuda; Dutch in 1800s—Saba, Curaçao, Aruba, St.Eustatius; Spanish and Venezuelan in 1800s—Isla Margarita; Danish in 1800s—St. Thomas, St. Johns; Swedish and French in 1800s—St. Barthélemy.

[a] This is an estimate of the period of highest influx of sugar planters and slaves and the highest rate of bringing new land under sugar cultivation. Where it was available, I have used the time at which the slave population was half of its stabilized value after sugar had filled its niche. For reasons discussed in the Author's Appendix, this ideal indicator was often not available, especially for the islands in the lower right of the table. Substitutes I used are discussed in the Author's Appendix.

[b] This is an estimate of the percentage of the labor force occupied by sugar workers and planters when the sugar labor force stabilized. Good data on this were less often available than early data on slave numbers. The substitutes I used where necessary were export data, agricultural land use data, slave/White ratios after stabilization of slave populations, or comments by travelers or residents. A low percentage indicates that there were other major agricultural crops, tree crops, or livestock, that there were relatively large urban populations, or that nothing much would grow on the island.

[c] Jamaica had considerable coffee cultivation and livestock raising, and a major entrepôt function, and may be misclassified.

especially when the core islands of the clouds of islands were physically smaller, less economically important, and had higher interisland distances. (This is also discussed in the Author's Appendix.)

Table 1 shows the differences in economic history of the islands that determined the degree of economic dominance of sugar planters and the timing of that dominance. It presents two causal variables; the degree of dominance of sugar in the island economy, and how long that dominance had lasted by the time I take my reading in the late eighteenth century. The dependent variable is implicit, and (as discussed above and in the Author's Appendix.) consists of the degree to which island governments devote themselves to interfering with many slave liberties in order to maximize the liberty of the slave owner to do what he or she wants with the slave. In Table 2 below, the results of Table 1 are combined with information on the form of island government to evaluate their joint effect on the dependent variable.

The length of time planters had been dominant in island government is estimated by estimating the "birth date" of planter economic dominance, which I call the "peak of the frontier period." When an island's land suitable for sugar had been devoted to other crops, converting it to sugar required an increase of the labor force of about five to ten times; this created a "frontier" period of very rapid immigration, combined with the rapid change of land from other crops or jungle to sugar. This frontier period could last from 40 years (e.g., Barbados) to 200 years (e.g., Cuba), depending on island size and heterogeneity, governmental restriction of development, and so on.

By the time half of the land to be devoted to sugar had been developed on islands that became quite completely sugar islands (and also occupied half of the labor force it would occupy at the end of the frontier period), about two-thirds of the population would be devoted to sugar, and around half of all sugar plantations would be run by people whose pioneering work was done and who thenceforth would be managing family estates. On islands that would end up having only half or less of their labor force devoted to sugar, such as Cuba, at the peak of the sugar frontier, only about one-quarter or less of the economic power of the island would be expected to be in the hands of planters who were managing family estates and were finished with their frontier pioneer work. During an island's frontier period, many of the planters were bachelor pioneers, many of the slaves were African, up to two-thirds of all slaves were adult males, fortunes were being made because the capital value of the plantation was being created, and in other ways an island showed demographic and economic signs of a rapid influx of people and resources characteristic of frontiers.

Thus, if the peak of the frontier, the time of most rapid influx and most rapid transition of land to sugar production, came before 1750, by the late eighteenth century about half of all the planters had been managing family estates instead of being pioneers for about two generations and could devote themselves to developing the political and social institutions of a slave society. Fewer of the rich and powerful would be sugar-family-estate planters if sugar would never occupy most of the island or if the last half of the frontier were just being settled and family sugar estates were newly stabilized. So sugar planter social and economic dominance in the late eighteenth century should be greatest in the upper left of Table 1 and weakest in the lower right. My prediction, borne out by the informal data (described in the Author's Appendix), is that islands in the upper left of Table 1 would be more likely to have intense slave societies by the late eighteenth century, and those in the lower right would have less intense slave societies at the level of the island as a whole.

Table 2 combines the results of Table 1 (the stub runs from strong sugar dominance at the top to weak sugar dominance at the bottom, corresponding to the diagonal dimension from upper left to lower right in Table 1) with the type of local government granted to or imposed on the island by its empire in the late eighteenth century (in the heading). My final prediction is that the island units in the upper left showed signs of being "slave societies" in the late eighteenth century, and those in the lower right had governments that devoted less effort to limiting slave freedom.

The Spanish islands (which, in order of importance to Spain, were: Cuba, Trinidad, Puerto Rico, Santo Domingo, and Isla Margarita) appear in the lower right of Table 1 or in the table note, and so appear near the bottom of the stub in Table 2. In 1750, for example, the entire sugar production of Cuba was about equal to that of the small Danish island of St. Croix (MacNeill 1985:126). Trinidad, which until 1800 was a Spanish island with an infusion of French planters, also would appear in the lower right cell if it hadn't been for vigorous British development of sugar planting there after 1800.

In the French empire, Guadeloupe and Martinique were developed earlier than Haiti. Haiti was at its maximum growth rate around 1750 or 1775, so that its planters in the late eighteenth century were much more often bachelors out trying to get rich, and its slaves were mostly born in Africa or were first-generation creoles (i.e., born in the Americas of African parents).

In the British islands, Barbados was the first island to be developed in sugar and was dominantly a sugar island, and the Leewards were developed soon after (the Author's Appendix indicates which islands in Table 1 are grouped under the British Leewards). British Jamaica had a lot of broken terrain good for raising cattle, tobacco, coffee, or other "non-plantation" crops, but much sugar land there was developed in the middle period, before and after 1750. Jamaica was declining as a sugar plantation island by the end of the eighteenth century. Trinidad, and islands that became the British Windwards[4] (Grenada, St.

---

[4] "Windward" islands are the small islands to the east and south, conventionally starting with Guadeloupe in the northwest, but excluding Trinidad and Barbados. Around 1780, the British Windards included Tobago, Grenada, St. Vincent, St. Lucia, and Dominica. Why I classify them together here is explained in the Author's Appendix. The Appendix also describes the British Lee-

Table 2. Factors Leading to High Planter Power: Planter Economic Dominance and Island Autonomy around 1780

Planter Social and Economic Dominance[b]	Planter Representation and Island Autonomy[a]		
	(1) Autonomous Assembly, Justice of Peace	(2) Governor-Chosen Council	(3) Urban Cabildos, Strong Bureaucracy
Settler sugar planters with few other crops	Barbados	Martinique Guadeloupe Br. Leewards	
Bachelor adventurer-planters or other crops prevalent	Jamaica Surinam	Br. Windwards Haiti Guyana St. Croix	Trinidad[c]
Few planters, many ranchers, peasants, merchants	Curaçao	Dominica Bahamas St. Thomas	Puerto Rico Santo Domingo Cuba[c]

Note: Planter power is highest in the upper left corner.

[a] Autonomy and control over administration of the law led to high planter power (column 1), urban representation and strong bureaucracy led to low planter power (column 3). The classification is impressionistic, and I have considered factors not mentioned in the table showing high island power in empire policy as applied to the island. (See the Author's Appendix.)

[b] When there were fewer planters and when they were birds of passage developing a frontier and did not form local families to use power consistently (when they were "bachelor adventurer planters") then planter power was low. If settler planters dominated the economy on the islands where they had great organizing capacity developed over historical time, they had greater power. See the discussion of Table 1, which is collapsed diagonally to produce the stub of this table. (See also the Author's Appendix.)

[c] Cuba, taken as a whole, was never dominated by sugar cultivation, and Trinidad was not dominated by sugar in 1780, but was being settled rapidly by French planters under a Spanish government. Both had politically powerful sections dominated by sugar in the nineteenth century, and most of the literature on slavery on those islands deals with that period. Cuba never had as much as 50 percent of the labor force in sugar. In the late eighteenth century, Cuba and Trinidad were not as different as indicated by their different placement in the table.

Lucia, St. Vincent, and Dominica), were developed later, as was Guyana, which was economically and socially an island, although not geographically so. Much of the development of these colonies took place after England abolished slavery and so much of the development of new sugar land was done by indentured East Indians rather than by slaves. But looking at these islands in the late eighteenth century, planters were not yet in a position to develop a thoroughgoing slave society, even if England had been in a mood to let them.

ward islands; the core ones are Antigua, St. Kitts, and Nevis. Exactly which islands were included in these groups varied historically for the reasons discussed there, and the groupings were different in different languages.

But planters could organize their economic power into class power if they had extensive ties with each other, had much time to shape institutions to their liking, had established households and were looking to the long-term health of their class and its wealth. The extensive apparatus of slave society that was imported into South Carolina from Barbados (Jordan 1969:84–85), or that is so beautifully documented in the legal studies of Goveia ([1965] 1980) on Antigua and the other British Leeward Islands, was not quite as developed in Jamaica and was much weaker in Trinidad, Grenada, or Guyana. Thus the result of Table 1 (summarized in the vertical dimension of Table 2) is that the island governments in the top row should be more dominated by sugar planters, and thus more devoted to restricting slave liberties.

Table 2, then, combines into one dimension the two sources of planter domination of the economy and society of each island in Table 1, and adds a dimension that measures the powers granted to local legislatures by the empire. Local authorities, dominated by planters when there were many of them, were more powerful when they had an assembly to which they were elected (column 1), rather than a cabinet of the governor to which the governor appointed the local rich (column 2), and these were more powerful than urban *cabildos* with small powers dominated by a strong Spanish bureaucracy (column 3). The same amount of planter dominance (in the stub of Table 2) translated into more governmental power (column 1) than occurred in column 2, in which a governor might consult with planters of his own choice, and into least power in column 3. In general in the English colonies with powerful legislatures, the main agents in small localities were volunteer "gentry" justices of the peace, so implementation of laws was in the hands of planters.

At the opposite extreme were the Spanish colonies (again Cuba, Trinidad, Santo Domingo, Puerto Rico, and Isla Margarita), where the local councils were the cabildos of cities, where most legislation governing the colonies was not passed by such cabildos, but instead by the Council of the Indies in Spain, and where the implementation of all laws was in the hands of civil servants, "peninsulares," sent out from Spain. Planters had to apply to the cabildo for permission to turn their cattle ranches into sugar plantations (Riverend 1972:111–12, 119–20; Marrero 1978:15).

Thus, in the upper left of Table 2 are the islands where both demography and the structure of local government in the empire maximized planter power. Barbados was the high point of planter power and had the fullest development of slave institutions, the greatest devotion to limiting slaves' liberties (and free colored liberties as well), and an inclination to defy the colonial office soberly and effectively, claiming imperial power for its own. Jamaica, Surinam, the British Leewards, Martinique, and Guadeloupe were close competitors with Barbados.

The lower right of the table is dominated by the Spanish islands that Klein (1967) used to illustrate the relative softness of Caribbean slavery,[5] the entrepôt islands of the Danes and the Dutch, and many miscellaneous small non-sugar islands that did not have much autonomy and that I left out of Table 2 as unimportant.

Slave societies, then, were created when the dominant people were those to whom slavery in its most extreme form was desirable; in the eighteenth-century Caribbean these were sugar planters. Three main factors made them dominant: (1) sugar as a large proportion of the economy, (2) a planter aristocracy with a solidary style of life in which managers of family estates had an interest in slave institutions, (3) and empires that let planters run island government. These served as multipliers of slave institutions, making them more elaborately oppressive. On the other hand, on the Spanish islands and Dutch islands (only Dutch Curaçao and the Dutch "island colony on the coast," Surinam, are important enough to mention in Table 2) there were very few records of and regulations about manumissions, but very many free colored and free Black people. Most records of free Blacks and free colored people on Spanish islands are apparently based on censuses that asked them whether or not they were free. This was very much a nonslave-society way of finding out who was free, and indicates a low level of government interest in pushing slavery to its extreme.

---

[5] Klein (1967) compared Virginia, dominated by tobacco rather than cotton and so one of the softest slave regimes in North America (but with well-organized planters), with Cuba, where the region around Havana and Santiago de Cuba had some of the tough slavery of resident planters in the sugar islands, and the other regions had the very soft slavery of peasant farming with little rural access to the levers of power in the empire. Knight (1970) tried to refute Klein, but he looked only at the small sugar part of the only serious Spanish sugar island and at the internal system within the plantations rather than at planter success in instituting governmental limitations on the options of blacks and free colored people. None of the places compared in this literature resembled Barbados, Antigua, or South Carolina in the degree to which they were slave societies. The political situation of Cuban planters was changing very rapidly in the late eighteenth century (Kuethe 1986). Scott (1985) offers a good overview of this whole debate.

SLAVERY IN THE EIGHTEENTH-CENTURY CARIBBEAN

EXPLAINING WITHIN-ISLAND VARIATIONS IN SLAVERY: SLAVES AS AGENTS

The liberty of planters to deal with slaves as they liked meant that slave owners could make whatever deals they liked with the slaves. Often owners used such liberty to make contracts that look like those made with free people, except that one of the rewards was sometimes manumission. Manumission was in some sense often a "career" reward, the last promotion for a faithful and loyal slave. Like many such rewards in bureaucratic organizations, one does not know whether one gets the final reward until near the end of one's career. One should then expect to find manumission in the same sorts of places in the economy we find bureaucratic promotions and generous pension schemes in modern society—where long and skilled service showing loyalty, discretion, and good faith were required by the economic task.

Other features of the agency contract of modern civil law appear in the lives of some slaves. Contracts in which the principal (the person who "hired" the slave, in this case the slave owner) monitored the outcomes of slave (agent) behavior rather than supervising the behavior closely, granted much discretion to agents, rewarded the agent in proportion to outcomes, and delayed agent reward until the overall results were in, were rather like those between a house owner and a real estate agent in modern society. According to modern agency theory, in situations where the agent has more information and more control over effort, enthusiasm, and intelligence than does the principal, such contracts are thought to achieve the principal's purposes better than would close supervision. I argue that treating slaves as almost free, and sometimes eventually as legally free, in the eighteenth-century Caribbean was usually an agency contract. Such contracts solve the problem of trust between slave and master better than coercion does. Except in the case of sexual relations, such agency contracts reduce supervision costs. In the extreme, when the slave owner would have had to be on the sea bottom watching the slave collect pearls off Isla Margarita, the cost of supervision would exceed the total value of the slave's labor. At the other extreme, supervising cane holing with a whip is cheap and effective.

Coercion, Norms, and Social Ties in the Formation of Race

Coercion was central to creating the slave population of the Caribbean and determining its racial composition. It was because coercion could be and was applied by White Europeans to Black populations in west Africa, and could not be or was not applied as intensively in Europe, that the labor demand in the Caribbean was translated into an African slave population. Coercion, rather than reward, dominated labor relations in the Caribbean, especially in the core of the slave system, the sugar plantation.

The totality of the definition of the coercive relation was greatest in the islands in the upper left of Table 2, least in the lower right. But people define the meaning of such larger coercive and normative structures, even like those in the upper left, in the course of daily activity. What owners wanted out of slaves depended on the activities they were trying to carry out by means of the slaves. The sexual tie was probably the most important one modifying slavery in the direction of freedom in the late eighteenth century. A number of other relations between powerful Whites and slaves modified the use of coercion and the use of class-conscious planter normative definitions in daily life.

Unfortunately, negotiations between slave and master rarely appear in the historical record. Slaves had no right to appeal in court, had few or no property rights defended by the courts, could not sign legally enforceable contracts, did not pay taxes, were maintained illiterate by social policy, and were not regarded as objects of religious institutions that kept records. So the problem here is quite different from finding out about variations in government action between islands, as I have done above; here I want to distinguish among individual slaves within an island society according to their ties with masters.

Five main conditions generated records bearing on the daily lives of slaves and slave-master relationships. One was manumission, the establishment of a former slave as free by a governmental act initiated by the owner. Documents about the conditions of such manumissions often tell something about the relations between slaves and masters under various conditions, although slave societies generated more manumission documents per free Black person than did societies whose central institution was not slavery. (This is discussed further in the Author's Appendix).

A second source was records of emancipation, the proposal by governments to treat slaves as free people whose rights needed to be documented to be defended. Closely related was the abolition of the slave trade, which created the category of illegally imported (and therefore legally free) slaves, who had to be distinguished from legitimate slaves. The documents telling which categories of the emancipation law which slaves fell under often tell something about the relations of slave to free.

A third source of records about slave-master relations was plantation accounting books and other plantation books of well-run plantations or other slave enterprises. The value of a slave depended in part on the nature of his or her activity, and so activities and special arrangements were recorded.

Fourth, governments had military or political reasons to treat some slaves (or former slaves) differently than others, especially if they had had military training and experience, had been to the mother polity and so had a claim to freedom, belonged to powerful maroon (runaway rebel) groups in the interior, or otherwise had a distinctive relation to the coercive or normative system defining slavery. A troop of Black soldiers obviously presented a different coercive problem than a gang of fieldhands.[6]

Finally, some churches administered some religious activities that bore on the daily lives of slaves, especially on their marriages, births, and deaths.

All of these sources are irregularly available. Religious records of marriages and baptisms are much more available in the Catholic empires (Spanish and French) than in the Protestant ones (English, Dutch, and Danish). Records generated by the enforcement of the abolition of the slave trade are primarily available for the English islands, forming, for example, the basis of Higman's marvelous demographic analysis of slavery in the early nineteenth century (1984), because only there was the imperial government really behind abolition of the trade.

In what follows, I develop a thesis about how the need for slave agency increased the freedom of slaves. I examine the five kinds of data above, especially looking at what groups of slaves were most likely to be manumitted. My thesis, by necessity, draws also on my speculations about the facts underlying these patterns; one might say that I follow the "theoretical method," discredited in the discipline of history, in which "theory" involves, in part, guessing at the facts. I go now to the theory, presented primarily as a theory of manumission rates.

Slaves had to form a relationship with owners or other powerful people in order to be freed. It was of course simplest to form a relationship with their own owners, and to persuade them (see note 3) to give them freedom, either as a gift or by testament on the owner's death. Sometimes ties to free people other than masters could become indirect ties to their owners, as when another free person bought slaves for the purpose of freeing them. Relationships to White employees of the owner, for example, fairly often resulted in freedom, with the employee buying the slave or being given the slave by the owner in appreciation for long service (Cauna 1987:134–35).

To understand why slave liberties might depend on the sort of tie between owner and slave, I must explain how planters' ties to colored and Black creole and African slaves varied: Creole slaves were freer and were more often manumitted. Ties also varied between small and large slave holdings: Slaves were more often freed on small holdings. Ties varied between city and plantation: Urban slave owners freed slaves more often. Ties varied between colonies in which sugar planting was rapidly expanding and older colonies where sugar had filled its niche: Older colonies had denser ties but more developed slave societies. Ties varied over time with the political situation: The French revolution and the abolition campaign in England, for example, substantially increased manumissions. Ties varied between

---

[6] See Geggus (1982:315–25) for details on how the British thought about Black and colored troops in Haiti during the British occupation.

empires: English planters were the least liberal in giving liberties to their slaves (even though they were least constrained by their home governments); the French were somewhat more likely to free slaves and to treat them as they treated free people; the Dutch and Danish were more liberal yet, though slave owners' liberties were very little restricted; and the Spanish (to exaggerate) used slavery mainly to recruit immigrants, and after immigration slaves were often informally freed and managed by those means the Spanish colonial government and powerful people used to manage "free" labor.

The evidence I examined showed that these are the variables that shaped rates of manumission, so they must have been the variables that determined the sorts of ties slave owners had with slaves. By extension, these variables must have shaped ties slave owners had with people they had just freed, and so determined the meaning in daily life of the boundary between slave and free.

Besides the powers of property, there were also powers of governments on the islands; ties of slaves to those powers could also result in freedom. When slaves rendered military service, especially during slave revolts, or against maroons or foreign invaders, the government often freed them for their loyal service and reimbursed their owners.

Both France and England in the seventeenth century had explicit arrangements that slave ownership could not be enforced in the mother polity. Thus, slaves automatically became free if they got to Europe. But, at least in France, these arrangements were substantially modified in practice over time so that slaves could be brought into the mother polity under various special dispensations that preserved their slave status while in France (Peytraud 1897:373–400).

Sometimes treaties with rebels in the colonies (either White rebels with slave recruits, or slave or maroon rebellions) granted freedom. Quite often owners were not reimbursed; the presumption must have been that if it required great state expense and activity to enforce slave ownership, reimbursement was not an obligation of the state.

The general point is that legal freeing of slaves required slave access to power, either the power of the owner or the power of the government. The power of property in slave society was particularly oppressive, but that oppressiveness gave property owners great discretion to define what ownership meant for particular slaves. No contract or law guaranteed equality of treatment, so some could be freed and others kept slave without violating property rights.

To understand what the boundary between slave and free meant socially, then, I interpret the data on manumission in terms of what sorts of ties could produce freedom. Manumission is the extreme form of stratification created by planter will within the slave community, which made some slaves able to make claims on (or against) White power holders and which left freed people who had "only barely" been freed unable to claim the full rights of citizenship (insofar as there were rights so universally available that they could be called "citizenship").

Four Forms of the Slave-Master Agency Tie

Slaves had four main ways to form ties with White people that might result in freedom: sexual and other intimate ties, agency in cooperative work (domestic work and and management), commerce, and politics.

Sexual and other intimate ties. Sexual ties between slave and free were mostly between White men and Black or colored slave women, especially young, creole, colored, domestic servants (of course, slaves might become domestic servants because of sexual selection, rather than be sexually selected because they were domestic servants). Peytraud (1897) quoted a letter from two island authorities about the ties between White male lovers and their Black or colored mistresses:

> If we did not take care to stop the manumission of slaves, there would be four times as many as there are, for here there is such great familiarity and liberty between masters and negresses, who are well formed, which results in a great quantity of mulattos, and the most usual recompense for their obliging compliance to the wishes of the masters is the promise of liberty which is so gratifying that, together with their sensuousness, the negresses determine to do everything their masters wish. (Peytraud 1897: 409)

Of course this causal analysis is a "Just So Story," because it does not explain what a White male needed the consent of his slave for. It is clear that in such a coercive relation as that of master to slave, rape could as easily be part of the daily routine as seduction. The master had to promise something only if he wanted something more than a rape relation to his sexual partner.

Like enthusiastic work, enthusiastic voluptuousness was not easily elicited by typical slave master coercion. The statistical fact that the above letter tried to explain—that mistresses of owners were much more likely to gain manumission—was, however, there to be explained.

In self-reproducing free colored populations the sex ratio (ratio of males to females) tended toward a normal one. However, in the slave societies in which most free colored had been manumitted, the free colored sex ratio showed a very high ratio of women to men. For instance, Laurence (1983:40) gives the sex ratio for Tobago in 1790 among free colored with origin among the British slaves as a little over 2 women to 1 man or about .5, among those with origin among the French slaves a little over 3.5 women to 1 man or about .3. Most of the free colored in Tobago at that time must have been created by manumission, because this was very early in sugar development there.

For most English colonies there are also more direct data on manumissions themselves.

> [Slaves manumitted] tended to be female, creole, young, and colored, and to work as domestics. In the sugar colonies females were roughly twice as likely to be manumitted as males in the period before 1820, but this difference narrowed significantly in many colonies as emancipation approached. . . . Females, however, more often obtained manumission through sexual relationships with Whites or freedmen, and such relationships were by no means confined to the towns. (Higman 1984:383)

The children of such unions had an indirect sexual tie with White power. The patriarchal and "blood" ideology of European families in the eighteenth and nineteenth centuries reinforced these indirect sexual ties, though that ideology also downgraded blood ties for resulting from sexual relations outside marriage, and even more for "miscegenation." Manumissions of Whites' own colored children were indirectly sexual, or "paternal."

Presumably sexual and paternal ties would have more effect when they lasted longer. Family-like relationships between settled Whites and their lovers and children would tend to produce more egalitarian relationships between the couple and in the paternal relationship, and thus to result in manumission. Sexual ties with domestic slaves and with slaves on small farms would then tend to result in more manumissions.

Clients of slave prostitutes would probably rarely be involved in their manumission, but ties between owners and their prostitute slaves might result in commercial manumission as discussed below. Note that the manumission of mistresses because one wants honest love is not strictly the sort of agency analyzed in modern economics, as was the incentive system under which the prostitute apparently worked for her owner. The reasons why one does not want to elicit affection with threats of whipping at each step and why rape of slaves did not create manumissions the way concubinage did are deeper than agency theory in economics explains. The causes of wanting slaves to make "free" affectional decisions and granting eventual freedom are apparently the same.

Some evidence indicates that the strength of sexual and paternal bonds may be higher when the owners and slaves are racially more homogeneous:

> In Bridgetown, where freedman slave ownership was concentrated, 2.6 percent of the slaves owned by freedmen were manumitted between 1817 and 1820, compared to only 1.0 percent of those belonging to Whites. In rural St. Michael these percentages were 1.2 and 0.2 respectively. Thus, slaves of freedmen were two to three times more likely to be manumitted than those of Whites, both in town and in the country . . . . [T]he highest manumission rates occurred where freedmen were already relatively numerous, for example in Trinidad, St. Lucia, the Virgin Islands, and the Bahamas. (Higman 1984:385)

Manumission of young women, then, can serve as a tracer of intimate relations between master and slave that tended toward love rather than rape or prostitution. Manumission of children of mistresses can serve as a tracer of those relations between master and slave children that tended toward paternity rather than breeding.

SLAVERY IN THE EIGHTEENTH-CENTURY CARIBBEAN

Agency in cooperative work. By agency in cooperative work, I mean domestic and managerial ties—ties that involved close continuing contact between a White slave owner (or owner's wife or agent) and a slave who must be trusted to achieve objectives that could not sensibly be monitored as "gang labor." Domestic servants who were not sexual partners were more likely to be manumitted than field hands. Slave drivers, mechanics, and stockmen were freed more often than were less skilled slaves. Slaves were selected into these groups by skill and loyalty.

These groups were disproportionately creole and colored. For example, "by 1834 at least 60 percent of slave domestics in Jamaica were colored, compared to 10 percent of the total slave population" (Higman 1983:126). Having been exposed to European culture, creole colored people could communicate effectively with the master and carry out the "agency" with an understanding of the owner's purpose. Such relations established an "unequal colleagueship" between master and slave, sentimentally and morally closer than that in a field gang. Agency relationships were based on cultural similarities that produced trust and fellow-feeling. Agency often required the owner to set up an incentive system more like an employment contract than a slave-master relation. Such contracts often led the owner to conceive of the slave as having rights to the reward promised, as well as obligations. Among those rights could be the right to freedom.

Domestic slaves were generally much more likely to be manumitted than were field slaves. For example,

> [I]n St. Lucia in 1815–19 . . . only 11 percent of the slaves manumitted were field laborers, although they accounted for 44 percent of the slave population. On the other hand, 52 percent of those manumitted were domestics (17 percent of the population) and 15 percent were tradesmen (5 percent of the population). (Higman 1984:384)

The ratio of the probability of manumission of domestics to the probability for field laborers was about 12 to 1, and about the same for tradesmen compared to field hands. Some of the higher probability of manumission for domestics was sexual, but a good deal of that advantage was preserved for domestic slaves of female owners.

Slaves on smaller rural holdings were also more likely to be manumitted. Furthermore, in those Spanish islands in which slaves often worked in large ranching enterprises before the sugar boom of the late eighteenth and nineteenth centuries (Cuba, Santo Domingo, and Puerto Rico) the rate of manumission was much higher, as reflected in a large free colored population (few manumission documents exist, as is explained in the Author's Appendix). This may be due to the impossibility of supervising cowhands in gangs and the damage that can be done to valuable animals by carelessness. If so, these conditions would have produced more "employment-like" relations between the rancher and his or her agents than was true on sugar islands (Boin and Serrule Ramúe [1979] 1985:61, 63).

Commerce. I define commercial ties as master-slave relations whose basic form was the exploitation of the slave by a formal contract with the slave, similar to the institution of *obrok* in Tsarist Russia, though such contracts were not legally valid. The contract was generally one in which the slave exploited commercial opportunities at his or her own discretion. Women carried on huckstering enterprises in the market; men hired themselves out for episodic transportation work on the docks; women were prostitutes; both sexes manufactured goods; women provided laundry services for hire. The commercial opportunities available to slaves were mostly located in urban areas.

These opportunities could not easily be monitored, so the owner needed a contract with the slave to encourage the slave to seek out opportunities. The better the owner's monitoring, the higher the owner's share of the return. Slave prostitutes were often owned by female entrepreneurs, often free colored women, and often presumably exploited the commercial opportunities in a house maintained by their owners (see the painting in Hoyos 1978:170).

According to agency theorists (e.g., Heady 1952), agents with control over information and effort required a contract in which the agent (slave) collected most of the marginal product of his or her exploitation of those opportunities. Further the contract had to give *rights* to the agent (the slave), so the owner could not change the terms and claim the whole product. On eighteenth- and early nineteenth-century plantations, the owner claimed the whole marginal product, which is why production had to be organized as highly monitored "gang labor."

Thus, the optimum contract in such circumstances (i.e., where continuous monitoring is difficult or expensive) is one in which the agent (slave) pays a fixed rent for the use of the asset (the farm in farm tenancy, the Russian serf in *obrok*, the slave in urban huckstering) and takes the whole product of the commercial activity. This way, the person who has the most information about opportunities and whose effort and attention determines the profitability of exploitation of those opportunities collects the full marginal product of the exploitation and therefore is strongly motivated, even in the absence of supervision (Heady 1952). Such a situation tends to create rights for the slaves that the owner feels bound to respect and that give slaves money to buy themselves out of slavery.

This explains why African slaves in cities had high rates of manumission, whereas in other locations Africans had the lowest manumission rates. Urban African slaves were disproportionately males on the docks, working in a system that must have been much like the "shape up" in longshoring on the American East Coast (Bell [1954] 1993). Stevedoring entrepreneurs or merchants or ship captains—the loading was apparently normally actually managed by the mate or *maître*—needed strong men for casual labor on an episodic basis. Urban male slaves were uniformly more likely to purchase their freedom than any other group (Higman 1984: 382). A similar mechanism of wanting intense work for a while and then to get rid of the worker might explain why houses of prostitution did not own many old women.

Before the twentieth century, commercial relations were much more dominant in cities than in the countryside. Furthermore, there was not much gang labor in simple tasks requiring little skill and initiative in cities in the late eighteenth century. Much manual labor in pre-modern cities was carried on by independent artisans, and much of the rest of it was casual wage labor or piecework labor in temporary jobs. Permanent relations between the people who wanted the work done and those who did the work were not the normal way of organizing work in cities in the eighteenth century. The same forces that produced free-labor-contract incentive systems for free urban manual laborers would have tended to produce the same conditions for slave laborers in cities as well.

Politics. Finally, the slave's political services that led to freedom were largely military and police services. The more sugar was dominant on an island, the fewer Whites there were to defend the island, and the more valuable it was to an empire. Islands largely devoted to sugar such as Haiti, Martinique, Guadeloupe, and Jamaica were therefore militarily vulnerable and commercially valuable in the frequent wars of the seventeenth and eighteenth centuries. Islands were less vulnerable if their governments could recruit colored people and Blacks to defend them.

For example, Guadeloupe was less conquerable than Martinique in the wars between England and France in the 1790s. Guadeloupe had freed its slaves and recruited both Black and colored troops into the militia, and after a precarious conquest the British failed to hold or reconquer it. Martinique, which had not freed slaves or recruited colored troops extensively, was fairly easily conquered. Napoleonic France did not actually reconquer Guadeloupe, but rather the colored general (Malgloire Pélage—see Bangou 1989) at the head of the troops switched allegiance to the empire government.

Sometimes treaties with organized rebel or runaway Blacks were forced on colonial governments. For example, after a war in Surinam between the Dutch and maroons ("bush negros"), the French in Guiana agreed with the organized Blacks that they could settle as free negroes ("*de les établer comme des nègres libres et les contenir sur ce pied*," literally, settle them as free Blacks and restrain them on that footing [Peytraud 1897: 358–59]).

The empires had a great deal of trouble with military operations in the Caribbean because troops from Europe quickly became too ill to fight. Planters tended to form militias that were not reliable servants of the empire, but instead formed alliances with whoever would best defend planter interests.

SLAVERY IN THE EIGHTEENTH-CENTURY CARIBBEAN

Table 3. Factors in Manumission Rates

Causal Process	Examples	Categories Most Likely to Be Manumitted
Sexual and other intimate ties	Sexual partners	Young, creole,[a] colored,[b] women domestics Women on small slaveholdings Women owned by free colored people
	Colored offspring	Creole, colored, domestics, young slaves Children in stable relationships
Agency in cooperative work	Slave drivers	Male, creole, middle-aged
	Skilled	Male craftsmen, mechanics, artisans
	Domestics	Women, household domestics, nannies
Commerce	Dock workers	Male, African
	Prostitutes	Colored urban women Slaves owned by females (often colored)
	Hucksters	Creole women
Politics	Military and police	Young males Militia members Maroons

[a] "Creole" here means born in the Americas, as it is used in the British islands. In Spanish the comparable word implies White race; in Louisiana it means of French origin.

[b] "Colored" sometimes referred to people of mixed (African and European) race, especially when speaking of slaves. However, in some situations a manumitted slave, whether of mixed descent or exclusively African, might be referred to as "free colored," as would the children of such a slave.

Planters also sponsored independence movements if it was proposed to tax them to support the defense of the empire (for Haiti, see Frostin 1975). The empire's military officers needed "seasoned" (i.e., immune to diseases prevalent in the Caribbean) troops from the islands themselves, but who would be more deployable than the local militias. Free colored and slaves were often used for building fortifications and other nonfighting military work, and sometimes for international fighting. Sometimes they were used as "intelligence agents" to find out about rebellions or to hunt down runaways. In any of these cases, they might be freed for their political services.

The Boundary between Slave and Free

In daily life, then, the most oppressive slavery occurred among field laborers on the large, highly class-conscious, and oppressive sugar plantations. Hardly any sugar plantation field workers were manumitted; few had intimate relations with Whites, though they sometimes got pregnant in nonintimate relations; few managed work on a collegial basis with the owner or owner's agents; few sought commercial opportunities with autonomy and discretion; few earned freedom from governments as a reward for loyalty and bravery; and all were subjected to the most class-conscious slave owners, those most interested in the "health" of the slave system as a whole.

As this sugar plantation core of slave society sloped off into slave mistresses or slaves owned by freedmen, creole slaves in domestic service, slaves in skilled work or first-line management, slaves in cities and especially in urban commerce, slaves in smaller enterprises, slaves of masters to whom the maintenance of the whole slave system was a secondary consideration, the master-slave relation became more like the relations among free unequal in eighteenth-century urban society. And that slope also led to the boundary between slave and free colored. A few people at the high-freedom end of these slopes in fact became free colored or Black freedmen. If they were women, they maintained the free colored population thenceforth, because the children of free colored women were also free. With the exception of reconquered Guadeloupe, there were no large movements of free colored people back into slavery.

Table 3 presents these patterns of manumission rates. I argue that the causal processes shown in the stub of the table were different kinds of ties between master and slave—processes that required discretion, loyalty, enthusiasm, skill, career training, or other aspects of agency relations. These causes, then, explain high manumission rates, and by inference, therefore, other ways in which slaves were treated more nearly as free.

CONCLUSION

The sociology of slavery and freedom has been crippled by not treating freedom as a continuous variable. Freedom is often thought of as a legal concept, as for example that defined in the Bill of Rights of the United States, so it is either guaranteed or not. The size of the set of possibilities among which a group of people chooses—the core idea of freedom here—is clear enough conceptually but hard to specify in practice because possibilities not chosen do not leave a historical record. My intellectual strategy has been to specify freedom by its causes, the causes of more and less restriction on slaves' choices in the late eighteenth-century Caribbean. These causes fall into two main groups: (1) the causes related to the power in island societies that was held by sugar planters, who had a great interest in restricting slaves' liberties, and (2) the causes related to the use by slave owners of their property rights in making agency contracts with their slaves. The scattered evidence of what slaves and their owners in fact chose (such as manumission of the slave), or of what slaves could choose (such as how to spend their wages), or of what property rights slaves had (such as having enough money to buy freedom), suggests the shape and size of the set of possibilities under different causal conditions.

What was generally distinctive of the eighteenth-century Caribbean colonies of all the empires (as of the American South at the same time) was the building of slave societies, societies whose principal governmental problem was holding slaves in bondage of varying degrees of restrictiveness. But the intensity of governmental effort to restrict possibilities, the degree of enforcement of oppressiveness of slavery, was greater where planters were more economically dominant, had better class unity, and were well represented in the system of government of the islands in the relevant empire.

But within a given level of slave society (in this case, low), an Isla Margarita pearl diver who had to risk his life under the water, where his owner could not monitor the work without risking his own life, presented a different control problem than did a gang worker digging holes for planting sugar cane. So within the Spanish empire, slavery in the region near Havana in Cuba resembled that in Jamaica, while slavery in Isla Margarita resembled that in the Bahamas. This occurred because fishing off the Bahamas was more nearly like the agency problem of pearl diving than like the agency problem of getting more dirt moved by a gang of recent African immigrant slaves in either Havana province or Jamaica.

As a practical matter, a thoroughgoing slave society was a utopian vision by planters, but in many situations they could not get from that vision what they wanted out of real slaves. The more their society resembled that in Barbados, the easier it was to get work done at low cost on their sugar plantations—but the harder it was to get the slaves to look after the livestock carefully or to harvest fish from the Caribbean, and the more salted fish they had to buy from New England. The more their society resembled that in Curaçao, the more easily they could send their slaves off as their agents on business or household matters. But in either kind of society, those slave owners who wanted commercial trustworthiness, initiative, courage, enthusiasm, or love, had to grant the slave enough freedom to be able to make deals with elements of equality and choice in them. Absolute power may have corrupted absolutely, but it had the additional disadvantage that it would not get the pearls off the bottom.

ARTHUR L. STINCHCOMBE is writing a book on The Political Economy of the Caribbean from 1775 to 1900. *He is thinking about retiring so he can get some work done; advice is welcome.*

REFERENCES

Bangou, Henri. 1989. *La revolution et l'esclavage à la Guadeloupe 1789–1802: Épopée noir et génocide* (Revolution and Slavery in Guadeloupe 1789–1802: Black Epic and Genocide). Paris, France: Messidor/Éditions sociales.

SLAVERY IN THE EIGHTEENTH-CENTURY CARIBBEAN

Bell, Daniel. [1954] 1993. *The Racket-Ridden Longshoremen*. New York, NY: Irvington.

Boin, Jacqueline, and José Serrule Ramúe. [1979] 1985. *Processode Desarrollo del Capitalismo en la República Dominicana 1844–1930* (Process of Development of Capitalism in the Dominican Republic 1844–1930). 3d ed. Santo Domingo, Dominican Republic: Graniel.

Cauna, Jacques. 1987. *Au temps des isles à sucre: Histoire d'une plantation de Saint-Domingue au XVIIIe siècle* (In the Time of the Sugar Islands: History of a Plantation of Haiti in the Eighteenth Century). Paris, France: Éditions Karthala et A.C.C.T.

Commons, John R. 1924. *Legal Foundations of Capitalism*. Madison, WI: University of Wisconsin Press.

Frostin, Charles. 1975. *Les révoltes blanches à Saint-Domingue aux XVIIe et XVIIIe siècles (Haòti avant 1789)* (The White Revolts in Saint-Domingue in the Seventeenth and Eighteenth Centuries [Haiti before 1789]). Paris, France: L'École.

Geggus, David Patrick. 1982. *Slavery, War, and Revolution: The British Occupation of Saint Domingue, 1793–1798*. Oxford, England: Clarendon.

Goveia, Elsa F. [1965] 1980. *Slave Society in the British Leeward Islands at the End of the Eighteenth Century*. Westport, CT: Greenwood.

Heady, Earl Orel. 1952. Economics of Agricultural Production and Resource Use. New York: Prentice-Hall.

Higman, Barry W., ed. 1984. *Slave Populations of the British Caribbean 1807–1834*. Baltimore, MD: Johns Hopkins University Press.

Holt, Thomas C. 1992. *The Problem of Freedom: Race, Labor, and Politics in Jamaica and Britain, 1832–1938*. Baltimore, MD: Johns Hopkins Press.

Hoyos, F. A. 1978. *Barbados: A History From the Amerindians to Independence*. London, England: Macmillan Education Limited.

Jordan, Winthrop D. 1969. *White Over Black: American Attitudes Toward the Negro, 1550–1812*. Baltimore, MD: Pelican.

Klein, Herbert S. 1967. *Slavery in the Americas: A Comparative Study of Virginia and Cuba*. Chicago, IL: University of Chicago.

Knight, Franklin W. 1970. *Slave Society in Cuba During the Ninteenth Century*. Madison, WI: University of Wisconsin Press.

Kuethe, Allan J. 1986. *Cuba, 1753–1815: Crown, Military, and Society*. Knoxville, TN: University of Tennessee Press.

Laurence, K. O. 1983. "Tobago and British Imperial Authority, 1793–1802." Pp. 39–56 in *Trade Government and Society in Caribbean History, 1700–1920*, edited by B. W. Higman. Kingston, Jamaica: Heineman.

MacNeill, John Robert. 1985. *Atlantic Empires of France and Spain: Louisbourg and Havana, 1700–1763*. Chapel Hill, NC: University of North Carolina Press.

———. 1978. *Cuba: Economía y Sociedad.* Vol. 7: *Del Monopolio Hacia la Libertad Comercial (1701–1763), del II* (Cuba: Economy and Society. Vol. 7: From Monopoly to Commercial Liberty [1701–1763] part II). Madrid, Spain: Editorial Playor.

Moreno Fraginals, Manuel. [1964] 1976. *The Sugarmill: The Socioeconomic Complex of Sugar in Cuba 1760–1860*. Translated by C. Belfrage. New York: Monthly Review Press.

Naipaul, V. S. [1969] 1984. *The Loss of Eldorado: A History*. New York: Vintage Books.

Patterson, Orlando. 1967. *The Sociology of Slavery: An Analysis of the Origins, Development, and Structure of Negro Slave Society in Jamaica*. London, England: MacGibbon and Kee.

———. 1991. *Freedom*. Vol. I: *Freedom in the Making of Western Culture*. New York: Basic Books.

Peytraud, Lucien. 1897. *L'Esclavage aux Antilles Françaises* (Slavery in the French Antilles). Paris, France: Librairie Hachette.

Riverend, Julio le. 1972. *Historia Económica de Cuba* (Economic History of Cuba). Barcelona, Spain: Ediciones Ariel.

Scott, Rebecca Jarvis. 1985. *Slave Emancipation in Cuba: The Transition to Free Labor, 1860–1899*. Princeton, NJ: Princeton University Press.

Tannenbaum, Frank. 1946. *Slave and Citizen: The Negro in the Americas*. New York: Vintage.

Williams, Eric. [1944] 1964. *Capitalism and Slavery*. London, England: Andre Deutsch.

12
Was the Plantation Slave a Proletarian?
Sidney W. Mintz

Between the beginnings of the African slave trade to the New World, shortly after 1500, and the abolition of slavery in the last New World territories where it had remained legal (Puerto Rico: 1873-1876; Cuba: 1886; Brazil: 1888), probably more than 9,000,000 enslaved Africans were shipped westward across the Atlantic.[1] The institution embodied in the capture, sale, transportation, and exploitation of African slaves in the western hemisphere thus lasted nearly four hundred years, and was legal for centuries, in large and much differentiated regions within the Americas. Many different European powers were involved in the sale, use and, often, resale of enslaved Africans. Local practices in these matters varied widely, and were usually subject to metropolitan codes of law and metropolitan bureaucracies (though these never were the last word in regulating the treatment, care, and defense of the enslaved). Hence to try to address generally the nature of slavery as it existed in the New World, or its common features

---

*First presented at a seminar of the Fernand Braudel Center, State University of New York at Binghamton, February 2, 1977, I am grateful to Professor Wallerstein for the opportunity to air my views and, indeed, for the choice of topic, to which he asked me to address myself.

[1] See Philip A. Curtin, *The Atlantic Slave Trade: A Census* (Madison: Univ. of Wisconsin Press, 1969).

as an institution in the New World setting, is a risky and frequently unprofitable undertaking. Not only was slavery different in the colonies of one power from what it was in those of another, but even within one imperial system, there were often significant differences in the slavery institution from colony to colony. Moreover, time and circumstance deeply affected the way slavery worked in particular milieux. Demography mattered; as did the prevailing form of work at which slaves were employed; whether the slaves were "creolized" — seasoned to the slavery regimen, or born into it, acculturated to the New World conditions, or caught up in the meaning and memories of a distant life — all these, and many other factors, much influenced what slavery was, and how it was experienced.

In this paper, I shall attempt to limit the geographical scope of my inquiry and, thereby, at least some part of the economic, political, and cultural variation with which I might otherwise have to struggle, were I attempting to look at the whole hemisphere. But I deliberately do not limit the time-span with which I deal, since one of my major concerns here is the significance of different time-periods (and what those differences entailed) for the question the paper means to address: the relationship between the terms and categories "proletarian" and "slave". Plainly, a number of fairly firm lines need to be drawn, to avoid drowning in generalities. The term "plantation slave", as I mean to use it here, refers to chattel slaves, persons purchased or inherited and owned as property, who were used as laborers on large agricultural estates producing commodities for (mainly) European markets, between the first decade of the sixteenth century and the ninth decade of the nineteenth. Nearly all, but by no means all, such slaves were born in Africa or were the descendants (at least in part) of people who were. By the "Caribbean region" I have in mind in particular the Greater and Lesser Antilles, with an important nod in the direction of the Guianas. I think that it would not be impossible (though it would entail extremely burdensome difficulties and a good deal more space) to extend the treatment to include Brazil, parts of Mexico and Central America, and even much of the United States South; I deliberately avoid such extensions, while recognizing that I have already taken on too much.

I am unable to limit myself similarly in time, as I have said; nor can I avoid the complications implicit in refining what I mean by "plantation". Just as slavery itself varied with place and with time, so, too, did the nature of the enterprises upon which slaves toiled. Plantations themselves also varied very widely, according to a great many environing conditions. Perhaps it is enough to say for the present that I am here particularly concerned with sugar-cane plantations, which were present throughout the four centuries that interest me, and were doubtless more important than any other type of Caribbean plantation, during this entire four-hundred year period.

I am not prepared to be so offhand in dealing with the term "proletarian", but I can state briefly, at least, what I have in mind by it. In the first volume of *Capital*, Karl Marx discusses the buying and selling of labor-power as an aspect of the capitalist mode of production,[2] wherein it becomes very clear that a "free" laborer is not thereby and automatically a member of the proletariat. Indeed, as

---

[2.] Karl Marx, *Capital* (New York: International Publ., 1939), I, 145 ff.

Was the Plantation Slave a Proletarian?

Marx employs the term "proletariat", it is bound up quite narrowly and specifically with the rise of capitalism, wherein "labour-power can appear upon the market as a commodity, only if, and so far as, its possessor, the individual whose labour-power it is, offers it for sale, or sells it, as a commodity."[3] Indeed, that is the first criterion of proletarian labor-power. Second, by Marx's reckoning, such a seller of labor-power as a commodity cannot sell himself, or sell his labor-power "once and for all," since by so doing he would become something other than a free seller of his own effort. Third, the seller must be obligated to sell his labor, by virtue of having nothing else either to sell, or by which to sustain himself; he has no choice but to sell his labor-power. That a free laborer has nothing to sell but his effort, that he sees and offers to sell that effort as a commodity to its prospective buyer, and that he has nothing but his labor-power to sell, all become parts of the definition of the proletarian.

"We have seen," Marx writes, "that the expropriation of the mass of the people from the soil forms the basis of the capitalist mode of production;"[4] and "so-called primitive accumulation . . . is nothing else than the historical process of divorcing the producer from the means of production."[5] What I refer to by "proletarian", then, consistent with these assertions, is the free but propertyless seller of his own labor-power as a commodity to a capitalist buyer of commodities, among them the commodity of labor-power, to undertake fresh production.

It was never Marx's sole or explicit intention, so far as I know, to draw an orderly contrast between slaves and proletarians in order to endow these terms with definitions that could become eternal verities. His concern was above all to understand and to reveal the inner nature of the capitalist system, and of the capitalist mode of production, as these typified the history of Europe. Well aware that he could not ignore or treat as irrelevant the activities of the Europeans outside the European heartland, he saw that the forms of labor exaction in different parts of the world in which the Europeans were active both arose from, and reacted back upon, developments in Europe itself:

> Freedom and slavery constitute an antagonism. . . . We are not dealing with indirect slavery, the slavery of the proletariat, but with direct slavery, the slavery of the black races in Surinam, in Brazil, in the southern states of North America. Direct slavery is as much the pivot of our industrialism today as machinery, credit, etc. Without slavery, no cotton; without cotton no modern industry. Slavery has given their value to the colonies; the colonies have created world trade; world trade is the necessary condition of large-scale machine industry. Before the traffic in Negroes began the colonies supplied the Old World with very few products and made no visible change in the face of the earth. Thus slavery is an economic category of the highest importance.[6]

But his interest throughout remained Europe, the pivot of what could be incited to happen elsewhere, the beating heart of capitalist endeavor. From that center,

---

[3]. *Ibid.*, I, 146.

[4]. *Ibid.*, I, 793.

[5]. *Ibid.*, I, 738.

[6]. Letter of Karl Marx to P. V. Annenkov, December 28, 1846, in *Karl Marx & Frederich Engels: Selected Works* (New York: International Publ., 1968), 13-14.

men, materials, and wealth flowed outward in order to integrate within the central design regions, populations, and resources that had lain outside and largely unaffected beforehand. Thus the expansion of European capitalism involved the assimilation to homeland — that is to say, to European metropolitan — objectives, of societies and peoples that were not yet part of the capitalist system. The ways in which this assimilation was set in motion, and the forms that it took were of course highly variable. They were not, they could not be, identical to those processes that had typified European economic growth; yet it was precisely European expansion itself that brought these external areas within the ambit of European power and economy, even if the forms of their integration differed radically from those familiar from Europe itself. In spite of his prevailing concern with Europe, Marx understood this well:

> The discovery of gold and silver in America, the extirpation, enslavement and entombment in mines of the aboriginal population, the beginning of the conquest and looting of the East Indies, the turning of Africa into a warren for the commercial hunting of black skins, signalized the rosy dawn of the era of capitalist production. These idyllic proceedings are the chief momenta of primitive accumulation. . . .
>
> The different momenta of primitive accumulation distribute themselves now, more or less in chronological order, particularly over Spain, Portugal, Holland, France and England. In England at the end of the 17th century, they arrive at a systematical combination, embracing the colonies, the national debt, the modern mode of taxation, and the protectionist system. These methods depend in part on brute force, *e.g.* the colonial system. But they all employ the power of the State, the concentrated and organized force of society, to hasten, hothouse fashion, the process of transformation of the feudal mode of production into the capitalist mode, and to shorten the transition. Force is the midwife of every society pregnant with the new one. It is itself an economic power. . . .
>
> Whilst the cotton industry introduced child-slavery in England, it gave in the United States a stimulus to the transformation of the earlier, more or less patriarchal slavery, into a system of commercial exploitation. In fact, the veiled slavery of the wage-workers in Europe needed, for its pedestal, slavery pure and simple in the new world.[7]

We see here that, in Marx's view, the looting of the world outside Europe contributed to European economic growth. (In spite of the spirited debates about *how much* it contributed, we have fortunately not yet reached that cliometric melting point where the non-European world will turn out miraculously to have been an economic burden upon Europe from the very beginning.) That growth in turn affected the new ways in which Europe continued its developmental efforts elsewhere. But in spite of the citations from Marx, it is not completely clear, at least to me, just how he envisioned slavery — and particularly plantation slavery, for the production of agricultural commodities for European markets — in his picture of world capitalism. I have suggested elsewhere[8] that Marx himself may not have been wholly satisfied with his own understanding of how "slavery pure and simple" fit within capitalism — as when he refers to

---

[7.] Marx, *Capital,* op. cit., I, 775, 776, 785.

[8.] Sidney W. Mintz, "The So-Called World System: Local Initiative and Local Response," *Dialectical Anthropology,* II, 4, Nov. 1977, 253-70.

Was the Plantation Slave a Proletarian?

plantation owners in America as capitalists who "exist as anomalies within a world market based on free labor"[9] — but I do not wish to pursue this exegetical problem further.

Indeed, my task as I understand it must be to concentrate on the Caribbean region, on the plantation system that developed within it, on the nature of slavery as the principal form of labor exaction over nearly four centuries, and on the linkages between slavery and other forms of labor in the same region. I will not, that is, seek to counterpose definitions of slaves and proletarians in some specified epoch, in order to see to what extent they are similar or different. Such an undertaking might be useful within narrow limits; but I would rather concentrate on the nature of slavery in certain specific historical instances to give some idea of its character and variation, against which notions of the proletariat and of proletarians might then be silhouetted.

In a recent essay,[10] I have hypothesized why slavery turned out to be so appropriate a solution to the labor problem in the Caribbean region, beginning as early as the dawn of the sixteenth century, and disappearing completely only in the dusk of the nineteenth. It is not necessary to repeat the argument here, but I do need to make several points in passing, to advance my wider presentation. First, the history of Caribbean slavery was usually marked by the accompanying presence of other forms of labor exaction, frequently in the same industry and even on the selfsame enterprises. That is, only for certain periods, and in certain colonies, did slavery function as the sole form of land-labor relationship on the plantations. Second, the other forms of labor exaction which accompanied slavery all seem to have involved varying degrees of coercion, though the laborers themselves were in most such cases "free" by conventional definition.

For present purposes, I would schematize Caribbean plantation and slave history as falling within five periods:
a) the first Hispanic sugar-cane plantations in the Caribbean, located on the Greater Antilles, *ca.* 1500-1580, manned with enslaved aborigines, and enslaved and imported Africans;
b) the first British and French sugar-cane plantations in the Caribbean, located in the Lesser Antilles, *ca.* 1640-1670, manned with enslaved aborigines, European indentured servants, and enslaved Africans;
c) British and French plantations based exclusively on enslaved African labor, at their apogee in English Jamaica (post-1655) and French St. Domingue (post-1697);
d) a new stage of Hispanic sugar-cane plantations, again on the Greater Antilles (now only Cuba and Puerto Rico), *ca.* 1770-1870, based on enslaved, "contracted" and coerced labor;
e) plantations based on free and "contracted" labor, successively throughout the sugar colonies after emancipation (post-1838, British; post-1848, French; post-1876, Puerto Rico; post-1886, Cuba, etc.).

This five-part schema could be carried forward into the present by the addition of at least two other stages (the emergence of a "genuine" rural proletariat, and

[9.] Karl Marx, *Grundrisse* (London: Penguin Books, 1973), 523.

[10.] Mintz, *op. cit.*

then its elimination with progressive mechanization); and of course, it should be elaborated and detailed far more fully. Its principal usefulness here, I believe (as in my review of Wallerstein's *Modern World System*, where I first proposed it),[11] is to indicate how labor forms other than slavery were usually combined with slavery itself, in practice.

These different forms of labor exaction, existing for the most part in combination in Caribbean history, were *not* interchangeable, each representing a variant response to labor needs; nor was it accidental or random that they usually occurred in combined form, answering needs for labor that could not be most conveniently or profitably met by using one or another form exclusively. Padgug has argued eloquently against the notion that such forms were freely interchangeable, though he concedes that some substitutability was possible:

> There can be no doubt that, to a certain degree, this view is correct. The postemancipation American systems, for example, were indeed able to convert to other systems of labor without losing their position in world markets. But that they were able to do this was not in fact a function of the absolute interchangeability of labor systems, but rather of the dominance of capitalism in the world, a dominance which created and kept in operation a major system of commodity production and exchange, and which could convert to its own use several more primitive systems of labor, which otherwise would have been by themselves incapable of sustaining a commodity system. . . .
>
> The apparent interchangeability of labor systems at particular historical moments paradoxically exists, therefore, only because of the peculiar nature of the *dominant* labor form, a form which in terms of dominance is not at all interchangeable with other forms. That this should be so ought not to be surprising. For slavery, like other modes of production, has particular characteristics and particular effects which differentiate it from all other modes. And at points where it is precisely those characteristics and effects which dominate the entire socio-economic formation or which are decisive for its functioning (as, for example, in the period when slavery in the Americas proved to be the only system capable of providing labor in sufficient quantities to enable the colonies to be tied to the world), it is not all interchangeable with other modes. It is true that Marx tends to lump slavery and serfdom together on occasion as if they were interchangeable, but this is only vis-à-vis wage-labor, and is only meant to demonstrate the vast differences which exist between all pre-capitalist labor relationships and the capitalist one.[12]

[11.] *Ibid.*

[12.] Robert A. Padgug, "Problems in the theory of slavery and slave society," *Science and Society*, XL, 1, Spr. 1976, 24-25. Padgug's use of the term "pre-capitalist", under which he places slavery and "other pre-capitalist formations... the real division [being] between capitalism and all earlier socio-economic formations," can be seriously questioned on several grounds. As Tomich points out in *Prelude to Emancipation: Sugar and Slavery in Martinique, 1830-1848*, (unpubl. Ph.D. dissertation, Univ. of Wisconsin, Madison, 1976) and in "Some Further Reflections on Class and Class-conflict in the World-Economy," (Seminar I, Working Papers, Fernand Braudel Center, Dec. 1, 1976, mimeo), plantation slavery in the New World was in no sense "pre-capitalist", but a very specific *product* of evolving capitalism.

"*Negro slavery* — which is besides incompatible with the development of bourgeois society and disappears with it, *presupposes* wage labor, and if other free states with wage labor did not exist alongside it, if, instead the Negro states were isolated, then all social conditions there would immediately turn into pre-civilized forms." (Marx, *Grundrisse, op. cit.*, 224). It is essential to draw analytical distinctions between different abstracted stages in the history of capitalism, and to explore the differences between so-called merchant capital and industrial capital. But it does not follow inevitably that slavery was coterminous with

Was the Plantation Slave a Proletarian?

Indeed, the history of Caribbean plantations does not show a clear break between a slave mode of production and a capitalist mode of production, but something quite different. The succession of different mixes of forms of labor exaction in specific instances reveals clearly how the plantation systems of different Caribbean societies developed as parts of a worldwide capitalism, each particular case indicating how variant means were employed to provide adequate labor, some successful and some not, all within an international division of labor transformed by capitalism, and to satisfy an international market created by that same capitalist system.

My division of Caribbean plantation labor history into five periods, except insofar as these can be vouchsafed by *legislative* (which is to say, politically documentable) stipulations as to the laws *intended* to regulate the employment and care of laborers of different categories, are quite arbitrary and imperfect. Yet they at least may suggest in some ways the progression of forms of labor exaction or, more precisely, the progression of mixtures of labor exaction, in certain selected cases. We move back and forth here, between some specific historical situation, more or less describable in terms of a dominant mode of production and certain subsidiary, complementary or subordinate but interdependent modes, and an abstract, ahistorical characterization, useful for helping us to understand all instances of concrete and the particular more fully. For my present purposes, it may be sufficient to defend this assertion with a sketchy comparison of two different cases.

Cuba and Puerto Rico, both Spanish colonies, began periods of renewed and rapid plantation expansion dating a few decades apart. In Cuba, the English occupation of Havana for nearly a full year (1762-63) marked the opening of a new epoch; in Puerto Rico, though stirrings of new developments predated the event, the "reforms" of 1809 were the turning-point. In Puerto Rico, the prime mover was legislative, not military; but the legislative process was forced by wider economic pressures, immediately following the loss of all Spanish power on the Latin American mainland, and soon after the Haitian Revolution had destroyed the world's greatest sugar-producing colony. In Cuba, the British set many local economic and political forces in motion by their invasion.[13] Cuba, which was more than ten times larger than Puerto Rico, richer and more populous, and with considerably greater influence in the metropolis, sought to solve its plantation labor problem with more enslaved Africans, and the importation rates in the decades following 1762 were horrifyingly high. Even after Spain had signed an accord with Britain not to import more slaves to its New World possessions, the importations continued, well past the middle of the nineteenth century.

---

one stage only in the world development of capitalism, and surely not that it was pre-capitalist in nature. Because Marxists approach the historical study of capitalism from an evolutionary perspective, it is understandable (but no less in error, I would argue) that they sometimes confuse *non*-capitalist with *pre*-capitalist social formations. Marx himself appears to have understood the difference clearly. The title of the book by Hindess and Hirst, *Pre-Capitalist Modes of Production* (London: Routledge & Kegan Paul, 1975), strikes me as being erroneous for the same reasons.

[13]. See Manuel Moreno Fraginals, *El Ingenio* (La Habana, 1964), 5 ff.

But enslaved African labor never sufficed for the Cuban planters of the times. To increase even more the available labor supply, they wrung from the Crown the right to import Chinese contract labor, and imported, during a period stretching out more than half a century, from the height of the plantation system to well after emancipation, perhaps as many as 135,000 Chinese. These "contract" laborers were not slaves, nor could they be said to have been entirely "free", though they were certainly free (as opposed to enslaved) by conventional standards of the time. Knight has cause, it seems to me, for claiming that "Chinese labor in Cuba in the nineteenth century was slavery in every social aspect except the name."[14] But the status of these laborers was not inherited; there were no international treaties against their importation; and there roles on the plantations were not at all precisely those of the slaves.[15] Aimes points out that the large estates of the mid-nineteenth century had mixed labor supplies of Chinese contract laborers and African slaves.[16] "Not one of the giant *ingenios* composed their stock entirely of negroes," he tells us. The gradual addition of Chinese contract laborers to the slave labor force played a particular part in "easing the transition" — to use the euphemism most common in describing this process in the Caribbean — from slavery to freedom. "The industries of Cuba," Aimes writes, "were in an evolutionary stage between slave labour and free labour, and in this change the great *ingenios* were taking the lead. Their first contribution was in the economy of labour effected through better organization and improved machinery, and their second, in replacing half of the slaves by coolies."[17] I shall not attempt here to detail the rationale for this particular process of modernization; suffice it to say that what occurred in Cuba was, on the one hand, consistent with the universal replacement of slave labor by free in the nineteenth century, and on the other, distinctively and uniquely Cuban in some regards.

Puerto Rico, the smaller, poorer, less influential island, entering into the renewed expansion of the sugar industry somewhat more tardily, had no luck in its efforts to influence the Spanish Crown to permit the importation of Asian contract labor. It possessed, however, another potential source of labor which it succeeded in tapping by legislative chicanery. The "reforms" of Don Ramón Power y Giralt achieved before the Cortés in 1809, made it possible for the Puerto Rican government to force onto the plantation freeborn but landless Puerto Ricans, on the elegant grounds that, being landless, they were "vagrants".[18] These measures approximately doubled the available labor force for

---

[14]. Franklin W. Knight, *Slave Society in Cuba During the Nineteenth Century* (Madison: Univ. of Wisconsin, 1970), 119.

[15]. See also Denise Helly, *Histoire des gens sans histoire: les Chinois Macao à Cuba* (in press).

[16]. See Hubert H. S. Aimes, *Slavery in Cuba, 1511-1868* (New York: G. P. Putnam's Sons, 1907), 212-13.

[17]. *Ibid.*, 213.

[18]. See my "The Role of Forced Labour in Nineteenth Century Puerto Rico," *Caribbean Historical Review*, II, 1951, 134-41; "Labor and Sugar in Puerto Rico and Jamaica, 1800-1850," *Comparative Studies in Society and History*, I, 3, Mar. 1959, 273-81; and *Caribbean Transformations* (Chicago: Aldine, 1974).

Was the Plantation Slave a Proletarian?

the plantations; and though Puerto Rico's nineteenth-century sugar industry was very modest, compared to Cuba's, in fact its regimented creole workers played a role neatly analogous to that played by the Chinese in Cuba.

In these two cases, we see at once the significance of the particular and specific, and the general rule each case substantiates. That rule is that forms of Caribbean plantation labor exaction were not interchangeable, and that slavery rarely occurred in absolutely pure form. It is my contention that findings of this sort throw some light on the general question as to whether the categories "slave" and "proletarian" can be viewed as the same, similar, or best understood only by contrast. I intend to enlarge on this general point at greater length in another publication, so that it need not be developed further here. Let me, then, return briefly to my "stages" to suggest something of the different character of each.

The first developments of the sugar industry in the Hispanic Greater Antilles involved early importations of enslaved Africans, who were used as laborers alongside enslaved Native Americans, on the plantations. These developments had no significant long-range implications for the European sugar market; indeed, the early plantations of this period disappeared in some cases, and exported declining quantities of sugar, for the most part, after the middle of the sixteenth century. Though we lack adequate details, it seems that the labor arrangements for enslaved Africans and Native Americans on these early estates were in fact quite different, American Indians were supposedly "commended" (*encomendados*), a status vaguely resembling enfeoffment, and based upon European practice as a source of legal status. In contrast, enslaved Africans were known to be, and recognized as, slaves, subject to different legal conceptions and laws. Granting that legal prescriptions are a poor guide to actual behavior, it is nonetheless the case that this first period of Caribbean plantation history does not seem to have been characterized by a uniform slave code for its labor force.

The development of more modern plantations in the Lesser Antilles by the British and French involved, first, the use of indentured Europeans, and later, the importation of ever-increasing numbers of enslaved Africans. (There were also some enslaved Native Americans used as labor on these plantations.) Once again, we find a mix of labor-exaction forms, subject to different usages and interpretations. Only after the middle of the seventeenth century does African slave labor begin to prevail; and thereafter indentured European labor plays an ever-declining role in the Lesser Antilles.

Only in the third period, when large-scale plantations were fully developed in Jamaica and French St. Domingue — which is to say, at the zenith of the slave-based system in the eighteenth century — did the plantation labor force (in these two colonies, at least) eventuate in being exclusively African and enslaved. It bears noting that in neither case was this for long the norm. Jamaica was redeveloped as a plantation colony by the English after its invasion in 1655, and became significant as such only well into the eighteenth century. Yet by the first decade of the nineteenth century, the Jamaican sugar industry was in some difficulty, and emancipation came in 1834-38. St. Domingue was developed by the French as a plantation colony even before the western third of the island of Española was ceded to them in 1697; but the plantation system did not reach its zenith there until the eighteenth century. And by the eighth decade of that century, the Revolution was ready to explode upon the colony. In other words, the epoch of "pure slavery" in these two colonies, the most lucrative in European history, was in each case less than a century in length.

In the fourth so-called period, Cuba and Puerto Rico developed their renewed sugar industries on a slave and forced-labor basis; since I have referred to these cases already, however, no more need be added here, except to underline once again the mixed character of the system of labor-exaction.

Finally, a word may be offered concerning the "transitional" period following formal emancipation. In the case of Cuba, as we have seen, Chinese contract labor "eased the transition" to freedom. But in many other instances, it was necessary to destroy the bargaining power of the newly freed in order to approximate conditions of coercion sufficiently continuous with slavery to make the plantation system worthwhile for those who underwrote it. Hence the period following the coming of formal freedom was, in many Caribbean cases, one of intensified chicanery, intimidation, and legislative coercion, reminiscent in its intent of the postbellum U.S. South, but never typified by the specific racist terrorism of the South. The taxes levied on Jamaican freedmen; the trickery used to facilitate the importation of Indian contract laborers to that country; the legislative devices developed to keep land out of the hands of the Guianese freedmen; the so-called "apprenticeship systems" employed to immobilize labor, ostensibly while laborers *learned how to be free*; the importation of Javanese to Surinam — indeed, the list of differentiated "solutions" to the "labor problem" typical of the post-emancipation Caribbean staggers the imagination, and numbs the reader's sense of ethics and fair play. It is only really in the closing decades of the nineteenth century and, in some cases, even later than that, when we are able to note the decline of legislative and other devices limiting in one regard or another the completely free movement of the laborer and the completely free sale of his labor as a commodity. One can argue, accordingly, that only when such a point arrives is it possible to speak of "true proletarians" — but I wish to defer that presumption, and what it brings in its wake.

Instead, I prefer to turn to a somewhat different subject at this point, having to do with slave labor-power, and its significance for the case I am seeking to make. I have already suggested that, like proletarians, slaves are separated from the means of production; but of course, it is not that they have nothing but their labor to sell. Rather, they are *themselves* commodities, their labor is not, under most circumstances, a commodity within the slave economy, but the products of their labor are, under most circumstances, commodities; they themselves appear to be a form of capital, though they are human beings.

The cost of labor, under these conditions:

> ... appears as a series of investments in fixed capital. ... Moreover, since the planter has to bear the costs of reproducing the slave, all of the slave's labor appears as unpaid surplus labor for the master.[19] The whole of the slave's product is the property of the

---

[19.] Rod Aya, in criticizing the analysis of slavery in *Pre-Capitalist Modes of Production* by Hindess and Hirst, shows how they have misunderstood Marx's treatment. (Review in *Theory and Society*, III, 4, Winter 1976, 623-29). Hindess & Hirst argue: "For the slave all labour is surplus-labour." (*Op cit.*, 132). But neither is true nor did Marx ever claim it. Indeed, he is very explicit: "The wage-form thus extinguishes every

Was the Plantation Slave a Proletarian?

master. Nonetheless, if the productive activity of the slave is examined, it is apparent that one part of his labor produces the value necessary for his subsistence and the other part produces a surplus. The production of this surplus is the basis of the slave economy, but the value of labor and the distinction between necessary and surplus labor are hidden by the property relation in slave society.[20]

Slaves differ from proletarians not only in that they appear as a form of capital while their labor is not a commodity, but also because they receive no wages, only receiving instead that portion of their labor-power that takes the form of necessary labor, so called. Accordingly, one could assert that they lie outside the commodity system within which they produce; they cannot generate internal demand; and they do not form a consumer market.

This is all very well, to the extent that it allows us to begin to characterize the slave mode of production. All that remains to be done, however, is to move from such postulates to the everyday realities of slave life on Caribbean plantations. In doing so, our grasp of the slave system inevitably becomes more complicated, even as it becomes more nuanced. The cost of slave labor appears, Tomich stresses, "as a series of investments in fixed capital (housing, food, clothing, etc.) . . . [while] all of the slave's labor appears as unpaid surplus labor for the master."[21] Maintenance during the effective productive period of the slave's life (and, indeed, often thereafter) represents a quite different cost from that represented by the original outlay — the purchase price — by which his owner acquires exclusive access to his labor-power.

Not calculated as a part of maintenance is the cost of coercion which, in my view, deserves mention not just because it was an important part of the reality of slave life, but also because I believe that it meshes with the problem of maintenance, and in curious ways. I would be inclined to argue that these two different sorts of running expense, maintenance on the one hand and coercion on the other, can cancel each other out, as it were, under certain conditions. The principal long-term supply-cost of maintaining the slave was, I believe, nutrition. In the slave codes of the Caribbean, slave nutrition usually figured importantly, codes often specifying the kinds and quantities of food with which slaves were supposed to be supplied. Indeed, the provision of adequate food was a prime preoccupation of Caribbean slave systems, and we need not look to altruism for explanation. Debbasch, in his monograph on *marronage* in St. Domingue, argues that inadequate food was probably a principal cause of slave flight from the

---

trace of the division of the working-day into necessary labour and surplus-labour, into paid and unpaid labour. All labour appears as paid labour. In the corvée, the labour of the worker for himself, and his compulsory labour for his lord, differ in space and time in the clearest possible way. In slave-labour, even that part of the working-day in which the slave is only replacing the value of his own means of existence, in which, therefore, in fact, he works for himself alone, appears as labour for his master. All the slave's labour appears as unpaid labour. In wage-labour, on the contrary, even surplus-labour, or unpaid labour, appear as paid. There the property relation conceals the labour of the slave for himself; here the money-relation conceals the unrequited labour of the wage-labourer." (*Capital, op. cit.*, I, 550).

[20.] Tomich, *Prelude to Emancipation, op. cit.*, 140-41.

[21.] *Ibid.*

plantations there.[22] Yet we immediately see here certain contradictions. The importation of food was always expensive. The slave systems (in their nature, it appears) tended to eliminate the local production of commodities other than those (sugar, coffee, indigo, or whatever) produced on the plantations for export. What is more, plantation systems also tended to eliminate free small-scale producers, as happened over and over again in the Lesser Antilles, as sugar-cane and slavery grew.

In many cases the planters, faced by these contradictions, sought to solve them by using some part of the slave labor force to produce food. Having dealt at length with this subject elsewhere,[23] I do not wish to dwell upon it here; but a few points in passing may be useful. First, it is noteworthy that the slave economies, both directly and indirectly, stimulated the exchange of food plants between the Old World and the New. The most famous particular case, by no means unique, was the commissioning of Capt. Bligh by the Jamaica Assembly to bring the breadfruit from Oceania to that island. Though mutiny thwarted his first attempt, Bligh was successful on his second, and the breadfruit did become an important source of slave subsistence. Second, it deserves note in passing that both the agriculture and the cuisine of the contemporary Caribbean region manifest the interblending of numerous different major traditions, among them African, Asian, European, and Native American; this contemporary picture is, however, centuries old, for the most part, and a byproduct of the economic and demographic history of the Caribbean region. Third, it needs to be stressed that a very substantial part of the slaves' subsistence was, in fact, produced by the slaves themselves, and that in many cases the slaves also produced a goodly measure of the subsistence of the free populations of plantation societies. It is to these latter two points that I wish to devote a little more attention.

In compelling or permitting the slaves to grow subsistence, planters had to balance the value of land put in sugar-cane against its value in food crops. Normally, upland or poorer tracts were used for subsistence cultivation, except on those islands so poor or dry that land could not be made available for such cultivation. It was necessary as well for the planters to balance the slave labor power used on the plantations against its yield if put into subsistence cultivation. Here, once again, the solution where possible was to use the very young and the very old, as well as the adult and ablebodied, and to confine such labor to the periods when work in the sugar-cane fields was less needed. In balancing labor use, a common solution was to leave the slave Sunday and an additional half-day, at least during the so-called "dead time", for the production of foodstuffs. Even this arrangement, however, contained contradictory elements within it.

That these were not without their comical side is suggested by the arguments of Mr. Edward Long, a pro-slavery figure as eloquent as he was virulent, in his two-volume *History of Jamaica*. Long's loyalties were at times confused by the

---

[22.] See Yvon Debbasch, "Le marronnage: essai sur la désertion de l'esclave antillais," *L'année sociologique*, 1961, 1-112; 1962, 117-95.

[23.] E.g., "Currency Problems in Eighteenth Century Jamaica and Gresham's Law," in R. Manners, ed., *Process and Pattern in Culture* (Chicago: Univ. of Chicago Press, 1964), 248-65; and *Caribbean Transformations, op. cit.*

Was the Plantation Slave a Proletarian?

circumstances of slave labor in Jamaica, where slaves were granted at the time each Sunday and another half-day off in order to work on their subsistence plots and go to the market. The slaves of Christians received 86 days per year free from plantation labor (except in cases of very urgent business), which included every Sunday and normally half of every Saturday. The slaves of Jews however, received at least 111 days per year for themselves, because the Jews had more holidays than the Christians.

Long calculated how these additional days not only improved the slaves' morale, but also increased significantly their ability to accumulate capital for themselves. But he recognized that few Christians were Christian enough to give their slaves two free days per week. At the same time, since markets had to be held on Sunday, the only day on which the majority of the slaves was entirely free, the larger part of the market trade was engrossed by Jewish merchants who could work on Sunday, unlike the Christian merchants. In order for Christian shopkeepers to compete with the Jews for the slaves' custom, the market day ought to have been changed to some day other than Sunday. Yet that would have meant a significant loss of labor to Christian slave owners. Long argues for adding Thursday as a free day, to Sunday, both to improve the slaves' morale and to afford Christian shopkeepers a firmer purchase on the buying power of the slaves. He even points out the potential value of religious education for the slaves on Sundays, quoting another writer. "On this day some pains should certainly be taken to instruct them to the best of their comprehension, especially the children, in some of the principles of religion and virtue — particularly humility, submission, and honesty which become their condition."[24] But Long sounds rather half-hearted here; perhaps he knew his Christians too well. The elimination of Sunday markets only came about in 1838, with total emancipation.

Though there was certainly an element of compulsion in the initiation of this form of work, in which slaves devoted a day and one-half per week to the cultivation of their own food plots, we nonetheless see very early in the history of both Jamaica and St. Domingue (the cases for which the information seems to be richest) that the institution soon became one which the slaves themselves preferred. I think this development was of great importance. It reveals simultaneously a whole series of contradictions or inconsistencies implicit, I think, in the slave mode of production, and points to some reservations that I feel about the concept itself. Let me try to enumerate some aspects of this contradiction, or inconsistency.

First of all, the development of food cultivation outside the slavery regimen ran entirely counter to the whole conception of how the slave mode of production was supposed to operate. It meant, above all, that slaves were able to work without supervision. Secondly, it made it possible, (and I believe that it was the only circumstance within the plantation framework in which this was true) for slaves to work in groups of their own choosing — normally in family groups, to judge by the descriptions we have. Thirdly, it permitted the slaves to make calculations — what they would grow, and how much — that not only nourished

---

[24]. Edward Long, *History of Jamaica* (London: T. Lowndes, 1774), I, 491-92.

their own sense of autonomy, but also must have permitted a demonstration within the slave group itself of individual differentiation — a differentiation that did not depend upon the whim of the master. Fourthly — and this comes out in the record, too, particularly in the reports of travellers — it dramatized the nature of the slave regimen, and the humanity of the slaves, to anyone intelligent enough to make the inferences. That these people, seemingly so sodden and stupid and dull, incapable of the simplest operations when cutting cane, could turn out to be lively, intelligent, and even happy when working on their own plots, amazed the planters. But foreigners — travellers — had no difficulty in understanding what the difference was. Moreover, subsistence cultivation by slaves had consequences of even wider significance. In both Jamaica and in Haiti, and in practically all Antillean societies where cultivation of this kind developed, this institution led to production that was not for direct use. Indeed, it led to more than simply the production of food which the producers themselves might consume. Thus the slaves were able to transform what had begun as a coercive form into something else: when a slave sold part of his own production, this meant a "radical breach" in the slave mode of production.[25] The concept of the mode depends, as does that of the capitalist mode, on the separation of the worker from the means of production. When the slave produces food for himself and his family he is adding direct-use production to the economic picture of his structural position. And when he adds the sale of his own product, he adds yet another, somewhat contradictory element to the reality of Antillean slavery. When he buys, with the money he earns by selling, he adds yet another element of a contradictory kind. And when — as was the case in these societies — he provisions the free classes within slave society, this adds yet another such element.[26]

[25.] The expression was apparently coined by T. Lepkowski, and appears in his *Haiti*, Vol. I (La Habana, 1968). It is also employed by Ciro F. S. Cardoso in his interesting paper, "La brecha en el sistema esclavista" (ms., 1977). But the idea that Caribbean slaves should not suffer the terminological confinement to which some scholars had consigned them goes back a good deal further; long before the twentieth century, observers noted that slaves and runaways both had done much to alter the nature of slavery itself, and to produce a reality the masters had neither intended nor calculated upon. I have treated this matter more fully in: *Caribbean Transformations, op. cit.*; "Toward an Afro-American history," *Cahiers d'Histoire Mondiale*, XIII, 2, 1971, 317-32); and, with Richard Price, in *An Anthropological Approach to the Afro-American Past: A Caribbean Perspective*, Ishi Occasional Papers in Social Change 2 (Philadelphia: Institute for the Study of Human Issues Press, 1976). Neither Cardoso nor Lepkowski, however, views these "breaches" in the slave system as requiring any revision in the concept of a slave mode of production. I remain a little unsure.

[26.] That one mode of production is dominant over other modes within the same formation; that the coexistence of such modes is entirely to be expected and that the concept of mode of production is not intended nor expected to be identical with any particular, on-the-ground reality, are assertions generally accepted by Marxist scholars, I believe. But it does not seem to me to be useful to treat particular historical instances as irrelevant to our understanding of what the ideal-type mode of production consists in, and represents. Nor do I find it useful to seek to explain what might mistakenly be perceived as exceptions, irregularities, or freak instances as being "transitional" phenomena. This part of the argument relates, on the one hand, to old-fashioned dispositions to describe concrete historical cases as examples of "feudal" or "slave" stages of evolution cut off from events elsewhere in the capitalist world, and, on the other, to ignore those very concrete particulars that enable us to grasp precisely what the term "contradiction" means, in understanding better how social formations, and their component modes of production, change over time.

Was the Plantation Slave a Proletarian?

One may say in response to this that, while the case complicates our understanding, it does not affect the nature of the mode of production, or our means for conceptualizing it. Nonetheless, I think we must try to specify what, precisely, is happening here. Moreau de St.-Méry, one of the most thorough observers of prerevolutionary St. Domingue, tells us in a beautiful passage that, in the marketplace of Clugny, in Cap François (today's Cap Haïtien), in the years immediately preceding the revolution, 15,000 slaves came each Sunday to buy and sell.[27] Again, in Jamaica, we know that the first marketplace was established in 1662, only seven years after the conquest of Jamaica by the British, and was followed by hundreds of others. Edward Long tells us that, in the late 18th century, 20% of the metallic currency in Jamaica at that time was in the hands of the slaves who sold to each other, to their masters, to the free population of the towns, and — a fact that would be funny if it were not so tragic — to the garrisons of British soldiers maintained in Jamaica to control the slaves.

Now, if one leaves aside the significance of cultivation and marketing for any elegant theory of mode of production, considering it only in terms of its everyday meaning, I think it leads to at least three points. First this institution puts in doubt any economic formulation that bases itself purely on commodity production in interpreting Antillean slave society. Second, it raises questions about any monolithic definition or explanation of what constitutes resistance. The way that I have couched this before — and one can think of other examples — the cook of the master's family, that faithful lady who prepared the meals three times a day, sometimes put ground glass in the food of her diners. But she had to become the cook before this option became available. What I mean to say, of course, is that the concept of resistance is really very complicated, ideological considerations aside. Third, the institutions of slave cultivation and marketing can help to throw light upon the historical sequences from slavery to other forms of labor exaction — though I believe that neither the research nor the thinking needed to reveal the full meaning of these institutions has taken place so far. "There is something in human history like retribution," Marx has written, "and it is a rule of historical retribution that its instrument be forged not by the offended but by the offender himself."[28] Nothing else during the history of Caribbean slavery was as important as marketing and provision cultivation in making it possible for the free person — in the case of Haiti, the successful revolutionary — to adapt to freedom without the blessings of the former masters.

But of course the process was in no sense a simple one, and both slaves and masters knew it:

> The practice which prevails in Jamaica of giving the Negroes lands to cultivate, from the produce of which they are expected to maintain themselves (except in times of scarcity, arising from hurricanes and droughts, when assistance is never denied them) is

---

[27] See Louis Moreau de St. Méry, *Description topographique, physique, civile, politique, et historique de la partie française de l'isle Saint-Domingue* (Paris: Sociéte d l'histoire des colonies françaises, 1958), I, 433.

[28] Karl Marx, "The Indian Revolt," *New York Tribune*, Sept. 16, 1857, in S. Avineri, ed., *Karl Marx on Colonialism and Modernization* (Garden City, N.Y.: Doubleday Anchor, 1969), 224.

universally allowed to be judicious and beneficial; producing a happy coalition of interests between the master and the slave. The negro who has acquired by his own labour a property in his master's land, has much to lose, and is therefore less inclined to desert his work. He earns a little money, by which he is enabled to indulge himself in fine clothes on holidays, and gratify his palate with salted meats and other provisions that otherwise he could not obtain; and the proprietor is eased, in a great measure, of the expense of feeding him.[29]

Bryan Edwards was too shrewd an observer of eighteenth-century Jamaica to have missed the mutual benefit flowing from these institutions — or to have failed to see how the short-term satisfactions of independent cultivation and sale might have dulled long-term dissatisfactions with the realities of slavery itself. All the same, the development of such institutions within the context of slavery suggests that our conceptions of freedom and unfreedom are probably too narrow and extreme.

Indeed, it is by this assertion that I return to the major aim of this paper: to consider in what ways, and to what extent, the categories "proletarian" and "slave" really approach each other in practice. "The proper role of a definition," Aya tells us, "is to focus attention on observables, to convert disputation over words into disagreement about what they stand for, and thereby open arguments to further inquiry, testing, and refutation. Taken by themselves," he continues, "definitions are arbitrary; they 'prove' nothing. At most they serve to demarcate the problem at issue, not to solve it. They are not subject to 'proof and demonstration' any more than you can 'prove' that a square is a rectangle with all four sides equal."[30] Starting from very meager definitional statements, I have sought to concentrate upon slaves, leaving aside any serious characterization of proletarians. Those slaves with whom I chose to deal were, as we have seen, disposing of some of their own labor-power independently, on the one hand, and often coexisting with representatives of other categories of labor exaction on the other. My aim, clearly, has not been to narrow what might be said about the slaves, so much as to broaden it. Thus, in certain regards it would be accurate to assert that I have touched on some of the ways in which the slaves participated in productive activities not conventionally associated with slavery, or not part of the slave mode of production.

If, on the one hand, I have sought to indicate some ways in which slave economic activities resembled those of free persons, it is also true that I would have liked to have shown how the activities of free persons, working alongside the slaves, were constrained by coercion and force. I have not really done so here; but the note taken of non-slave categories of labor in the Caribbean plantation context was intended to make this general point. Just as slaves were not completely encapsulated by the state of servitude, so those who, technically free, labored at their side were not in fact completely unshackled.[31] The con-

---

[29]. Bryan Edwards, *The History, Civil and Commercial, of the British Colonies in the West Indies* (London: J. Stockdale, 1793), II, 131.

[30]. Aya, Review of Hindess and Hirst, *op. cit.*, 625.

[31]. F. H. Cardoso, in criticizing very helpfully an early draft of my review of Wallerstein (*Dialectical Anthropology, op. cit.*), writes: "On the one hand, it does not seem to me that these new 'indentured

Was the Plantation Slave a Proletarian?

trast between free and slave, when drawn as Marx drew it in order to dramatize the distinctive nature of European capitalism, is not incorrect, but extreme, and does not — could not — take account of specific historical conditions in every case. As Tomich has asserted, "while Marx stressed the importance of the capitalist world-economy for understanding New World slavery, he never explicitly developed a theory of slave economies, and the question of the social forms of slave production is not systematically treated in his work."[32] Padgug makes a different, but related, point when he writes: "It is true that Marx tends to lump slavery and serfdom together on occasion as if they were interchangeable, but this is only vis-à-vis wage labor, and is only meant to demonstrate the vast differences between all pre-capitalist [sic] labor relationships and the capitalist one."[33]

I do not mean to suggest by these citations that I believe the fundamental economic difference between Caribbean plantation slaves and European factory proletarians can be abandoned by simple recourse to the theme of the global world-economy. I do believe, however, that Wallerstein's insistence is justifiable, that local forms of labor can be made analytically more comprehensible by prior reference to the world-economy:

> The point is that the "relations of productions" that define a system are the "relations of production" of the whole system, and the system at this point in time [the sixteenth century] is the European world-economy. Free labor is indeed a defining feature of capitalism, but not free labor throughout the productive enterprises. Free labor is the form of labor control used for skilled work in core countries whereas coerced labor is used for less skilled work in peripheral areas. The *combination* thereof is the *essence* of capitalism.[34]

Put otherwise, it is not analytically most useful to define either "proletarian" or "slave" in isolation, since these two vast categories of toiler were actually linked intimately by the world economy that had, as it were, given birth to them both, in their modern form. I have not aimed here at assimilating either category

---

servants' from China, India or Java could be thought of as free by anyone making a considered judgement. On the other, abolition did not mean to anyone the passage to a typical capitalist system in regard to productive relations, since slavery was replaced by sharecropping and similar arrangements, which represented a high level of personal dependence, including extra-economic coercion. I believe this is one of the clearest cases of the formal subjection of non-capitalist forms of labor to a clearly capitalist process, thereby preventing internal opportunities for preexisting structures — productive forces, forms and levels of accumulation, and a whole historical context — from responding differently to new influences of the world market.... For me... this reveals the necessity of analyzing, in transitions of this sort, the contradictions [arising from] the confluence of external and internal forces" (personal correspondence, my translation).

While I agree entirely, I believe that these contradictions must raise continuing questions of a theoretical nature about the categories themselves ("proletarian", "slave") and the adjectives ("free", "unfree") we use to describe them. The contradictions are both a cause and an outcome of specific and particular circumstances that *should* affect the nature of our categories. It is the categories which are *abstract*.

32. Tomich, *Prelude to Emancipation,* op. cit., 138.

33. Padgug, op. cit., 24-25.

34. Immanuel Wallerstein, *The Modern World-System: Capitalist Agriculture and the Origins of the European World-Economy in the Sixteenth Century* (New York: Academic Press, 1974), 27.

to the other, but at suggesting instead why a purely definitional approach leaves something to be desired. I shall not attempt to broach a related theme — the specific economic linkages between European proletarians and Caribbean slaves through the products of their labor — which deserves separate and detailed treatment in its own right. But it may be appropriate to conclude by suggesting that both the similarities and differences between these abstract categories will become much clearer, once those linkages have become fully exposed.

Index

Please note page numbers which appear in italics are references to tables or illustrations.

Acre, 6
Acuña, González de, bishop, 150, 152
Afonso, Diogo, Portuguese captain, 27
Afonso I, Henriques, king of Portugal, 12
Afonso V, king of Portugal, 20–1, 25, 29
Africa
  Portuguese in, 20–9, 103–4
  slave trade in, 103–4; economic causes of, 37–8
Africans, European attitudes toward, 250–1
Agga, slave in Mount Airy plantation, 221–2
agriculture; *see also* plantations
  Native American, in Brazil, 39–40
  use of slaves in, origins of, 5, 9, 28
Alcaçovas, Treaty of (1479), 33
aldeias, Jesuit villages in Brazil, 45–51, *47, 50*, 71
Alexander III, pope, 12
Alexander the Great, king of Macedonia, 4
alhorra, blight, 144–5, 147
Amazon basin; *see also* Brazil
  agricultural potential of, 96–8, 100, 104
  disease in, 117–19
  forest collecting industry in, 106–8, 111, 113, 119
  Portuguese plantations in, 96, 98
  products of, 104–5
America(s)/New World
  colonization of, and global economy, 37–8
  gold in, 34
  silver in, 34, 147, 155
Andeiro, João Fernandez, count, 13
Anderson, Alexander, Virginia resident, 181
Antigua, 210
Antilles, plantations in, 309, 313
Aquilena, Francisco de, 141–2
Arrington, Thomas, planter, 181
Arruda, Manuel Monteiro Velho, historian, 26
Ashton, Henry, planter, 181
Augustus, slave in Mesopotamia plantation, 221–2
Aviz dynasty, 45
Azores, 24–6
  plantations in, 2–3, 26–7; slaves and, 29

Bahia, Brazil, 42, 45, 48–53, 58–9, 100–1
Baker, John, planter, 181
Baños y Sotomayor, *see* Sotomayor, Baños y
Barbados, 210, 239, 252, 287–8, *290–1*, 292, *293*
Barcelos, count of, 19
Barham, Henry, physician, 218
Barham, Joseph Foster I, 219
Barham, Joseph Foster II, planter, 215, 218–20, 234, 236, 247
Bayon de Libertat, *see* Libertat, Antoine François Bayon de
Beatriz, princess of Portugal, 13
Belém, *see* Pará
Bermuda Hundred, Virginia, 157
Blanco Ponte family, *see* Ponte
Braganza, duke of, 21
Brazil, 32–3, *97; see also* Amazon basin, Bahia, Balém, Pará
  African slaves in, 67–72, 74–98; disease and, 117–19; economic effects of, 95–6, 99, 101–14; flight of, 114–16; geographic distribution of, 100–1, 106, 108, 119–21, *120, 122*; historiographic considerations, 120–1; introduction of, 98–102; number of, 101, 111, 119–21, *120, 122*; occupations of, 100, 104, 108; price of, 69–72, 99–100, 105–7; resistance of, 114, 116–17
  cattle in, 74–5
  demography of, 59–60, 65, 74–94, 119–21
  economy of, 48–51, 54, 57, 95–6, 101, 109–14; capital and, 102, 106–7, 110; credit and, 106–7, 112; free trade and, 111; labour and, 38, 44–5, 49–51, 54–8, 66–73, 95–6, 99, 101–9, 114; peasantry in, 44–5, 48–9; sugar and, 45, 48, 51–4, 57–8, 69, 72, 95–6
  famine in, 53–4
  Native Americans in: agriculture of, 39–40, 48–9; Christianity and, 44–8, 62–5; demography of, 59–60, 65; disease and, 52, 54, 69, 117–19; economy of, 40–4; ethnography of, 38–40, 46–7, 58–9; historiographic considerations, 55, 57–8; interaction with Portuguese, 38–52, *43*; mortality rates of, 61; names of, 58, 62; resistance to forced labour by, 53–4, 116; as slaves, 42, 44–5, 49–69, 98; terminology for, 55–6

Brazil continued
  plantations in: African slaves in, 37–8, 51, 54, 62, 65–72, *68*, 95–6; cultural aspects of, 3, 35, 44–52, 59–66; introduction of, 3; Native American labourers in, 42, 44–5, 49–70, *68*; Portuguese planning of, 95–6, 100–9, 119; racial composition of, 65–72, *68*; structure of, 54–7, 60–1, 67, *68*, 71, 96; sugar, 3, 34, 45, 48, 50–65, 100
  political structure of, 34, 45–6, 100, 102–3, 119–20
  Portuguese in, 34, 38; *see also* Brazil, plantations in; immigration program for, 110; interaction with Native Americans, 38–52, *43*; slave raids by, 51
  slave prices in, 69–72
  slaves in, demography of, 74–94
  taxes in, 101, 111–13
brazilwood, 33, 41–2
Bréda, comte de, planter, 268
Bréda plantations (Haiti), 265–8; *see also* Haut-du-Cap; Plaine-du-Nord
  slaves in, 270; African, 272–5; age distribution of, 270; birth rates of, 271, 278; creole language and, 273; fertility of, 275–7; health of, 271–2; mortality of, 271; number of, 270; occupations of, 274–5; racial composition of, 270, 272–5; sex ratios of, 270; treatment of, 277–9
  structure of, 270
  sugar production in, 265, 277
Britain/British, plantations of
  legal aspects of, 286, 294
  in North America, 202
Byrd, William III, 168

Cabral, Gonçalo Velho, Portuguese captain, 25–6
Cabral, Pedro Alvares, navigator, 32–3
cacao
  in Amazon basin, 104
  in Caracas, 130–1
Cachaça, sugar cane brandy, 100
Cadamasto, Luis de, chronicler, 18, 27
Caldas, João Pereira, Portuguese governor, 109
Camara, Rui Gonçalves da, Portuguese captain, 26
Caminha, Pero Vaz da, scribe, 33
Caminho, Alvaro de, Portuguese captain, 30
Canary Islands
  Portuguese in, 16–18
  as source of slaves, 2, 17–19, 28
Cape Verdes, 20, 25, 27, 104
  plantations in, 2–3, 28–30
capital
  and Brazilian economy, 102, 106–7, 110
  slaves as, in Caribbean region, 315
capitalism
  and colonization of New World, 37–8
  and global economy, historiographic considerations, 37–8, 321–2
  labour and, 305–22; slave versus proletarian forms of, 307–22
Caracas, 129–55
  African slaves in, 130, 133, 140, 142–3, 145, 151–5; demography of, 149–50, *150*, 153–4; population of, 149, 152
  cacao in, 130–1, 144; *alhorra* blight and, 144–5, 147; origins of, 131–2; terminology for, 131
  cacao trade of: African slaves and, 133, 140, 142–3, 145, 148–51, 153–4; with Dutch merchants, 148; economic aspects of, 138, 140–9, 151; *encomienda* labour and, 133–4, 140, 145; expansion of, 132–3; external competition and, 146–7; with Mexico, 133–4, 146–7; natural disasters and, 145–7; prices in, *144*, 145–7, *146*; profits in, *144*, 144–5, 151; supply and, 145–6; trends in, 132–3, *144*, *146*, 149
  encomiendas in, 129–55; abolition of, 152–3; geographic distribution of, 134–8, *135–6*, 153–4; number of, 135, 153; number of labourers in, 135, *136*, 137–8, 153; organizational aspects of, 134–5, 141; profits in, 137–8; *rentas* in, 134, *136*, 136–8; source material on, 133–4; types of, 134
  exports of, 129–30, *130*, 143
  imports of, 130
  slave trade in, 133, 141–3, 148
  social structure of, 138–42, 148, 151–5
  Spanish in, 129
  taxes in, 152–3
Caribbean region
  aboriginal cultures in, 1
  colonization of, cultural impact of, 1
  plantations in: British, 292, 297, 309, 313; classification of, 309; Danish, *see* St. John, plantations in; Dutch, 35, 297; economic aspects of, 309–22; French, 292, 297, 309, 313; introduction of, 1,

Caribbean region continued
35; labour systems in, 306, 309–22; social aspects of, 1, *290–1*; Spanish, 292, 297, 309, 313; structure of, 201, 203–4; sugar, 34–5, 45, 50, 204, 217, 239, 250, 316; *see also* Brazil, plantations in, sugar; Cuba, sugar production in; Haiti, sugar production in
  silver currency in, 147, 155
  slaves in, 306; as capital, 315; coercion and, 295–7; comparison with proletarians, 306–22; cultivation of provisions by, 316–20; maintenance of, 315–16; racial composition of, 295–7; relationships with owners, 295–302; source material on, 295–6; trade networks of, 316–17
  slave trade in, 133, 140–3, 148, 209–11, 226
Carrington, Paul, attorney and slave owner, 157–9
Carrington, Paul, Jr., slave owner, 159
Carstens, John Lorentz, planter, 250–1, 258
Carter, Robert, planter, 169, 213
Carvalho e Melo, *see* Melo, Sebastião José de Carvalho e
cassava, *see* manioc
Castile, *see* Spain
Castillo, Suárez del, governor, 142
Castro, Manoel Bernardo de Mello e, Portuguese governor, 107
Catalão, Antonio, Portuguese colonist, 26
Catholicism
  in Brazilian sugar plantations, 63–4; godparents and, 63–5, *64*; Native American practice of, 62–4
  and Portuguese policies toward Native Americans, 44–7
Cayenne, 116
  runaway slaves in, 115–16
Ceuta, 14, 19
Chesapeake region; *see also* Virginia
  slaves in, reproduction of, 168, 181–3, 212–13, 223, *224*, 226, 233, 240–8, *241*
  tobacco in, 162–3, 179, 183–4, 204, 218
Christians; *see also* Catholicism
  conflict with Muslims, 5–6, 12
  enslavement of pagans by, 9
  in Holy Land, Italian trade and, 7
Chuao, 148–9, *150*
Cintro, Gonçalo de, Portuguese captain, 20
Clement VI, pope, 16

Coale, Ansley, demographer, 225
Coelho, Duarte, Portuguese captain, 34–5
coercion, and definition of freedom, 287
coffee, in Amazon basin, 104
Cohn, Bernard, historian, 249
Columbus, Christopher, explorer, 1
Commons, John R., 285
Companhia de Cacheu e Cabo Verde, 99
Companhia de Comercio do Maranhão, 99
Companhia Geral do Grão Pará e Maranhão, 102, 104–11, 117–18
Côrtes, Portuguese governmental body, 13, 20
corvée labour, in Mediterranean sugar plantations, 8–9
Costa, Duarte da, Portuguese governor-general, 51
cotton
  in Amazon basin, 104, 110, 113, 119
  in Atlantic region, 29
  in Caracas, 129, *130*
  in St. John, 250
Coutinho, Francisco de Sousa, Portuguese governor, 111–12, 114
Covington, Leonard, slave owner, 173
Craven, Wesley Frank, historian, 212
credit, and Brazilian economy, 106–7, 112
Crete, sugar plantations in, 2, 9–10
Crusades, 5–6, 21
Cuba, 292, 302, 309, 313–14
  slave trade in, 211
  sugar production in, *290–1*, 292, *293*
Cunha, Gaspar da, plantation administrator, 72
Cunha, Nuno da, Portuguese captain, 20
Curaçao
  Dutch in, 148, 294
  slavery in, 287–8
Curtiba, Brazil, 97
Curtin, Philip D., historian, 214
Cyprus, sugar plantations in, 2, 8–10
  corvée labour in, 8–9
  slave labour in, 9

Danish, Caribbean plantations of, *see* St. John, plantations in
Debien, Gabriel, historian, 265–6, 268, 277
Demeny, Paul, demographer, 225
demography
  of African slaves, in Caracas, 149–50, *150*, 153–4
  of Brazil, 59–60, 65, 74–94, 119–21
  of Mesopotamia plantation, 223–5, *224*
  of Mount Airy plantation, 223–5, *224*
  of Native Americans, in Brazil, 59–60, 65
Désir, Philippe Jasmin, free black in Haiti, 267

Dias, Dinis, navigator, 27
disease
   among African slaves: in Brazil, 117–19; in Jamaica, 231–2, 245–7
   among Native Americans: contact with Africans and, 117–19; contact with Europeans and, 52, 69
   in cacao trees, 144–5, 147
   slave trade and, 117–19
Dominican Republic (Santo Domingo), 287, *290–1*, 292, *293*
Duarte, king of Portugal, 19–20
Duncan, T. Bentley, historian, 24–5, 28
Dutch
   in Curaçao, 148
   plantations of, in West Indies, 35
   trade with Caracas, 148

Edwards, Bryan, historian, 319–20
Egmont, Earl of, 202–3, 205
Elkins, Stanley, historian, 247
encomienda, labour grant system, 37, 56, 313
   in Caracas, *see* Caracas, *encomiendas* in
Engenho Santana, 58–62, 67
Engenho Sergipe, 50, 56–72
Engerman, Stanley, historian
Escobar, Pero, navigator, 30
Escovedo, Baltasar de, merchant, 143
Europe
   capitalism in, labour and, 307–8, 321–2
   slave plantations in, 1–2
European expansion, 12
   Crusades and, 5–6
   economic aspects of, 307–22
   effects on Native American cultures, 38
   and plantations, 2–3, 11, 14, 19, 23–5, 308–22
Europeans, attitudes toward Africans, 250–1

Farías, Eduardo Arcila, historian, 132
farinha, 40–1
feitorias, Portuguese factories, 3, 20, 22, 29, 33
Ferdinand the Catholic, king of Castile, 33
Fernandes, Valentim, chronicler, 28
Fernando I, king of Portugal, 12
Feudalism, in Holy Land, 6
Florida, plantations in, 202
   slaves and, 202
Fogel, Robert, historian, 248
forest collecting industry, in Amazon basin, 106–8, 111, 113, 119
Fortaleza, *97*
Foster, John, planter, 218

Foster, Thomas, planter, 218
Fouchard, Jean, historian, 266, 268
Fraginals, Moreno, 286
France, competition with Portugal, in New World, 34
Frederick II, holy Roman emperor, 10
freedom
   definition of, 285–7
   relation to slavery, sociological considerations, 285–302
French Revolution, 277, 296
Furtado, Francisco Xavier de Mendonça, Portuguese governor, 102, 104

Gama, Vasco da, explorer, 22, 32
Gámez, Francisca, 142
Gandavo, Magahães de, historian, 72
Genoa/Genoese
   in slave trade, 2
   trade in Levant, 7
Genovese, Eugene D., historian, 199–200, 248
Gibraltar, Straits of, 12
Glen, James, British governor, 205
global economy
   capitalism and: historiographic considerations, 37–8, 321–2; labour systems and, 305–22
   emergence of, and colonization of New World, 37–8
Godinho, Vitorino Magalhães, historian, 26
Gomes, Diogo, explorer, 27
Gomes, Fernão, Portuguese captain, 29
Gonçalo Velho, *see* Cabral, Gonçalo Velho
Gonçalves, Antão, 28
Grant, James, British governor, 202–3, 205, 207
Great Schism, 13–14
Guipuzcoana Company, 151
Gutman, Herbert, historian, 248

Haagensen, Reinert, planter, 250
Haiti (Saint Domingue), 265, 313, 390
   plantations in, *see* Bréda plantations; Haut-du-Cap; Plaine-du-Nord
   slave revolt in, 265, 277–9
   slaves in, economic status of, 317–19
   sugar production in, 265, 277, *290–1*, 292, *293*, 300
Hanson, Richard, 157
Hapsburg dynasty, 45
Harris, P.M.G., historian, 176
Haut-du-Cap, 265–70
   slaves in: age distribution of, 270; birth rates of, 271; fertility of, 275–7;

Haut-du-Cap continued
    mortality of, 271; number of, 270; occupations of, 274–5; racial composition of, 270, 272–5; sex ratios of, 270; treatment of, 277–9
Havana, see Cuba
Henry of Burgundy, count of Portugal, 12
Henry (Henrique), 'the Navigator', prince of Portugal, 15, 17–20, 23, 25, 27
Herrera, Antonio de, historian, 141
hides, in Caracas, 130, *130*
Holt, Thomas C., sociologist, 288
Holy Land, sugar cane in, 6
Hundred Years' War, 14

Ibarra, Juan de, merchant, 131, 140
India, Portuguese in, 33
indigo, in South Carolina, 204, 206
Innocent XI, pope, 152
Inquisition, Mexican, 133, 147
irrigation, in Madeira, 16, 18
Isabella, queen of Castile, 33
Islam; see also Muslims
    spread of, and slavery, 4–5
islands, slavery on, sociological considerations, 290–302, *291*
Italy, see Genoa; Venice

Jamaica, *290–1*, *293*; see also Mesopotamia plantation
    geography of, 292
    plantations in, 215, 309, 313–14; structure of, 217
    slaves in, 300, 302; economic status of, 215, 316–20; emancipation of, 288
    slave trade in, 226
Jerusalem, 6
Jesuits, in Brazil, 40, 44–6, 49, 52, 74–5, 88
    expulsion of, 74
    relations with colonists, 44–5, 48–51, 54, 56
    relations with Native Americans, 44–51, *47*, 54–6
Jews
    in Jamaica, 317
    in Portuguese plantations, 31
João I, king of Portugal, 13, 19
João II, king of Portugal, 21–2, 24, 30
João III, king of Portugal, 34
Juan I, king of Castile, 13

Kate, slave in piedmont Virginia, 158–61, 183–4, 191–3
Klein, Herbert S., historian, 213, 286–7

Kulikoff, Allan, historian, 212

Laurens, Henry, planter, 208, 212
Leanor of Aragon, 20
Leite, Duarte, historian, 26
Leonor Teles, queen-mother of Portugal, 13
Levant, 6, 10
Levi-Strauss, Claude, philosopher, 47
Lewis, 'Monk', novelist, 219
Libertat, Antoine François Bayon de, attorney, 265–6, 268–73, 277–9
liberty
    definition of, 285–6
    of slaves, 285–9
Liendo, Domingo de, planter, 143–4, 151
Liendo family, planters, 131
Liendo, Pedro de, *encomendero*, 139, 142
Liendo, Santiago de, planter, 147
Linhares, count of, planter, 56
Long, Edward, historian, 316–17, 319
Louverture, François Dominique Toussaint, governor of Saint Domingue, 265–70, 273, 279
Lusignan, Guy de, planter, 8

Macapá, 107
MacLeod, Murdo J., historian, 132, 146
Madeira, 14–17
    plantations in, 1–3, 11, 18; irrigation and, 16, 18; slaves and, 17–19, 28–9, 67; sugar cane and, 18–19, 24
Magalhães Godinho, see Godinho, Vitorino Magalhães
Manaus, *97*
manioc, 39–42, 48–9, 70
Manuel I, king of Portugal, 32–4
manumission, 295–302, *301*
Maranhão, Brazil 98–105, 109–21, *122*
Marchant, Alexander, historian, 42–3
Marques, Oliveira, historian, 13, 19
Marx, Karl, economist, 306–8
Maryland, slaves in, reproduction of, 212–13
Mattos, Pedro Gonçalves de, planter, 66
Mawe, John, traveler, 88
McCarty, Daniel, planter, 181
measles, effect on Native Americans, in Brazil, 52
Mediterranean region, slave plantations in, 1–2, 8–10
Mello e Castro, see Castro, Manoel Bernardo de Mello e
Melo, Sebastião José de Carvalho e, Marquis of Pombal, 101–2, 120
Mem de Sá, see Sá, Mem de

Menard, Russell R., historian, 181, 212
Meneses, Diogo de, Portuguese governor, 41
Menezes, José Narciso de Magalhaes,
    Portuguese governor, 118–19
Menier, Marie-Antoinette, historian, 266, 268
Mesopotamia plantation (Jamaica), 215–48
    slaves in: African, 243–4; age distribution of, 228, *228*, 243; demography of, 223–5, *224*; diet of, 236, 245–6; disease among, 231–2, 245–7; family structure of, 230–2; historiographic considerations, 222–3; invalid, 231, 246; mortality of, *224*, 225, 243, 245–6; occupations of, 220, *235*, 236–9, 244, 246; population changes of, 223, *224*, 240; population of, 216–17; racial composition of, 231–2, 237; religious instruction of, 219, 247; reproduction of, 223–6, 228, 240–7, *241*; sex ratios of, *228*, 228–9, *235*, 237, 243; sexual exploitation of, 231, 246; source material on, 215–16, 221–2, 230
    structure of, 217, 219–20
    sugar in, 217, 220
Mexía, Catalina, 142
Mexican Inquisition, 133, 147
Mexico, 155
    trade with Caracas, 133–4, 146–7
monopolies
    Portuguese, 3, 33–4
    Spanish, 33–4
Moors, 10, 12, 32–3
Morgan, Edmund S., historian, 204
Morgan, William Henry, buccaneer, 148
Mount Airy plantation (Virginia), 215–48
    corn in, 220, 238
    slaves in: age distribution of, 228, *228*, 243; demography of, 223–5, *224*; family structure of, 230, 232–3; historiographic considerations, 222–3; kinship network of, 233; mortality of, *224*, 225; occupations of, 219–22, 234–40, *235*; population changes of, 223, *224*, 240; population of, 216–17; prices of, 229; racial composition of, 232; reproduction of, 223, *224*, 226, 233, 240–4, *241*, 246–8; sex ratios of, *228*, 229, *235*, 237, 243; source material on, 215–16, 221–2
    structure of, 219
    tobacco in, 218
    wheat in, 220, 238
mucambos, jungle encampments, 115
mulattos
    in African trade, 29–30
    in Brazil, 119
    in Jamaican plantations, 231–2, 237
    in Portuguese plantations, 29–30, 49
    in St. John, 261
    in Virginia plantations, 232
Mullin, Gerald W., historian, 200–1
Muslims; *see also* Islam; Moors
    conflict with Christians, 5–6, 12
    in Iberia, 12, 32–3
    sugar cane cultivation by, 4–5

Naipaul, V.S., writer, 288
nation-state, properties of, 10–11
Native Americans
    in Brazil: agriculture of, 39–40; Christianity and, 44–8; economy of, 40–1; ethnography of, 38–40; geographic distribution of, 59; sex ratios of, 60
    in Caracas, *see* Caracas, *encomiendas* in
    as forced labourers, 38; in Brazil, 42, 44–5, 49–69; occupations of, *68*; prices of, 71–2; resistance of, 53–4, 116
    as free labourers: in Brazil, 50, 54, 57, 69–70, 106; wages of, 69–70
    interaction with Africans: demographic impact of, 117; disease and, 117–19
    interaction with Europeans, 38–41; cultural impact of, 38, 41–52; demographic impact of, 45, 52; disease and, 52, 69
Navarro, Juan de Azpilcueta, S.J., missionary, 46
negro (term), 55
Newton plantation (Barbados), 239–40
New World, *see* America(s)
Nóbrega, Manoel da, S.J., missionary, 40
Noé, comte de, planter, 266
Noli, Antonio da, navigator, 27
North America; *see also* Chesapeake region; Florida; South Carolina; Virginia
    plantations in: agricultural differences among, 204; crops grown in, 204; historiographic considerations, 216; industrial, 201, 203; paternalistic, 201, 207–8, 212, 247; sex ratios of slaves in, 201–2, 209, 212–14; structure of, 200–3, 206–7, 213, 217
    slaves in: family structure of, 208; historiographic considerations, 199–200; price of, 205; profitability of, 204–5; reproduction of, 199–203, 205, 208–9, 212, 214; sex ratios of, 201–2, 209, 212–14
    slave trade in, 209

INDEX

Novais, Fernando, historian, 37

Oldendorp, C.G.A., chronicler, 251, 259
Oswald, Richard, 202–3
Ottoman empire, conflicts of, effect on European trade, 7–8, 10
Ovalle, Diego de, *encomendero*, 131–2, 138–41

Paiva, João de, Portuguese captain, 30
Pará, Brazil, 96–9, *97*, 102, 104–21, *120*
Parreira, Amorim, 28
paternalism, in North American plantations, 201, 207–8, 212, 247
Patterson, Orlando, sociologist, 254, 285–6
peça de India, unit of measurement of slaves, 120
Pedro I, king of Portugal, 12
Pedro, prince regent of Portugal, 19–21, 24
Pereira, Agustín, merchant, 143
Pereira, João, Portuguese captain, 30
Pereira, Nuno Alvarez, Portuguese constable, 13, 19
Perestrelo, Bartolomeu, Portuguese captain, 15, 18
Pernambuco, 34–5, 42, 45, 51, 59, 61, 95, 100–1, 113
Phipps, James, shipping agent, 213
Pimentel, José Rengifo, *encomendero*, 139, 142
Pitta, Sebastião da Rocha, historian and planter, 72
Plaine-du-Nord, 265, 271–2, 274, 276, 278
plantations
  in Atlantic region, 1–3, 18, 24–8; cotton, 29; political aspects of, 11; racial composition of, 29–30; slaves and, 17–19, 24, 28–9; structure of, 29; sugar, 11, 18, 24, 26–32
  in Azores, 2–3, 26–7; slaves and, 29
  in Brazil, *see* Brazil, plantations in
  British: legal aspects of, 286, 294; in North America, 202
  in Caribbean region, *see* Caribbean region, plantations in
  European expansion and, 2–3, 11, 14, 19, 23–35, 308–22
  European origins of, 1–2, 8–9
  in Haiti, *see* Bréda plantations
  legal aspects of, 286, 288, 294
  in Madeira, *see* Madeira, plantations in
  in Mediterranean region, 1–2; origins of, 8–10; slaves and, 9–10; sugar, 2, 8–9
  Mesopotamia, *see* Mesopotamia plantation
  Mount Airy, *see* Mount Airy plantation
  in New World: degrees of freedom in, 287–9; introduction of, 1, 34–5; Native American labourers and, 38, 44–5; racial composition of, 35, 49, 231–2, 237; slaves and, *see* slavery, in New World; social aspects of, 1; structure of, 3–4, 201–4, 206–8, 217; sugar, 34–5, 45, 50, 239, 250, 316
  in St. John, *see* St. John, plantations in
Polly, ship, 157–8
Pombal, Marquis of, *see* Melo, Sebastião José de Carvalho e
Ponte, Alejandro Blanco, *encomendero*, 139, 142
Ponte, Pedro Blanco, 139
Pope, Nathaniel, planter, 181–2
Pôrto Alegre, *97*
Portugal/Portuguese
  in Africa, 20–9, 103–4
  in Brazil, *see* Amazon basin, Portuguese plantations in; Brazil, Portuguese in
  in Canary Islands, 16–18
  French competition with, in New World, 34
  in India, 33
  monopoly powers of, 3, 33–4
  navigation by: to Asia, 22–3, 32–3; in Atlantic, 22–3, 25, 30, 32–3, 103–4; dispute with Castile and, 33; and slave trade, 103–4, 113–14
  plantations of: Atlantic, 2–3, 11, 18; in Brazil, *see* Amazon basin, Portuguese plantations in; Brazil, plantations in; Jews in, 31; political expansion and, 2–3, 11, 14, 19, 23–35
  political structure of, 12–14, 19–20
  slave laws of, 44–6, 56
  in slave trade, *see* slave trade, Portuguese in
  socioeconomic structure of, 12, 14
  sugar cane and, transmission of, 11, 18–19, 24, 26–8, 31–2, 34–5
  war with Castile, 13, 32–3
Prawer, Joshua, historian, 6–7
Principe, plantations in, 3
proletarian
  comparison with slave, 305–22
  definition of, 306–7
prostitutes, slaves as, 298
Puerto Rico, 287, *290–1*, 292, *293*, 309, 313–14

Queiróz, João de São José, bishop, 107

Randolph, Peyton, planter, 169

Read, Clement, 158
Recife, *97*, 98
Reconguista, 12
repartimiento, draft labour system, 56
resgate, slave trade, 56
rice
  in Amazon basin, 104–5, 107
  in South Carolina, 204, 206
Riley-Smith, Jonathan, historian, 8
Rio de Janeiro, *97*
Rocha, Martin da, S.J., missionary, 40
Rojas, Juan Vásquez de, 139
Roman empire, slavery in, 5
Royal African Company, 213

Sahlins, Marshall, historian, 43
St. Croix, 258–9, *290–1*, *293*
Saint Domingue, *see* Haiti
St. John
  African slaves in, 250; Afro-Caribbean culture and, 253–6, 262–3; cultivation of provisions by, 252–3, 255, 257; Danish attitudes toward, 250–1, 257–63; family structure of, 253–5; kinship among, 253–4; laws concerning, 255–7; mortality of, 254; sexual exploitation of, 259–60; trade networks of, 253, 255, 257; treatment of, 251, 257
  Danish in, 250; attitudes toward African slaves, 250–1, 257–63; cultural aspects of, 257–61; gender roles of, 259–61
  maroons in, 252
  mulattos in, 261
  plantations in: cotton, 250; cultural aspects of, 249–63; economic structure of, 252; social structure of, 250, 255–61; sugar, 250
  trade with St. Thomas, 253
St. Thomas, 261
  trade with St. John, 253
Salvador, *97*
Sá, Mem de, Portuguese governor-general, 48, 51
Sancho IV, king of Castile-Leon, 12
Sanford, John, planter, 182
Sanford, Joseph, planter, 181
Santa Cruz (Brazilian estate), 74–94
Santarem, João de, navigator, 30
Santiago (Cape Verdes), 27–30
Santidade, resistance movement, 53
Santo Domingo, *see* Dominican Republic
Santos, Alonso Rodríguez, *encomendero*, 143
São Luis, 99–100; *see also* Maranhão

São Miguel, 24–6
São Paulo, *97*
São Tomé, 2, 30–1
  plantations in, 3, 31–2; Jews in, 31; racial composition of, 31; slaves and, 31, 67
sarsaparilla, 113, 129, *130*
Senegal, 25
sesmarias, land grants, 48–9
sexual exploitation, of slaves, 176, 231, 246, 259–60, 297–8
Shenandoah Valley, *161*
  slaves in, *164*, 169
Sicily, sugar cane in, 10
silver currency, in Caribbean region, 147, 155
slave raids
  Portuguese, in Brazil, 51
  Spanish, in Caracas, 137
slavery
  economic aspects of, 37, 305–22
  legal aspects of, 44–7, 56, 255–7, 285–302, 305
  in New World, 35, 38, 202; African, 37–8, 51, 54, 66–72; historiographic considerations, 305–6; Native American, 38; Portuguese legislation and, 44–7, 56
  origins of, 4–5, 9, 28
  relation to freedom, sociological considerations, 285–302
  resistance to: by Africans, 114–17, 265, 277–9; by Native Americans, 53–4, 116
slaves
  African: in Brazil, 37–8, 51, 62, 65–72, 74–98; occupations of, 67–71, *68*; price of, 69–72, 99–100, 105–7; reproduction of, 199–214; source regions of, 213, 272–4
  as agents, 295–302
  in Azores, 29
  in Brazil: breeding of, 77–8; demography of, 74–94; family structure of, 74–5, 78–84, *78–84*, 89, *92–4*; handicapped, 86–7, 87; historiographic considerations, 74, 87–9; occupations of, 84–6, *85*; sex ratios of, 75–7, *76–7*
  in Bréda plantations, *see* Bréda plantations, slaves in
  in Canary Islands, 2, 17–19, 28
  in Caracas, 130, 133, 140, 142–3, 145, 151–5; demography of, 149–50, *150*, 153–4; population of, 149, 152
  comparison with proletarians, 305–22
  disease among: in Brazil, 117–19; in Jamaica, 231–2, 245–7

INDEX 331

slaves continued
  emancipation of, 219, 285, 288, 296, 314
  in Haiti: economic status of, 317–19; revolt of, 265, 277–9
  in Haut-du-Cap, *see* Haut-du-Cap, slaves in
  legal liberties of, sociological aspects of, 285–93, 295–7
  in Madeira, 17–19, 28–9, 67
  manumission of, 295–302, *301*
  in Mesopotamia plantation, *see* Mesopotamia plantation, slaves in
  military service of, 297, 300
  in Mount Airy plantation, *see* Mount Airy plantation, slaves in
  Native American, *see* Native Americans, as forced labourers
  in North America, *see* North America, slaves in
  as prisoners of war, 9, 17, 45
  as prostitutes, 298
  relationships with owners, 295–302
  reproduction of, 199–214; in Brazil, 77–8; in Chesapeake region, 168, 181–3, 212–13, 223, *224*, 226, 233, 240–8, *241*; in Haiti, 271, 278; in Jamaica, 223–6, 228, 240–7, *241*; in South Carolina, 208–9, 212
  sexual exploitation of, 176, 231, 246, 259–60, 297–8
  in Shenandoah Valley, *164*, 169
  in South Carolina: family structure of, 208; profitability of, 204–5, 207; reproduction of, 208–9, 212
  in St. John, *see* St. John, slaves in
  in Virginia, *see* Virginia, slaves in
slave society
  definition of, 289
  sociological aspects of, 294, 302
slave trade
  abolition of, 296
  British in, 209–11, 213
  in Caribbean region, 133, 140–3, 148, 209–11, 226
  demand in, 99–100, 114
  disease and, 117–19
  Dutch in, 148
  economic causes of, 37–8
  Genoese in, 2
  geographic distribution in, 100, 106, 113–14
  in Jamaica, 226
  mortality in, 148–9
  Portuguese in, 28–9, 67–9, 99–109, 111–14, 210; credit and, 106–7; government involvement and, 101–5, 110–12; and Spanish American cacao trade, 133, 141–3
  profits in, 105
  routes of, 100, 113–14
  sex ratios in, 209–13
  Spanish in, 211
  structure of, 111
  taxes in, 101, 111–13
  textiles and, 29
  in Virginia, *see* Virginia, slave trade in
  volume of, 111, 119–21, *120*, *122*, 305
smallpox
  effect on Native Americans, in Brazil, 52, 117–19
  in Jamaica, 247
Smyth, John, Virginia resident, 181
Soares, Francisco, S.J., missionary, 71–2
Soares, João, Portuguese captain, 26
Society of Jesus, *see* Jesuits
Sotomayor, Baños y, bishop, 152
Sousa Coutinho, *see* Coutinho, Francisco de Sousa
Sousa, Tomé de, Portuguese governor, 49, 51
South Carolina
  indigo in, 204, 206
  plantations in, 204; structure of, 206–8
  rice in, 204, 206
  slaves in: family structure of, 208; profitability of, 204–5, 207; reproduction of, 208–9, 212
  slave trade in, 209, 211
sovereignty, definition of, 10–11
Spain (Castile)
  navigation by: in Atlantic, 32–3; dispute with Portugal and, 33
  plantations of, legal aspects of, 286, 288, 294
  war with Portugal, 13, 33
Speed, John, Jr., slave owner, 182
Steuart, Charles, merchant, 166–7
sugar cane
  in Atlantic region, 18–19, 24
  in Brazil, 3, 34, 45, 48, 50–65, 69, 72, 95–6, 100
  in Caracas, 129, *130*, 154–5
  in Caribbean region, 34–5, 45, 50, 204, 217, 239, 250, 316; *see also* Cuba, sugar production in; Haiti, sugar production in; sugar cane, in Brazil
  distillation of, in Amazon basin, 100
  domination of island economies by, sociological considerations, 290–4, *291*
  geographic origins of, 4

sugar cane continued
  in Mediterranean region, 2, 6–10
  transmission of: comparison with spices, 10; Crusades and, 5–6; Italian merchants and, 9–10; Portuguese and, 11, 18–19, 24, 26–8, 31–2, 34–5

Tabb, Thomas, merchant, 159
Tannenbaum, Frank, historian, 199–200
Tarasia, countess, 12
taxes
  in Brazil, 101, 111–13
  in Caracas, 101, 111–13
Tayloe, John III, planter, 215–19, 224, 226, 238–9, 247–8
Tayloe, John I, planter, 218
Tayloe, William, planter, 218
Taylor, William, captain, 167
Taylor, William Henry, planter, 233
Teixeira, Ruy, plantation administrator, 56
Teixeira, Tristão Vaz, Portuguese captain, 15, 18
Templars, 8
textile industry, in Atlantic region, 29
tobacco
  in Caracas, 129, *130*
  in Chesapeake region, 204
  types of, 162
Tomich, historian, 314–15
Tordesillas, Treaty of (1494), 33–4
Toussaint Louverture, *see* Louverture, François Dominique Toussaint
Tovar, Mauro de, bishop, 146
triangle trade, Caracas in, 133
Trinidad, 292
Tristão, Nuno, Portuguese captain, 20, 28
Tupinambá, Native Americans, 39–46, 58–9

Urban II, pope, 5, 12

Valle, Leonardo do, S.J., missionary, 52
Valsemey, plantation manager, 278
Vasconcellos, Simão de, S.J., missionary, 41
Vasconcelo, Francisco de, Portuguese governor, 141–2
Velho Arruda, *see* Arruda, Manuel Monteiro Velho, historian
Venezuela, *see* Caracas
Venice/Venetians
  in Levantine trade, 7
  in sugar trade, 7, 9
Verlinden, Charles, historian, 1–2, 6
Villanueva, Diego de, Spanish royal treasurer, 129–30

Villegas, Juan Maritínez de, *encomendero*, 137
Villevaleix, attorney, 278–9
Virginia; *see also* Chesapeake region; Mount Airy plantation
  economy of, 162–3
  mortality rates in, 204
  piedmont region of, 157–97, *160*; settlement of, 161–2; slaves in, *see* Virginia, slaves in
  plantations in: size of, 216–17; slave labour and, 157–9, 168–9, 179, 183–4, *185–7*; structure of, 217
  population of, geographic distribution of, *161*, 161–2
  slaves in, 157–9, 163, 168–9, 179, 183–4, *185–7*; adult-child ratios in, 165–6, *166*, *169*, 173–4, *175*, 182–3; African, 163–9, 173–4, 179, 183, 192–7, 227; ages of, *169*, 173, 193–7; American-born, 167–8, 173, 226–7; family structure of, 185–7; female-child ratios in, 170, *171–2*, 183–4; geographic distribution of, *164*; migration patterns of, 168–74; occupations of, 184; population of, 163–5, *185–7*, *188–90*, 197, *197*; quarters of, 191–2, *192*; relations with owners, 191; reproduction of, 168, 181–3, 212; sex ratios of, 167–81, *168*, *171–2*, *175*, *178*, *180*, 187, *188–90*; sexual exploitation of, 176
  slave trade in, 157–8; adults in, 165–6, *166*; auctions in, 157; children in, 163–5, *169*, 193–7, *194–6*; prices in, 157, 176; volume of, 164–5, 197, *197*
  tidewater region of, 161–2, *162*; agriculture in, 179, 217; slaves in, *see* Virginia, slaves in
  tobacco in, 218; price of, 162; slave labour and, 163, 179, 183–4

Wallerstein, Imanuel, historian, 37
Watson, William, planter, 182
West, Hans, school master, 251–2, 258, 261
West Indies, *see* Caribbean region
wheat flour, in Caracas, 129, *130*, 143, 154–5
White, Joseph, slave owner, 181
Williams, Eric, 286–7
Wimbish, Benjamin, slave owner, 182
Wolf, Eric R., historian, 201
Wood, Peter H., historian, 208–9

yaws, disease, 231–2

Young, Crawford, political scientist, 10

Zarco, João Gonçalves, Portuguese captain, 15, 18, 26

Zurara, chronicler, 18, 26

HD 1471 .A3 P54 1997

PLANTATION SOCIETIES IN THE
ERA OF EUROPEAN EXPANSION